BIOINORGANIC CHEMISTRY

BIOINORGANIC CHEMISTRY

A Short Course

ROSETTE M. ROAT-MALONE
Washington College
Chestertown, MD

WILEY-INTERSCIENCE
A JOHN WILEY & SONS, INC., PUBLICATION

Published by John Wiley & Sons, Inc., Hoboken, New Jersey.
Published simultaneously in Canada.

For general information on our other products and services please contact our Customer Care Department within the U.S. at 877-762-2974, outside the U.S. at 317-572-3993 or fax 317-572-4002.

Wiley also publishes its books in a variety of electronic formats. Some content that appears in print, however, may not be available in electronic format.

Library of Congress Cataloging-in-Publication Data is available.

ISBN: 0-471-15976-X

Printed in the United States of America.

10 9 8 7 6 5 4 3 2 1

To Michael

CONTENTS

PREFACE

When I began teaching a bioinorganic course in the mid-1980s, Robert Hay's short text (Hay, R. W. *Bio-Inorganic Chemistry*, Ellis Horwood Limited, Halsted Press, New York, 1984, 210 pp.) addressed most topics in the area. Coverage of the entire bioinorganic field in such a text now would be impossible because the number of topics and their complexity have grown tremendously. In all recent iterations of my bioinorganic courses, I have followed a philosophy upon which this "short" text is based:

1. Review enough introductory material in inorganic (Chapter 1: Inorganic Chemistry Essentials) and biochemistry (Chapter 2: Biochemistry Fundamentals) to bring all students to more or less the same level. I try to take the students' backgrounds into account: Are their principal knowledge and interest areas chemistry, biology, biochemistry, physics, spectroscopy, pharmaceuticals, and so on? Do they intend to become synthetic chemists, theoretical chemists, molecular biologists, physicians, pharmacologists, neuroscientists, and so on?
2. Introduce instrumental techniques used in analysis of the bioinorganic systems I will lecture on (Chapter 3: Instrumental and Computer-Based Methods). Typically, these would be electron paramagnetic resonance (EPR) and Mössbauer spectroscopies not often covered in undergraduate instrumental analysis courses plus X-ray diffraction and NMR techniques used for structural analyses of metalloproteins and their small molecule model compounds.
3. Present one or two bioinorganic topics of my choice. I follow the paradigms the reader will note in Chapters 4, 5, and 6: Use the structural information gained through X-ray crystallographic and NMR solution studies to try to understand the metalloprotein's function and mechanism of activity and then design small

molecule compounds to confirm mechanism, extend the catalytic activity for industrial purposes, or diagnose and cure the disease states caused by the metalloprotein's malfunction.

4. Encourage students to choose their own bioinorganic topics for individual study and class discussion in the remaining course time. I find that undergraduate and graduate student fascination and enthusiasm for the subject area always energizes them (and me) for the study of diverse and interesting bioinorganic topics. Certainly the subject area provides great opportunities for introducing the use of primary literature sources and the application of computer- and internet-based searching, visualization, and modeling techniques.

In keeping with this philosophy I believe the text is appropriate for a one-semester bioinorganic chemistry course offered to fourth-year undergraduate chemistry, biochemistry, and biology majors and first-year graduate students concentrating in inorganic and biochemical subject areas. The text presents review chapters on inorganic chemistry (Chapter 1: Inorganic Chemistry Essentials) and biochemistry (Chapter 2: Biochemistry Fundamentals), although ideally students will have previously taken undergraduate one-semester advanced inorganic and biochemistry courses. Chapter 3 (Instrumental and Computer-Based Methods) provides an introduction to some important instrumental techniques for bioinorganic chemistry; the computer section includes basic information about computer hardware and software available in late 2001. I hope I have passed along my enthusiasm for computer-aided modeling and visualization programs with an emphasis on free software and databases available via the World Wide Web. The bioinorganic subjects covered in Chapter 4 (Iron-Containing Oxygen Carriers and Their Synthetic Models), Chapter 5 (Copper Enzymes), and Chapter 6 (The Enzyme Nitrogenase) emphasize the use of primary sources. Every effort has been made to include up-to-date references in each subject area. Following (or instead of) presentation of these chapters, teachers and learners are encouraged to find bioinorganic subject areas of special interest to them. My special interest area is introduced in Chapter 7 (Metals in Medicine).

Some important bioinorganic areas not discussed in this text are mentioned in the following paragraphs, hopefully to jump-start student investigations of areas of interest to them. Other texts and sources—references 3–7 in Chapter 1, for instance—will be helpful in choosing bioinorganic topics for further discussion.

The use of alkali and alkaline earth group metal ions, especially those of sodium, potassium, magnesium, and calcium, for maintenance of electrolyte balance and for signaling and promotion of enzyme activity and protein function are not discussed in this text. Many of these ions, used for signaling purposes in the exciting area of neuroscience, are of great interest. In ribozymes, RNAs with catalytic activity, solvated magnesium ions stabilize complex secondary and tertiary molecular structure. Telomeres, sequences of DNA at the ends of chromosomes that are implicated in cell death or immortalization, require potassium ions for structural stabilization.

Hemoglobin's protoporphyrin IX, discussed in Chapter 4, is only one of many important porphyrin hemes involved in bioinorganic and catalytic science.

Cytochromes, catalases, and peroxidases all contain iron–heme centers. Nitrite and sulfite reductases, involved in N–O and S–O reductive cleavage reactions to NH_3 and HS^-, contain iron–heme centers coupled to $[Fe_4S_4]$ iron–sulfur clusters. Photosynthetic reaction center complexes contain porphyrins that are implicated in the photoinitiated electron transfers carried out by the complexes.

To access information about the structure and function of metalloproteins and enzymes, I recommend a visit to the Research Collaboratory for Structural Bioinformatics (RCSB) Protein Data Bank (PDB) website at www.rcsb.org/pdb. Begin by entering the enzyme's name into the search routine. If an X-ray crystallographic or NMR structural study has been published, the hits displayed will provide journal references. A recent PDB search on the term *cytochrome* yielded 478 hits including 22 referring to cytochrome P-450, a membrane-bound monooxygenase that catalyzes incorporation of one oxygen atom from O_2 into substrate molecules. Sixteen citations described cytochrome c oxidase, a membrane-bound electron transfer protein that couples electron donor NADH with electron acceptor O_2, releasing large amounts of energy. Fifteen citations result from a search on aconitase, an enzyme that contains an $[Fe_4S_4]^{2+}$ cluster and catalyzes the isomerization of citrate and isocitrate. The isomerization is a key reaction in the Kreb's cycle. In aconitase, one cluster iron binds citrate at the active site during the catalytic cycle. Liver alcohol dehydrogenase (LADH) and carbonic anyhydrase are two interesting zinc-containing metalloenzymes. The growing zinc-finger family of similarly shaped metalloproteins constitutes an important motif for binding nucleic acids. This last topic is discussed in Section 2.4 (Zinc-Finger Proteins) of Chapter 2.

A very brief introduction to the important topic of bioinorganic electron transfer mechanisms has been included in Section 1.8 (Electron Transfer) of Chapter 1. Discussions of Marcus theory for protein–protein electron transfer and electron or nuclear tunneling are included in the texts mentioned in Chapter 1 (references 3–7). A definitive explanation of the underlying theory is found in the article entitled "Electron-Transfer in Chemistry and Biology," written by R. A. Marcus and N. Sutin and published in *Biochem. Biophys. Acta*, 1985, **811**, 265–322.

Bioinorganic chemical knowledge grows more interesting and more complex with each passing year. As more details about the usage and utility of metals in biological species and more mechanistic and structural information becomes available about bioinorganic molecules, more biologists, chemists, and physicists will become interested in the field. Senior-level undergraduates in chemistry, biology, and physics, as well as beginning graduate students, will do well to educate themselves in this diverse and fascinating chemical area. I hope this introductory text will be catalytic for students, whetting their appetites for more information and encouraging them to join research groups engaged in the search for new knowledge in the continually expanding bioinorganic field.

ROSETTE M. ROAT-MALONE
Washington College

ACKNOWLEDGMENTS

Many groups of people contributed to the creation and realization of this book. Many thanks go out to former bioinorganic chemistry students whose enthusiasm for the subject material inspired the book's format. Professional colleagues at Washington College and other universities worldwide helped in may ways–as critical readers, as advisers on important subject areas to be included, with bibliographic assistance, and even with a last-minute delivery service when deadlines loomed. The book would not exist without the expert assistance of Wiley editors—Darla Henderson, Amy Romano, Danielle Lacourciere, Bob Golden, and Dean Gonzalez. Lastly, I express heartfelt gratitude to family and friends for their patience during the many months of gestation.

1

INORGANIC CHEMISTRY ESSENTIALS

1.1 INTRODUCTION

Bioinorganic chemistry involves the study of metal species in biological systems. As an introduction to basic inorganic chemistry needed for understanding bioinorganic topics, this chapter will discuss the essential chemical elements, the occurrences and purposes of metal centers in biological species, the geometries of ligand fields surrounding these metal centers, and the ionic states preferred by the metals. Important considerations include equilibria between metal centers and their ligands and a basic understanding of the kinetics of biological metal–ligand systems. The occurrence of organometallic complexes and clusters in metalloproteins will be discussed briefly, and an introduction to electron transfer in coordination complexes will be presented. Because the metal centers under consideration are found in a biochemical milieu, basic biochemical concepts, including a discussion of proteins and nucleic acids, are presented in Chapter 2.

1.2 ESSENTIAL CHEMICAL ELEMENTS

Chemical elements essential to life forms can be broken down into four major categories: (1) bulk elements (H, C, N, O, P, S); (2) macrominerals and ions (Na, K, Mg, Ca, Cl, PO_4^{3-}, SO_4^{2-}); (3) trace elements (Fe, Zn, Cu); and (4) ultratrace elements, comprised of nonmetals (F, I, Se, Si, As, B) and metals (Mn, Mo, Co, Cr, V, Ni, Cd, Sn, Pb, Li). The identities of essential elements are based on historical work and that done by Klaus Schwarz in the 1970s.[1] Other essential elements may be present in various biological species. Essentiality has been defined by certain

Table 1.1 Percentage Composition of Selected Elements in the Human Body

Element	Percentage (by weight)	Element	Percentage (by weight)
Oxygen	53.6	Silicon, Magnesium	0.04
Carbon	16.0	Iron, fluorine	0.005
Hydrogen	13.4	Zinc	0.003
Nitrogen	2.4	Copper, bromine	$2.\times 10^{-4}$
Sodium, potassium, sulfur	0.10	Selenium, manganese, arsenic, nickel	$2.\times 10^{-5}$
Chlorine	0.09	Lead, cobalt	$9.\times 10^{-6}$

Source: Adapted from reference 2.

criteria: (1) A physiological deficiency appears when the element is removed from the diet; (2) the deficiency is relieved by the addition of that element to the diet; and (3) a specific biological function is associated with the element.[2] Table 1.1 indicates the approximate percentages by weight of selected essential elements for an adult human.

Every essential element follows a dose–response curve, shown in Figure 1.1, as adapted from reference 2. At lowest dosages the organism does not survive, whereas in deficiency regions the organism exists with less than optimal function.

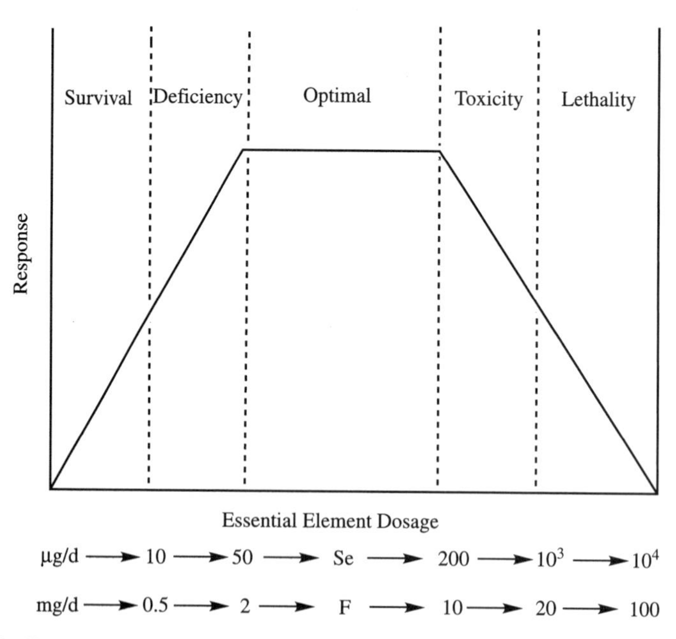

Figure 1.1 Dose–response curve for an essential element. (Adapted with permission from Figure 3 of Frieden, E. *J. Chem. Ed.*, 1985, **62**(11), 917–923. Copyright 1985, Division of Chemical Education, Inc.)

After the concentration plateau of the optimal dosage region, higher dosages cause toxic effects in the organism, eventually leading to lethality. Specific daily requirements of essential elements may range from microgram to gram quantities as shown for two representative elements in Figure 1.1.[2]

Considering the content of earth's contemporary waters and atmospheres, many questions arise as to the choice of essential elements at the time of life's origins 3.5 billion or more years ago. Certainly, sufficient quantities of the bulk elements were available in primordial oceans and at shorelines. However, the concentrations of essential trace metals in modern oceans may differ considerably from those found in prebiotic times. Iron's current approximate 10^{-4} mM concentration in sea water, for instance, may not reflect accurately its pre-life-forms availability. If one assumes a mostly reducing atmosphere contemporary with the beginnings of biological life, the availability of the more soluble iron(II) ion in primordial oceans must have been much higher. Thus the essentiality of iron(II) at a concentration of 0.02 mM in the blood plasma heme (hemoglobin) and muscle tissue heme (myoglobin) may be explained. Beside the availability factor, many chemical and physical properties of elements and their ions are responsible for their inclusion in biological systems. These include: ionic charge, ionic radius, ligand preferences, preferred coordination geometries, spin pairings, systemic kinetic control, and the chemical reactivity of the ions in solution. These factors are discussed in detail by Frausto da Silva and Williams.[3]

1.3 METALS IN BIOLOGICAL SYSTEMS: A SURVEY

Metals in biological systems function in a number of different ways. Group 1 and 2 metals operate as structural elements or in the maintenance of charge and osmotic balance (Table 1.2). Transition metal ions that exist in single oxidation states, such as zinc(II), function as structural elements in superoxide dismutase and zinc fingers, or, as an example from main group +2 ions, as triggers for protein activity—that is, calcium ions in calmodulin or troponin C (Table 1.3). Transition metals that exist in multiple oxidation states serve as electron carriers—that is, iron ions in cytochromes or in the iron–sulfur clusters of the enzyme nitrogenase or copper ions in

Table 1.2 Metals in Biological Systems: Charge Carriers

Metal	Coordination Number, Geometry	Preferred Ligands	Functions and Examples
Sodium, Na^+	6, octahedral	*O*-Ether, hydroxyl, carboxylate	Charge carrier, osmotic balance, nerve impulses
Potassium, K^+	6–8, flexible	*O*-Ether, hydroxyl, carboxylate	Charge carrier, osmotic balance, nerve impulses

Table 1.3 Metals in Biological Systems: Structural, Triggers

Metal	Coordination Number, Geometry	Preferred Ligands	Functions and Examples
Magnesium, Mg^{2+}	6, octahedral	*O*-Carboxylate, phosphate	Structure in hydrolases, isomerases, phosphate transfer, trigger reactions
Calcium, Ca^{2+}	6–8, flexible	*O*-Carboxylate, carbonyl, phosphate	Structure, charge carrier, phosphate transfer, trigger reactions
Zinc, Zn^{2+} (d^{10})	4, tetrahedral	*O*-Carboxylate, carbonyl, *S*-thiolate, *N*-imidazole	Structure in zinc fingers, gene regulation, anhydrases, dehydrogenases
Zinc, Zn^{2+} (d^{10})	5, square pyramid	*O*-Carboxylate, carbonyl, *N*-imidazole	Structure in hydrolases, peptidases
Manganese, Mn^{2+} (d^5)	6, octahedral	*O*-Carboxylate, phosphate, *N*-imidazole	Structure in oxidases, photosynthesis
Manganese, Mn^{3+} (d^4)	6, tetragonal	*O*-Carboxylate, phosphate, hydroxide	Structure in oxidases, photosynthesis

Table 1.4 Metals in Biological Systems: Electron Transfer

Metal	Coordination Number, Geometry	Preferred Ligands	Functions and Examples
Iron, Fe^{2+} (d^6) Iron, Fe^{2+} (d^6)	4, tetrahedral 6, octahedral	*S*-Thiolate *O*-Carboxylate, alkoxide, oxide, phenolate	Electron transfer, nitrogen fixation in nitrogenases, electron transfer in oxidases
Iron, Fe^{3+}(d^5) Iron, Fe^{3+}(d^5)	4, tetrahedral 6, octahedral	*S*-Thiolate *O*-Carboxylate, alkoxide, oxide, phenolate	Electron transfer, nitrogen fixation in nitrogenases, electron transfer in oxidases
Copper, Cu^{+}(d^{10}), Cu^{2+} (d^9)	4, tetrahedral	*S*-Thiolate, thioether, *N*-imidazole	Electron transfer in Type I blue copper proteins

Table 1.5 Metals in Biological Systems: Dioxygen Transport

Metal	Coordination Number, Geometry	Preferred Ligands	Functions and Examples
Copper, Cu^{2+} (d^9)	5, square pyramid 6, tetragonal	*O*-Carboxylate *N*-Imidazole	Type II copper oxidases, hydoxylases Type III copper hydroxylases, dioxygen transport in hemocyanin
Iron, Fe^{2+} (d^6)	6, octahedral	*N*-Imidazole, porphyrin	Dioxygen transport in hemoglobin and myoglobin

azurin and plastocyanin (Table 1.4); as facilitators of oxygen transport—that is, iron ions in hemoglobin or copper ions in hemocyanin (Table 1.5); and as sites at which enzyme catalysis occurs—that is, copper ions in superoxide dismutase or iron and molybdenum ions in nitrogenase (Table 1.6). Metal ions may serve multiple functions, depending on their location within the biological system, so that the classifications in the Tables 1.2 to 1.6 are somewhat arbitrary and/or overlapping.[4,5]

Table 1.6 Metals in Biological Systems: Enzyme Catalysis

Metal	Coordination Number, Geometry	Preferred Ligands	Functions and Examples
Copper, Cu^{2+} (d^9)	4, square planar	*O*-Carboxylate, *N*-imidazole	Type II copper in oxidases
Cobalt, Co^{2+} (d^7)	4, tetrahedral	*S*-Thiolate, thioether, *N*-imidazole	Alkyl group transfer, oxidases
Cobalt, Co^{3+} (d^6)	6, octahedral	*O*-Carboxylate, *N*-imidazole	Alkyl group transfer in vitamin B_{12} (cyanocobalamin)
Cobalt, Co^{2+} (d^7)	6, octahedral	*O*-Carboxylate, *N*-imidazole	Alkyl group transfer in Vitamin B_{12r}
Cobalt, Co^{+} (d^8)	6, octahedral, usually missing the 6th ligand	*O*-Carboxylate, *N*-imidazole	Alkyl group transfer in vitamin B_{12s}
Nickel, Ni^{2+} (d^8)	4, square planar	*S*-Thiolate, thioether, *N*-imidazole, polypyrrole	Hydrogenases, hydrolases
Nickel, Ni^{2+} (d^8)	6, octahedral		Uncommon
Molybdenum, Mo^{4+}(d^2), Mo^{5+}(d^1), Mo^{6+}(d^0)	6, octahedral	*O*-Oxide, carboxylate, phenolate, *S*-sulfide, thiolate	Nitrogen fixation in nitrogenases, oxo transfer in oxidases

1.4 INORGANIC CHEMISTRY BASICS

Ligand preference and possible coordination geometries of the metal center are important bioinorganic principles. Metal ligand preference is closely related to the hard–soft acid–base nature of metals and their preferred ligands. These are listed in Table 1.7.[6]

In general, hard metal cations form their most stable compounds with hard ligands, whereas soft metal cations form their most stable compounds with soft ligands. Hard cations can be thought of as small dense cores of positive charge, whereas hard ligands are usually the small highly electronegative elements or ligand atoms within a hard polyatomic ion—that is, oxygen ligands in $(RO)_2PO_2^-$ or in $CH_3CO_2^-$. Crown ethers are hard ligands that have cavities suitable for encapsulating hard metal ions. The [18]-crown-6 ether shown in Figure 1.2 with its 2.6- to 3.2-Å hole provides a good fit for the potassium ion, which has a radius of 2.88 Å.[6]

Table 1.7 Hard–Soft Acid–Base Classification of Metal Ions and Ligands

Metals, Ions, Molecules	Ligands
HARD	HARD
H^+, Na^+, K^+ Mg^{2+}, Ca^{2+}, Mn^{2+}, VO^{2+} Al^{3+}, Co^{3+}, Cr^{3+}, Ga^{3+}, Fe^{3+}, Tl^{3+}, Ln^{3+}, MoO^{3+} SO_3, CO_2	Oxygen ligands in H_2O, CO_3^{2-}, NO_3^-, PO_4^{3-} $ROPO_3^{2-}$, $(RO)_2PO_2^-$, CH_3COO^-, OH^-, RO^-, R_2O, and crown ethers Nitrogen ligands in NH_3, N_2H_4, RNH_2, Cl^-
INTERMEDIATE	INTERMEDIATE
Fe^{2+}, Ni^{2+}, Zn^{2+}, Co^{2+}, Cu^{2+}, Pb^{2+}, Sn^{2+}, Ru^{2+}, Au^{3+}, SO_2, NO^+	Br^-, SO_3^{2-}, nitrogen ligands in NO_2^-, N_3^-, N_2, HN N NH_2
SOFT	SOFT
Cu^+, Au^+, Tl^+, Ag^+, Hg_2^{2+} Pt^{2+}, Pb^{2+}, Hg^{2+}, Cd^{2+}, Pd^{2+} Pt^{4+}	Sulfur ligands in RSH, RS^-, R_2S, R_3P, RNC, CN^-, CO, R^-, H^-, I^-, $S_2O_3^{2-}$, $(RS)_2PO_2^-$, $(RO)_2P(O)S^-$

Source: Adapted from references 4 and 6.

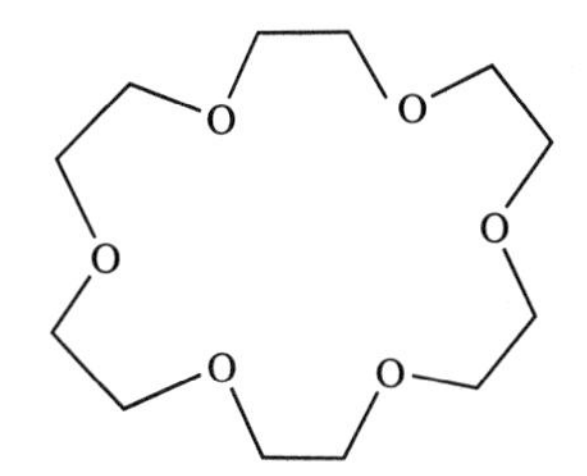

Figure 1.2 [18]-crown-6 ether.

It is possible to modify a hard nitrogen ligand toward an intermediate softness by increasing the polarizability of its substituents or the π electron cloud about it, an example being the imidazole nitrogen of the amino acid histidine. Increasing the softness of phosphate ion substituents can transform the hard oxygen ligand of $(RO)_2PO_2^-$ to a soft state in $(RS)_2PO_2^-$. Soft cations and anions are those with highly polarizable, large electron clouds—that is, Hg^{2+}, sulfur ligands as sulfides or thiolates, and iodide ions.

1.5 BIOLOGICAL METAL ION COMPLEXATION

1.5.1 Thermodynamics

The thermodynamic stability of metal ions are denoted by stepwise formation constants as shown in equations 1.1–1.3 (charges omitted for simplicity).

$$\mathrm{M + L \Leftrightarrow ML} \qquad K_1 = \frac{[\mathrm{ML}]}{[\mathrm{M}][\mathrm{L}]} \tag{1.1}$$

$$\mathrm{ML + L \Leftrightarrow ML_2} \qquad K_2 = \frac{[\mathrm{ML_2}]}{[\mathrm{ML}][\mathrm{L}]} \tag{1.2}$$

$$\mathrm{ML_2 + L \Leftrightarrow ML_3} \qquad K_3 = \frac{[\mathrm{ML_3}]}{[\mathrm{ML_2}][\mathrm{L}]} \tag{1.3}$$

Alternately, they are indicated by overall stability constants as shown in equations 1.4–1.6:

$$\mathrm{M + L \Leftrightarrow ML} \qquad \beta_1 = \frac{[\mathrm{ML}]}{[\mathrm{M}][\mathrm{L}]} \tag{1.4}$$

$$\mathrm{M + 2L \Leftrightarrow ML_2} \qquad \beta_2 = \frac{[\mathrm{ML}]}{[\mathrm{M}][\mathrm{L}]^2} \tag{1.5}$$

$$\mathrm{M + 3L \Leftrightarrow ML_3} \qquad \beta_3 = \frac{[\mathrm{ML}]}{[\mathrm{M}][\mathrm{L}]^3} \tag{1.6}$$

The equation relating the stepwise and overall stability constants is indicated by equation 1.7:

$$\beta_n = K_1 K_2 \ldots K_n \tag{1.7}$$

In biological systems, many factors affect metal–ligand complex formation. Hard–soft acid–base considerations have already been mentioned. Concentrations of the metal and ligand at the site of complexation are determined locally through concentration gradients, membrane permeability to metals and ligands, and other factors. Various competing equilibria—solubility products, complexation, and/or acid–base equilibrium constants—sometimes referred to as "metal ion speciation," all affect complex formation. Ion size and charge, preferred metal coordination geometry, and ligand chlelation effects all affect metal uptake. To better measure biological metal–ligand interactions, an "uptake factor" is defined as $K_{ML} \times [M]$, where K_{ML} is the stability constant K_1 and [M] is the concentration of metal ion.[3] Because naturally occurring aqueous systems have metal ion concentration varying roughly as

K^+, Na^+	Ca^{2+}, Mg^{2+}	Zn^{2+}	Cu^{2+}	Fe^{2+}
10^{-1} M	$\sim 10^{-3}$ M	$<10^{-9}$ M	$<10^{-12}$ M	$\sim 10^{-17}$ M

great selectivity for metal species is necessary to concentrate the necessary ions at sites where they are needed. Differentiating ligands are those preferred by the cation in question. A much more detailed discussion takes place in reference 3. Table 1.8 is adapted from this source.

1.5.2 Kinetics

In biological systems, as in all others, metal ions exist in an inner coordination sphere, an ordered array of ligands binding directly to the metal. Surrounding this is

Table 1.8 K_{ML} and $K_{ML} \times$ [M] for Some Cations and Their Differentiating Ligands

	K^+, Na^+	Ca^{2+}, Mg^{2+}	Zn^{2+}, Cu^{2+}	Differentiating Ligand
K^+, Na^+				*O*-Macrocycles such as crown ethers, cryptates, and naturally occurring macrocyclic antibiotics such as nonactin and valinomycin
K_{ML}	>10	$<10^2$	$<10^6$	
$K_{ML} \times$ [M]	>1.0	<0.1	<0.1	
Ca^{2+}, Mg^{2+}				Oxygen donors such as di- or tricarboxylates
K_{ML}	1.0	$<10^3$	$<10^6$	
$K_{ML} \times$ [M]	<0.1	>1.0	<0.1	
Zn^{2+}, Cu^{2+}				Nitrogen and sulfur ligands
K_{ML}	0.1	$<10^2$	$>10^6$	
$K_{ML} \times$ [M]	<0.1	<0.1	>1.0	

Source: Adapted from reference 3.

the outer coordination sphere consisting of other ligands, counterions, and solvent molecules. In stoichiometric mechanisms where one can distinguish an intermediate, substitution within the metals inner coordination sphere may take place through an associative (A), S_N2 process as shown in equations 1.8 (for six-coordinate complexes) and 1.9 (for four-coordinate complexes) or a dissociative (D), S_N1 mechanism as shown in equation 1.10 (RDS = rate determining step):

$$ML_6^{n+} + L' \xrightarrow{RDS} ML_6L'^{n+} \longrightarrow ML_5L'^{n+} + L \quad (1.8)$$

$$ML_4^{n+} + L' \xrightarrow{RDS} ML_4L'^{n+} \longrightarrow ML_3L'^{n+} + L \quad (1.9)$$

$$ML_6^{n+} \xrightarrow{RDS} ML_5L^{n+} + L' \xrightarrow{fast} ML_5L'^{n+} + L \quad (1.10)$$

Associative mechanisms for metals in octahedral fields are difficult stereochemically (due to ligand crowding); therefore, they are rare for all but the largest metal ion centers. The associative mechanism is well known and preferred for four-coordinate square-planar complexes. Pure dissociative mechanisms are rare as well. When an intermediate cannot be detected by kinetic, stereochemical, or product distribution studies, the so-called interchange mechanisms (I) are invoked. Associative interchange (I_A) mechanisms have rates dependent on the nature of the entering group, whereas dissociative interchange (I_D) mechanisms do not.

The simplest reactions to study, those of coordination complexes with solvent, are used to classify metal ions as labile or inert. Factors affecting metal ion lability include size, charge, electron configuration, and coordination number. Solvents can by classified as to their size, polarity, and the nature of the donor atom. Using the water exchange reaction for the aqua ion $[M(H_2O)_n]^{m+}$, metal ions are divided by Cotton, Wilkinson, and Gaus[7] into four classes:

Class I. Rate constants for water exchange exceed $10^8\ s^{-1}$, essentially diffusion-controlled. These are classified as the labile species.

Class II. Rate constants for water exchange are in the range 10^4–$10^8\ s^{-1}$.

Class III. Rate constants for water exchange are in the range 1–$10^4\ s^{-1}$.

Class IV. Rate constants for water exchange are in the range 10^{-3}–$10^{-6}\ s^{-1}$. These ions are classified as inert.

Labile species are usually main group metal ions with the exception of Cr^{2+} and Cu^{2+}, whose lability can be ascribed to Jahn–Teller effects. Transition metals of classes II and III are species with small ligand field stabilization energies, whereas the inert species have high ligand field stabilization energies (LFSE). Examples include Cr^{3+} ($3d^3$) and Co^{3+} ($3d^6$). Jahn–Teller effects and LFSE are discussed in Section 1.6. Table 1.9 reports rate constant values for some aqueous solvent exchange reactions.[8]

Outer-sphere (OS) reaction rates and rate laws can be defined for solvolysis of a given complex. Complex formation is defined as the reverse reaction—that is, replacement of solvent (S) by another ligand (L′). Following the arguments of

Table 1.9 Rate Constants for Water Exchange in Metal Aqua Ions

Class	Metal Ions	Rates $\log k\,(s^{-1})$
I	Group IA (1), Group IIA (2) except Be and Mg, Group IIB (12) except Zn^{2+} ($3d^{10}$), Cr^{2+} ($3d^4$), Cu^{2+} ($3d^9$)	8–9
II	Zn^{2+} ($3d^{10}$)	7.6
	Mn^{2+} ($3d^5$)	6.8
	Fe^{2+} ($3d^6$)	6.3
	Co^{2+} ($3d^7$)	5.7
	Ni^{2+} ($3d^8$)	4.3
III	Ga^{3+}	3.0
	Be^{2+}	2.0
	V^{2+} ($3d^3$)	2.0
	Al^{3+}	<0.1
IV	Cr^{3+} ($3d^3$), Co^{3+} ($3d^6$), Rh^{3+} ($3d^6$), Ir^{3+} ($3d^6$), Pt^{2+} ($3d^8$)	−3 to −6

Source: Adapted from references 7 and 8.

Tobe,[9] in aqueous solution the general rate law for complex formation (eliminating charge for simplicity),

$$[M(L_n)(S)] + L' \rightarrow [M(L_n)(L')] + S \tag{1.11}$$

takes the second-order form shown in equation 1.12:

$$-d\frac{[M(L_n)(S)]}{dt} = k'[M(L_n)(S)][L'] \tag{1.12}$$

The rate law frequently may be more complex and given as equation 1.13:

$$-d\frac{[M(L_n)(S)]}{dt} = \frac{k'K[M(L_n)(S)][L']}{(1 + K[L'])} \tag{1.13}$$

Equation 1.13 reduces to the second-order rate law, shown in equation 1.12, when $K[L'] \lll 1$ and to a first-order rate law, equation 1.14,

$$-d\frac{[M(L_n)(S)]}{dt} = k'[M(L_n)(S)] \tag{1.14}$$

when $K[L'] \ggg 1$.

Interchange mechanisms (I_A or I_D) in a preformed OS complex will generate the following observed rate laws (which cannot distinguish I_A from I_D) with the equilibrium constant $= K_{OS}$ (equation 1.15) and $k = k_i$ (equation 1.16):

$$[M(L_n)(S)] + L' \Leftrightarrow [M(L_n)(S)]\cdots L' \qquad K_{OS} \tag{1.15}$$

$$[M(L_n)(S)]\cdots L' \rightarrow [M(L_n)(L')]\cdots S \qquad k_i \tag{1.16}$$

The dissociative (D or S_N1) mechanism, for which the intermediate is long-lived enough to be detected, will yield equations 1.17 and 1.18, where $k = k_1$ and $K = k_2/(k_{-1}[S])$:

$$[M(L_n)(S)] \Leftrightarrow [M(L_n)] + S \qquad k_1 k_{-1} \tag{1.17}$$

$$[M(L_n)] + L' \rightarrow [M(L_n)(L')] \qquad k_2 \tag{1.18}$$

The associative (A or S_N2) will give the simple second-order rate law shown in equations 1.19 and 1.20 if the higher coordination number intermediate concentration remains small, resulting in the rate dependence shown in equation 1.21.

$$[M(L_n)(S)] + L' \Leftrightarrow [M(L_n)(S)(L')] \qquad k_a k_{-a} \tag{1.19}$$

$$[M(L_n)(S)(L')] \rightarrow [M(L_n)(L')] + S \qquad k_b \tag{1.20}$$

$$\frac{d[M(L_n)(S)]}{dt} = \frac{k_a\, k_b}{k_{-a} + k_b}[M(L_n)(S)][L'] \tag{1.21}$$

In all cases the key to assigning mechanism is the ability to detect and measure the equilibrium constant K. The equilibrium constant K_{OS} can be estimated through the Fuoss–Eigen equation,[10] as shown in equation 1.22. Usually, K_{OS} is ignored in the case of $L' = \text{solvent}$.

$$K_{OS} = \frac{4\pi N_A a^3}{3000}(e^{-V/kT}) \tag{1.22}$$

where a is the distance of closest approach of the oppositely charged ions (~5 Å), N_A is Avogadro's number, and V is the electrostatic potential at that distance (equation 1.23).

$$V = \frac{Z_1 Z_2 e^2}{4\pi\varepsilon_0\varepsilon_R a} \tag{1.23}$$

As the above discussion indicates, assigning mechanisms to simple anation reactions of transition metal complexes is not simple. The situation becomes even more difficult for a complex enzyme system containing a metal cofactor at an active site. Methods developed to study the kinetics of enzymatic reactions according to the Michaelis–Menten model will be discussed in Section 2.2.4.

1.6 ELECTRONIC AND GEOMETRIC STRUCTURES OF METALS IN BIOLOGICAL SYSTEMS

Tables 1.2–1.6 list some of the important geometries assumed by metal ions in biological systems. Common geometries adopted by transition metal ions that will

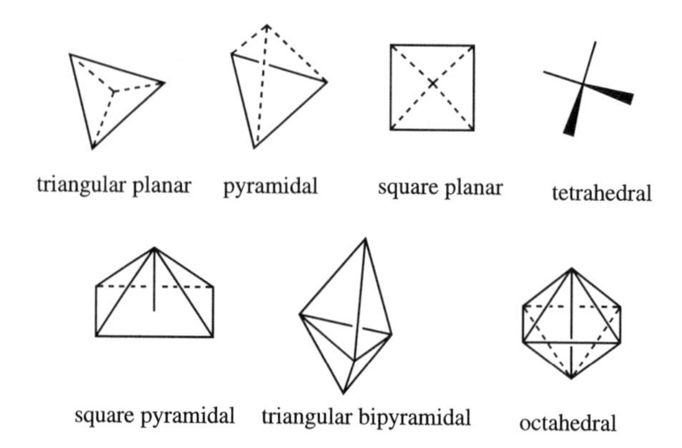

Figure 1.3 Common transition metal coordination geometries.

be of most concern to readers of this text are illustrated in Figure 1.3. In biological systems these geometries are usually distorted in both bond length and bond angle.

Transition metal ions play special roles in biological systems, with all elements from the first transition series except titanium (Ti) and scandium (Sc) occurring with great variety in thousands of diverse metalloproteins. Metals determine the geometry of enzymatic active sites, act as centers for enzyme reactivity, and act as biological oxidation–reduction facilitators. Molybdenum (Mo) appears to be the only transition element in the second transition series with a similar role. Vanadium (V), technetium (Tc), platinum (Pt), ruthenium (Ru), and gold (Au) compounds, as well as gadolinium (Gd) and other lanthanide complexes, are extremely important in medicinal chemistry as will be discussed in Chapter 7. Tables 1.2–1.6 list the *d*-electron configuration for transition metal ions common to biological systems. To find the number of *d* electrons for any transition metal ion, the following is a useful formula:

$$\text{Number of } d \text{ electrons} = \text{Atomic number for the element(Z)} - \text{oxidation state of the element's ion} - \text{Z for the preceding noble-gas element}^{4}$$

Examples:

$$\text{Fe(II)}: \quad 26 - 2 - 18 \text{ (argon)} = 6$$
$$\text{Mo(V)}: \quad 42 - 5 - 36 \text{ (krypton)} = 1$$

As a consequence of their partially filled *d* orbitals, transition metals exhibit variable oxidation states and a rich variety of coordination geometries and ligand spheres. Although a free metal ion would exhibit degenerate *d*-electron energy levels, ligand field theory describes the observed splitting of these *d*-electrons for metal ions in various ligand environments. In all cases the amount of stabilization or destabilization of *d*-electron energy levels centers about the so-called barycenter of unsplit *d*-electron energy levels. The most important of these for bioinorganic

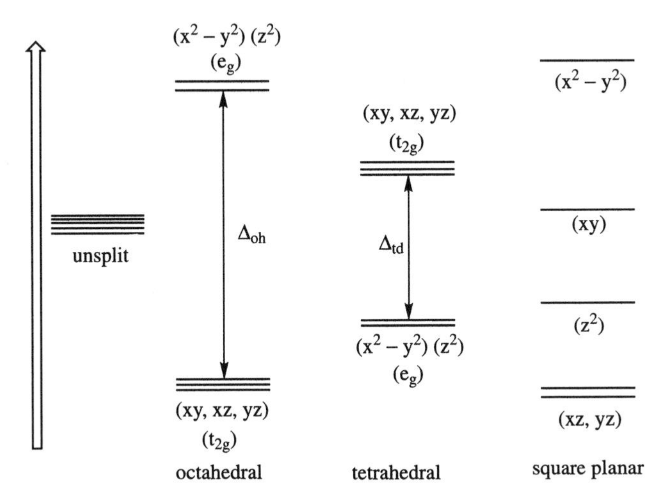

Figure 1.4 Approximate energy levels for *d* electrons in octahedral, tetrahedral, and square-planar fields.

applications are shown in Figure 1.4 for octahedral, tetrahedral, and square-planar ligand fields. The $t_{2g}(d_{xy}$, d_{yz}, and $d_{xz})$ and $e_g(d_{x^2-y^2}$ and $d_{z^2})$ energy level designations identify symmetry properties of the *d* orbitals and are often used to indicate the degenerate energy levels under discussion. (See LFSE discussion below.) Generally, the energy gap for tetrahedral fields is approximately one-half that for octahedral fields, and that for square-planar fields is approximately $1.2\Delta_{oh}$. Many thermodynamic and kinetic properties of transition metal coordination complexes can be predicted by knowing the magnitude of Δ. Measurement of ultraviolet and visible absorption spectra of transition metal complexes that arise from these quantum mechanically forbidden *d–d* transitions provide a measure of Δ.

To describe the *d*-orbital splitting effect for the octahedral field, one should imagine ligand spheres of electron density approaching along the *x*, *y*, and *z* axes, where the $d_{x^2-y^2}$ and d_{z^2} lobes of electron density point. Figure 1.5 illustrates representations of high-probability electron orbit surfaces for the five *d* orbitals.

For octahedral (O_h) geometry the repelling effect of like charge approach of the ligand electrons toward regions of high *d* electron density along the *x*, *y*, and *z* axes elevates the energy of the e_g ($d_{x^2-y^2}$ and d_{z^2}) orbitals while the t_{2g} (d_{xy}, d_{yz}, and d_{xz}) orbitals are proportionally lowered in energy. For the tetrahedral (T_d) case, ligands approach between the *x*, *y*, and *z* axes, thereby stabilizing ($d_{x^2-y^2}$ and d_{z^2}) and destabilizing d_{xy}, d_{yz}, and d_{xz} orbital energy levels. For the square-planar case, ligands will approach along the *x* and *y* axes. Distorted octahedral and tetrahedral geometries are quite common in biological systems. Square-planar geometries, less common, are found for d^8 transition metal ions, especially for gold(III), iridium(I), palladium(II), and platinum(II) and for nickel(II) species in strong ligand fields. The platinum anticancer agent, *cis*-dichlorodiammineplatinum(II), shown in

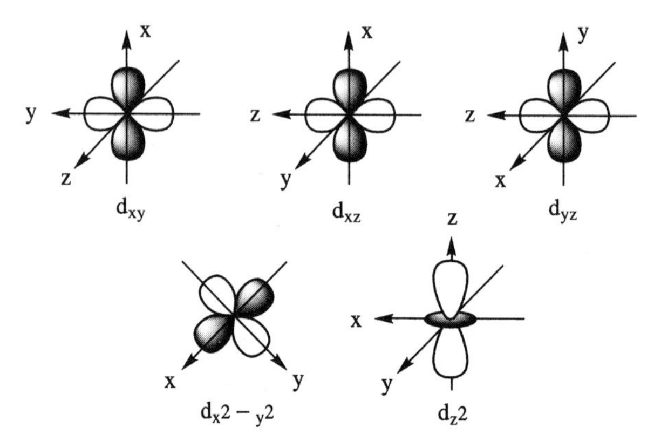

Figure 1.5 Representations of the five *d* orbitals along *x*, *y*, and *z* axes.

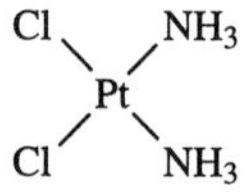

Cl NH$_3$
Pt
Cl NH$_3$

cis-dichlorodiammineplatinum(II)
cisplatin, cisDDP

Figure 1.6 The antitumor active platinum coumpound *cis*-dichlorodiammineplatinum(II).

Figure 1.6, has a square-planar geometry all important for its utilization as an antitumor agent. See Section 7.2.3 for further discussion of this drug molecule.

The strength of the ligand field at a metal center is strongly dependent on the character of the ligand's electronic field and leads to the classification of ligands according to a "spectrochemical series" arranged below in order from weak field (halides, sulfides, hydroxides) to strong field (cyanide and carbon monoxide):

$$\begin{aligned} &I^- < Br^- < S^{2-} < Cl^- < NO_3^- < OH^- \sim RCOO^- < H_2O \sim RS^- < NH_3 \\ &\sim \text{imidazole} < \text{en (ethylenediamine or diaminoethane)} \\ &< \text{bpy(2, 2}'\text{-bipyridine)} < CN^- < CO \end{aligned}$$

Ligand field strength may determine coordination geometry. For example, $NiCl_4^{2-}$ occurs as a tetrahedral complex (small splitting—small Δ_{td}), whereas $Ni(CN)_4^{2-}$ occurs in the square-planar geometry (large energy gap—large Δ_{sp}). In octahedral fields, ligand field strength can determine the magnetic properties of metal ions because for d^4 through d^7 electronic configurations both high-spin (maximum unpairing of electron spins) and low-spin (maximum pairing of electron spins) complexes are possible. Possible configurations are shown in Figure 1.7. In general, weak field ligands form high-spin complexes (small Δ_{oh}) and strong field ligands

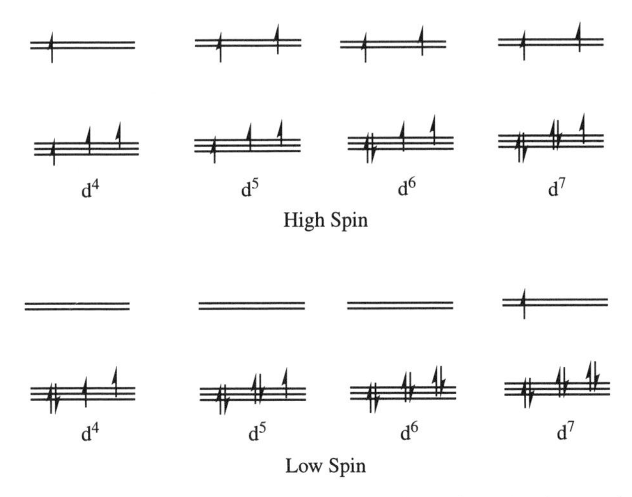

Figure 1.7 High-spin and low-spin *d*-electron configurations for the octahedral field.

form low-spin complexes (large Δ_{oh}). Detection of paramagnetism (unpaired electrons) and diamagnetism (all electrons paired) in bioinorganic ligand fields can help determine coordination geometry at active sites in enzymes. In the case of hemoglobin, for example, the d^6 iron(II) center cycles between high-spin and low-spin configurations affecting the placement of the iron center in or out of the plane of its porphyrin ligand. See Section 4.3 for further discussion. In Type III copper enzymes, two d^9 copper(II) centers become antiferromagnetically coupled resulting in a loss of the expected paramagnetism. See Sections 5.2.4 and 5.3.4.

The sum of the *d*-electron contributions to LFSE can be calculated with the formula shown in equation 1.24 for octahedral complexes:

$$\text{LFSE} = -\tfrac{2}{5}(\#e^{-}\text{in } t_{2g})\Delta_{oh} + \tfrac{3}{5}(\#e^{-}\text{in } e_g)\Delta_{oh} \tag{1.24}$$

where $\#e^-$ is the number of *d* electrons.

The 2/5 stabilization (negative energy values) and 3/5 destabilization (positive energy values) modifiers arise from the displacement of three *d* orbitals to lower energy versus two *d* orbitals to higher energy from the unsplit degenerate *d*-orbital state before imposition of the ligand field. Splitting values for *d*-orbital energy levels, based on $\Delta_{oh} = 10$, have been adapted from reference 7 and appear in Table 1.10.

The Jahn–Teller effect arises in cases where removal of degeneracy of a *d*-electron energy level is caused by partial occupation of a degenerate level. Two common examples are those of Cu(II) and Cr(II) as shown in Figure 1.8. Electrons in the e_g level could be placed in either the $d_{x^2-y^2}$ or d_{z^2} orbitals. Placing the odd electron in either orbital destroys the degeneracy of the e_g orbitals and usually has the effect of moving the ligands on one axis in or out. For Cu(II) complexes, this

Table 1.10 Splitting Values for *d* Orbitals in Common Geometries

C. N.[a]	Geometry	$d_{x^2-y^2}$	d_{z^2}	d_{xy}	d_{xz}	d_{yz}
4	Tetrahedral	−2.67	−2.67	1.78	1.78	1.78
4	Square planar[b]	12.28	−4.28	2.28	−5.14	−5.14
5	Square pyramidal[c]	9.14	0.86	−0.86	−4.57	−4.57
6	Octahedral	6.00	6.00	−4.00	−4.00	−4.00

[a] C. N. stands for coordination number.
[b] Bonds in *xy* plane.
[c] Pyramidal base in *xy* plane.

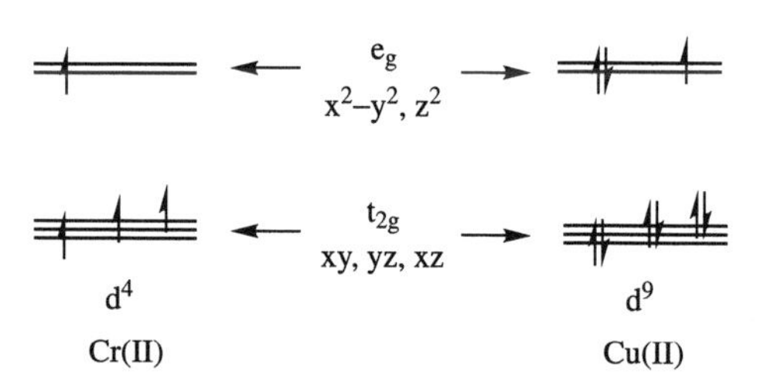

Figure 1.8 Electron configurations for Cr(II) and Cu(II).

effect is very common, resulting in longer bond lengths on what is usually taken as the complex's *z* axis. The effect is also seen for high-spin d^4 Mn(III) and for low-spin d^7 Co(II) and Ni(III).

1.7 BIOORGANOMETALLIC CHEMISTRY

The eighteen (18-e) and sixteen (16-e) electron rules for organometallic complexes may also be applied to bioinorganic systems. In this system, the valence electrons of transition metals are considered to be filling the 4*s*, 3*d*, 4*p* or 5*s*, 4*d*, 5*p* shells. The most stabilized filled shell is determined to be eighteen electrons—s^2, d^{10}, p^6, differing by the 10 electrons of the filled *d* shell from main group element compounds stabilized by electron octets. Compounds or complexes fulfill the 18-e rule by addition of metal valence electrons and electron contributions from ligands. Metal valence shell electrons may be counted as if the metal, in its 0, +1, +2 oxidation states, combine with ligand electrons counted according to Table 1.11. Many stable coordination complexes can be counted as having 16 electrons (16-e rule), especially those having square planar geometry and those bonded to aromatic rings through their π electronic systems. Two of these complexes, belonging to a group of compounds called metallocenes, bind to DNA and are antitumor agents. The anticancer agent cisplatin, *cis*-dichlorodiammineplatinum(II), also obeys the 16-e rule. These complexes will be discussed in Chapter 7. Several illustrations of these molecules are shown in Figure 1.9.

Table 1.11 Ligand Contributions to the 18-Electron Rule

Ligand	Number of Electrons
Hydrogen $H^{\bullet}$, chloride radical $Cl^{\bullet}$	1
Alkyl or acyl groups	1
Carbonyl group	2
Nitrosyl group, linear	3
Lewis bases Cl^-, O^{2-}, S^{2-}, NH_3, PR_3	2
Alkenes	2 per double bond
Benzene	6 per ring (π donation)
cyclopentadienyl (cp) $C_5H_5^{\bullet}$	5 per ring

Many clusters contain metal–metal bonds, and these are counted by contributing one electron to each metal connected. Some simple examples in Figure 1.9 illustrate application of the rules.

Iron–sulfur clusters, such as those discussed in Chapter 6, cannot be treated using the 16-e or 18-e rules. Other frameworks exist to treat large metal clusters, and these have some utility in treating $[Fe_4S_4]^{n+}$ clusters. One method treats the

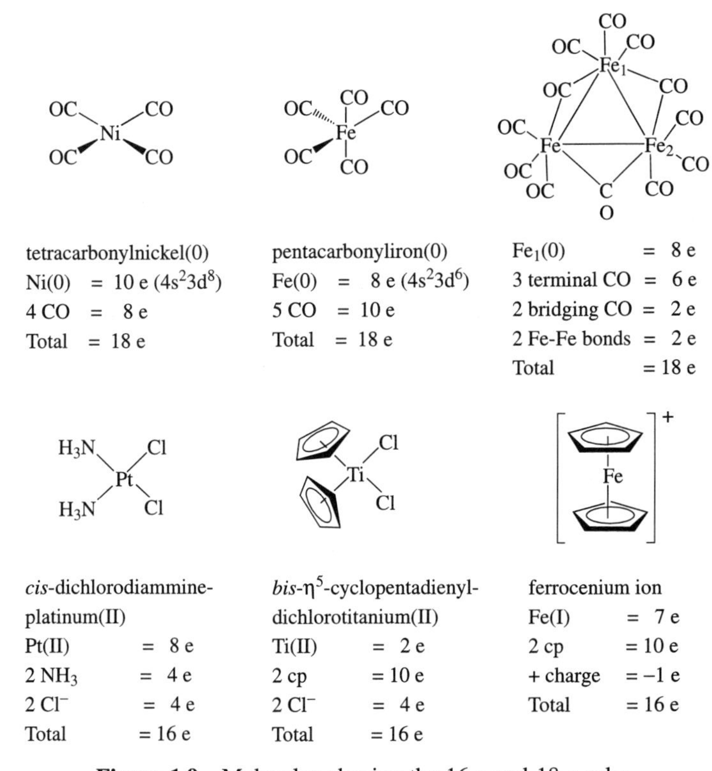

Figure 1.9 Molecules obeying the 16-e and 18-e rules.

number of metal atoms and the metal–metal bonds in a cluster according to the following formula[11]:

$$\sum \text{Valence electrons} = \text{no. of cluster Me atoms} \times 18 - \text{no. of metal–metal bonds} \times 2$$

Applying this formula to the cubane $[(Fe(II))_4(\eta^5\text{-}C_5H_5)_4(\mu_3\text{-}S)_4]$ shown in Figure 1.10A results in the following electron count:

$$\sum \text{Valence electrons} = 4 \times 18 - 6 \times 2 = 60 \text{ electrons}$$

the so-called "magic number" for four metal atoms in a cluster.

If one applies the same procedure to Figure 1.10B, an iron–sulfur cluster often used as a model for those in biological systems, the same magic number of 60 would be obtained. Cluster magic numbers would occur as 48 e for a triangular clusters, 60 e for tetrahedral, 72 e for trigonal bipyramidal, 74 e for square pyramidal, 86 e for octahedral, 90 e for trigonal prisms, and 120 e for cubic structures.

For biological systems such as ferredoxins, problems arise when counting electrons by the valence electron method. This system assumes six Fe–Fe bonds within the tetrahedral iron–sulfur clusters, but Fe–Fe bond distances within biological iron sulfur clusters as found by X-ray crystallography do not often indicate Fe–Fe bonds. As discussed in reference 11, it is known that for the Fe_4S_4 cubane found in biological systems, oxidations are accompanied by increasing distortion of the cubane frame. Also, ^{57}Fe Mössbauer spectra indicate that the four iron atoms remain equivalent, suggesting delocalization within the Fe–S framework. Nitrogenase iron–sulfur clusters (discussed in Chapter 6) deviate substantially from the rules for $[Fe_4S_4]$ clusters discussed here.

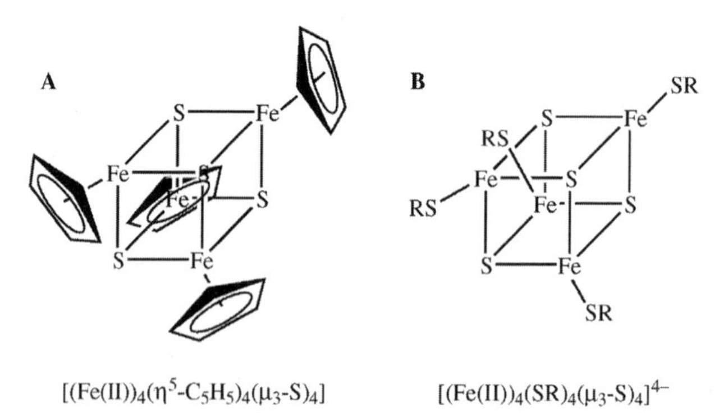

Figure 1.10 Cubanes (A) $[(Fe(II))_4(\eta^5\text{-}C_5H_5)_4(\mu_3\text{-}S)_4]$ and (B) $[(Fe(II))_4(SR)_4(\mu_3\text{-}S)_4]^{4-}$.

1.8 ELECTRON TRANSFER

Many reactions catalyzed by metalloenzymes involve electron transfer. On the simplest level, one can consider electron transfer reactions to be complementary when there are equal numbers of oxidants and reductants and the metals transfer equal numbers of electrons as shown in equation 1.25:

$$\mathrm{Fe(III)(CN)_6^{3-} + Ru(II)(NH_3)_6^{2+} \rightarrow Fe(II)(CN)_6^{4-} + Ru(III)(NH_3)_6^{3+}} \quad (1.25)$$

Noncomplementary reactions, as shown in equation 1.26, involve unequal numbers of oxidants and reductants because the number of electrons gained or lost by each metal differs.[6] Noncomplementary reactions, especially for large biomolecules, must proceed by a number of bimolecular steps because the possibility of termolecular or higher-order collisions is very small.

$$\mathrm{Mn^{7+} + 5Fe^{2+} \rightarrow Mn^{2+} + 5Fe^{3+}} \quad (1.26)$$

Two types of electron transfer mechanisms are defined for transition metal species. Outer-sphere electron transfer occurs when the outer, or solvent, coordination spheres of the metal centers is involved in transferring electrons. No reorganization of the inner coordination sphere of either reactant takes place during electron transfer. A reaction example is depicted in equation 1.27:

$$\mathrm{Fe(II)(CN)_6^{4-} + Rh(IV)Cl_6^{2-} \rightarrow Fe(III)(CN)_6^{3-} + Rh(III)Cl_6^{3-}} \quad (1.27)$$

Inner-sphere electron transfers involve the inner coordination sphere of the metal complexes and usually take place through a bridging ligand. The classic example, typical of those studied and explained by H. Taube,[12] is illustrated by Figure 1.11's

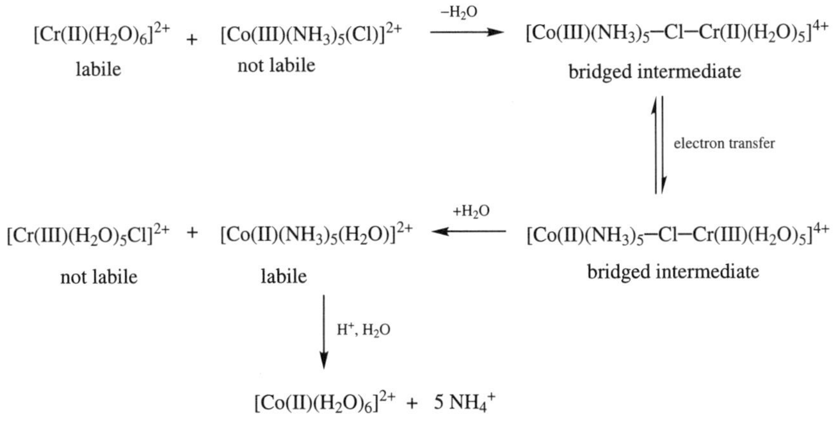

$[Cr(II)(H_2O)_6]^{2+}$ (labile) + $[Co(III)(NH_3)_5(Cl)]^{2+}$ (not labile) $\xrightarrow{-H_2O}$ $[Co(III)(NH_3)_5\text{–Cl–}Cr(II)(H_2O)_5]^{4+}$ (bridged intermediate)

$\rightleftarrows$ electron transfer

$[Co(II)(NH_3)_5\text{–Cl–}Cr(III)(H_2O)_5]^{4+}$ (bridged intermediate) $\xrightarrow{+H_2O}$ $[Cr(III)(H_2O)_5Cl]^{2+}$ (not labile) + $[Co(II)(NH_3)_5(H_2O)]^{2+}$ (labile)

$\downarrow$ H^+, H_2O

$[Co(II)(H_2O)_6]^{2+}$ + 5 NH_4^+

Figure 1.11 An inner-sphere electron transfer reaction sequence. (Adapted from reference 7.)

reaction sequence adapted from reference 7. In this reaction sequence, production of $[Cr(III)(H_2O)_5Cl]^{2+}$ implies that electron transfer through the bridged intermediate from Cr(II) to Co(III) and Cl^- transfer from Co to Cr are mutually interdependent acts.

Harry B. Gray and Walther Ellis,[13] writing in Chapter 6 of reference 13, describe three types of oxidation–reduction centers found in biological systems. The first of these, protein side chains, may undergo oxidation–reduction reactions such as the transformation of two cysteine residues to form the cystine dimer as shown in equation 1.28:

$$2R{-}SH \rightarrow R{-}S{-}S{-}R \qquad (1.28)$$

The second type of biological electron transfer involves a variety of small molecules, both organic and inorganic. Examples of these are (a) nicotinamide adenine dinucleotide (NAD) and nicotinamide adenine dinucleotide phosphate (NADP) as two electron carriers and (b) quinones and flavin mononucleotide (FMN), which may transfer one or two electrons. The structure of NAD and its reduced counterpart NADH are shown in Figure 1.12.

The third type of biological electron transfer involves metalloproteins themselves. These may be electron carriers (i.e., azurin) or proteins involved in the transport or activation of small molecules (i.e., nitrogenase). These so-called electron transferases have some or all of the following characteristics: (1) a suitable cofactor, such as NAD^+/NADH, acting as an electron source or sink; (2) geometry that allows the cofactor close enough to the protein surface for the transfer of electrons; (3) a hydrophobic shell on the protein surface around or near the cofactor; and (4) architecture that permits changes in protein conformation to facilitate electron transfer. These last changes should be small.[14] Electron transferases that will be discussed in this text include the blue copper proteins such as azurin and plastocyanin (Chapter 5) and nitrogenase, one of the iron–sulfur proteins

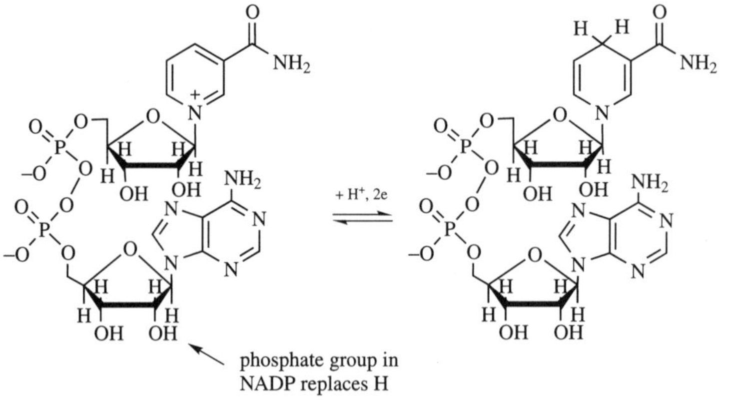

Figure 1.12 Electron transfer cofactors NAD^+ or $NADP^+$.

(Chapter 6). Other iron–sulfur proteins, so named because they contain iron sulfur clusters of various sizes, include the rubredoxins and ferredoxins. Rubredoxins are found in anaerobic bacteria and contain iron ligated to four cysteine sulfurs. Ferredoxins are found in plant chloroplasts and mammalian tissue and contain spin-coupled [2Fe–2S] clusters. Cytochromes comprise several large classes of electron transfer metalloproteins widespread in nature. At least four cytochromes are involved in the mitochondrial electron transfer chain, which reduces oxygen to water according to equation 1.29. Further discussion of these proteins can be found in Chapters 6 and 7 of reference 13.

$$\tfrac{1}{2}O_2 + NADH + H^+ \rightarrow H_2O + NAD^+ \tag{1.29}$$

The simplest electron transfer reactions are outer sphere. The Franck–Condon principle states that during an electronic transition, electronic motion is so rapid that the metal nuclei, the metal ligands, and solvent molecules do not have time to move. In a self-exchange example,

$$A_{ox^*} + A_{red} \rightarrow A_{red^*} + A_{ox} \tag{1.30}$$

the energies of donor and acceptor orbitals as well as bond lengths and bond angles remain the same during efficient electron transfer. In a cross reaction between two different species, one can write the following set of equilibrium statements (K) and rate equations (k_{et}):

$$A_{ox} + B_{red} \Leftrightarrow [A_{ox}B_{red}]\text{(precursor complex)} \tag{1.31}$$

$$[A_{ox}B_{red}] \xrightarrow{k_{et}} [A_{red}B_{ox}]\text{(successor complex)} \tag{1.32}$$

$$[A_{red}B_{ox}] \xrightarrow{\text{fast}} A_{red} + B_{ox} \tag{1.33}$$

Electron transfer theory is further explained by Marcus using potential energy diagrams to describe electron transfer processes.[15] In the diagrams such as shown in Figure 1.13, electron donors and acceptors behave as collections of harmonic oscillators. The diagram expresses donor and acceptor in a single surface representing the precursor complex and one representing the successor complex. Point S represents the activated complex, and E_R and E_P are the reactant and product surfaces, respectively.

It is beyond the scope of this text to continue the discussion of Marcus theory. Qualitatively, the student should understand that electrons must find a path through the protein from the donor species to the acceptor. This may take place through bonds as outlined above or through electron tunneling events in which electrons travel through space between orbitals of the donor species to the acceptor species. Chapter 6 of reference 13 presents a clear explanation for further reading.

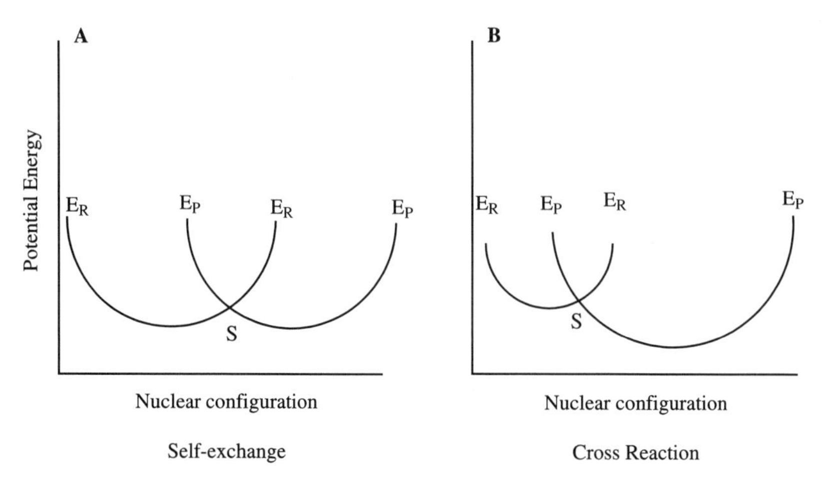

Figure 1.13 Potential energy diagrams describing electron transfer processes according to Marcus theory. (A) Self–exchange (B) Cross Reaction.

1.9 CONCLUSIONS

The preceding brief review of inorganic chemistry has been oriented toward questions that will arise in the following discussion of several bioinorganic systems. The inorganic and bioinorganic chemistry texts referenced in this chapter are good sources for answering the additional questions sure to arise in studying the behavior of metals in biological systems. It is important to keep in mind that metal behavior in the biological milieu will be influenced greatly by the surroundings. Metal–ligand systems existing in thermodynamic equilibrium and slow to react to changing cellular or noncellular dynamics will not long endure. Therefore, most of the metalloenzyme systems to be described in the following chapters contain metals in distorted and changeable ligand fields. These systems will continue to challenge the ingenuity of inorganic and bioinorganic chemists attempting to understand, modify, model, or design synthetic substitutes for them.

REFERENCES

1. Schwarz, K. *Ged. Proc.*, 1974, **33**, 1748–1757.
2. Frieden, E. *J. Chem. Ed.*, 1985, **62**(11), 917–923.
3. Frausto da Silva, J. J. R.; Williams, R. J. P. *The Biological Chemistry of the Elements: The Inorganic Chemistry of Life*, Clarendon Press, New York, 1991.
4. Lippard, S. J.; Berg, J. M. *Principles of Bioinorganic Chemistry*, University Science Books, Mill Valley, CA, 1994.

5. Hay, R. W. *Bio-Inorganic Chemistry*, Ellis Horwood Limited, Halsted Press, New York, 1984.
6. Cowan, J. A. *Inorganic Biochemistry, An Introduction*, 2nd ed., Wiley-VCH, New York, 1997.
7. Cotton, F. A.; Wilkinson, G.; Gaus, P. L. *Basic Inorganic Chemistry*, 3rd ed., John Wiley & Sons, New York, 1995, pp 192–194.
8. (a) Eigen, M. *Pure Appl. Chem.* 1963, **6**, 105. (b) Bennetto, H. P.; Caldin, E. F. *J. Chem. Soc. A*, 1971, 2198.
9. Tobe, M. L. Substitution Reactions, in *Comprehensive Coordination Chemistry*, Wilkinson, G., ed., Pergamon Press, Oxford, 1987, pp. 281–329.
10. Shriver, D. F.; Atkins, P. W.; Langford, C. H. *Inorganic Chemistry*, Oxford University Press, Oxford, 1990, pp. 477–478.
11. Elschenbroich, C. *Organometallics: A Concise Introduction*, VCH, New York, 1992.
12. Taube, H. *Electron Transfer Reactions of Complex Ions in Solution*, Academic Press, New York, 1970.
13. Gray, H. B.; Ellis, W. R., in Bertini, I.; Gray, H. B.; Lippard, S. J.; Valentine, J. S. *Bioinorganic Chemistry*, University Science Books, Mill Valley, CA, 1994, pp. 315–363.
14. Adman, E. T. *Biochim. Biophys. Acta*, 1979, **549**, 107–144.
15. Marcus, R. A. *Annu. Rev. Phys. Chem.*, 1964, **15**, 155–196.

2

BIOCHEMISTRY FUNDAMENTALS

2.1 INTRODUCTION

Biochemistry concerns itself with the study of life at the molecular level. The subject area encompasses a huge body of information that grows and changes rapidly in the hands of thousands of capable teachers, researchers, and writers. An introduction to the biochemistry topics of proteins, protein kinetics, nucleic acids, and genetics will be presented in this chapter. In addition, a discussion of genomics and proteomics introduces topics of current relevance. The last section of the chapter discusses zinc-finger proteins, a topic that combines information introduced in this chapter on proteins, protein analysis, DNA, DNA binding, and cloning. Other important biochemical topic areas such as bioenergetics, saccharides and polysaccharides, lipids and lipoproteins, membrane compositions and dynamics, vitamins, hormones, biochemical pathways, and many others are not covered here. Students are referred to biochemistry texts[1] and, for the most up-to-date information, to the primary literature and to the internet.

In introductory biochemistry, one becomes familiar with amino acids (aa) and how they combine (polymerize) to become peptides and proteins. Proteins fold into three-dimensional shapes and become enzymes, the catalysts of biochemical reactions, or structural materials in muscles and tendons, or small molecule carriers such as the dioxygen-carrying metalloproteins myoglobin and hemoglobin. At least one-third of all proteins and enzymes contain metals and are known as metalloproteins and metalloenzymes. Some of these will be the bioinorganic chemistry topics treated in Chapters 4, 5, and 6 of this text. *Proteomics*, a recently coined term, refers to the study of a large collection of proteins that occur and function together in a particular biochemical entity such as a cell or other organelle. Important interactions among proteins in a proteome distinguish the proteome's behavior as being more complex than that of a single protein studied individually.

A second major area of biochemical importance concerns study of nucleotide polymerization to produce ribonucleic acids (RNA) and deoxyribonucleic acids (DNA). Genes, the basis for inherited characteristics, are contained in DNA double helical sections incorporated into coiled and supercoiled DNA structures. Genomics, the study of the total genetic assemblage of any species, is now a well-known topic to all, especially with the announcement of the sequencing of the human genome in 2001. A basic understanding of the biochemical processes undertaken by proteins and nucleic acids is necessary for students, teachers, and researchers wishing to understand and replicate the catalytic activity of metalloenzymes or to design compounds (many of them containing inorganic atoms) that will be able to detect, diagnose, and treat disease.

2.2 PROTEINS

2.2.1 Amino Acid Building Blocks

Polypeptides are formed through the polymerization of any combination of the 20 naturally occurring amino acids (aa). In humans, 10 of these amino acids are essential (cannot be synthesized by the body and must be ingested in the diet). The 10 essential amino acids (see Figures 2.1–2.4 for full names and structures) are: arg, his, ile, leu, lys, met, phe, thr, trp, and val. Relatively short amino acid chains, called polypeptides, have important hormonal (control) functions in biological species. Proteins are classified as polypeptide chains exceeding 50 amino acids in length, whereas enzyme molecules usually contain more than 100 amino acid residues. Amino acids contain a central carbon, called the α carbon, to which four substituent groups are attached: the amine group ($—NH_2$), a carboxylic acid group (—COOH), a hydrogen atom (—H), and a side chain (—R) group unique to each amino acid. The structure of these basic building blocks are grouped in Figures 2.1 to 2.4 to indicate the neutral, polar, acidic, or basic characteristics of their R groups. The three- and one-letter common abbreviations for amino acids are shown in the figures. The amino acids are shown in their zwitterion form at pH 7 in which they have their COOH group ($pK_a = 2.35$) in the COO^- form and their NH_2 group ($pK_a = 9.69$) in the NH_3^+ form.[2] Side-chain R groups will be protonated or deprotonated based on the pK value of the side chain. All amino acids found in proteins are called α-amino acids because the amine group is bonded to the α carbon. All amino acids found in proteins are L-stereoisomers with respect to the α carbon (except glycine whose R group is H), although D-stereoisomers are found in bacterial cell walls and some peptide antibiotics.

In nomenclature for amino acids as ligands, the atoms in the R group are labeled with Greek letters starting with β for the first atom attached to the α carbon followed by γ and δ and ε. The common bioinorganic ligands histidine, cysteine, and aspartic acid have their atoms labeled in the manner shown in Figure 2.5. The

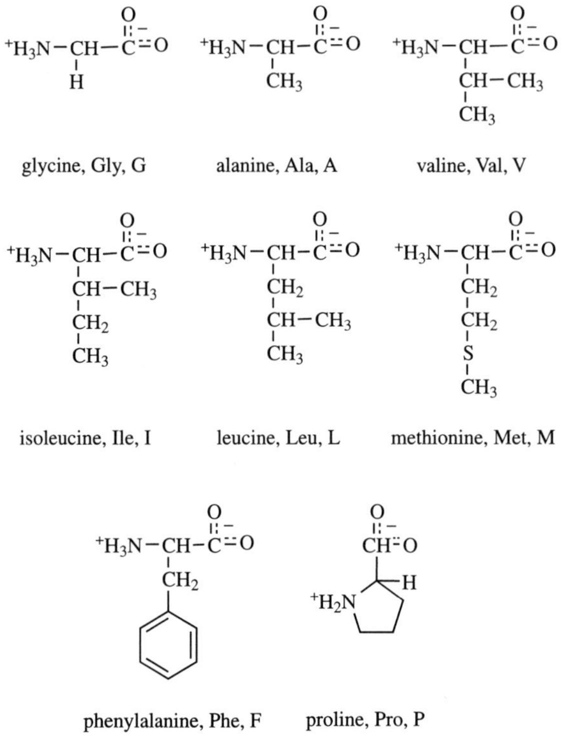

glycine, Gly, G
alanine, Ala, A
valine, Val, V
isoleucine, Ile, I
leucine, Leu, L
methionine, Met, M
phenylalanine, Phe, F
proline, Pro, P

Figure 2.1 Zwitterions of nonpolar hydrophobic amino acids at physiological pH.

asparagine, Asn, N
tryptophan, Trp, W
glutamine, Gln, Q
Serine, Ser, S
threonine, Thr, T
cysteine, cys, C

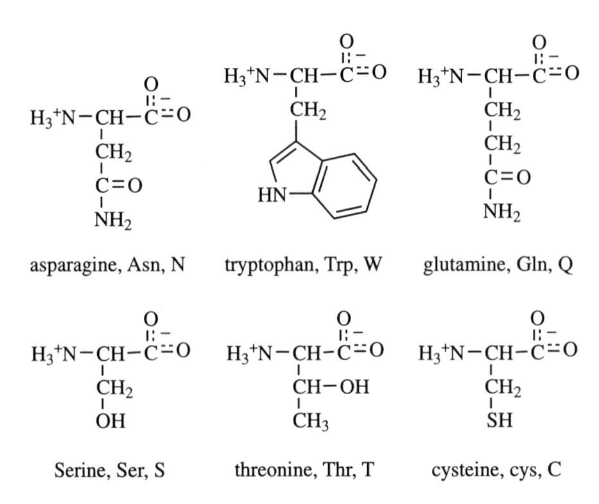

Figure 2.2 Zwitterions of polar neutral amino acids at physiological pH.

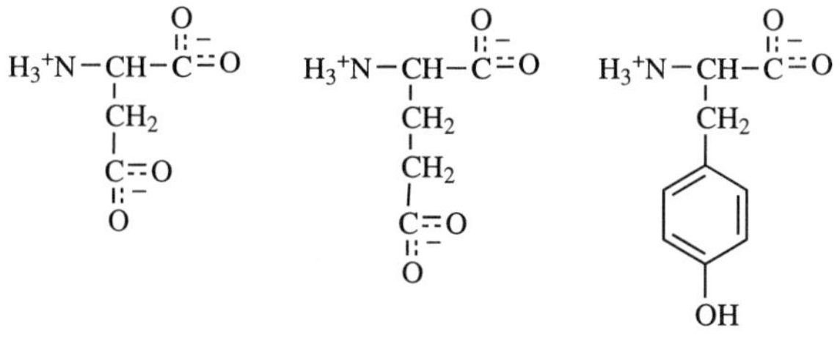

aspartic acid, Asp, D | glutamic acid, Glu, E | tyrosine, Tyr, Y

Figure 2.3 Zwitterions of acidic amino acids at physiological pH.

histidine, His, H | arginine, Arg, R | lysine, Lys, K

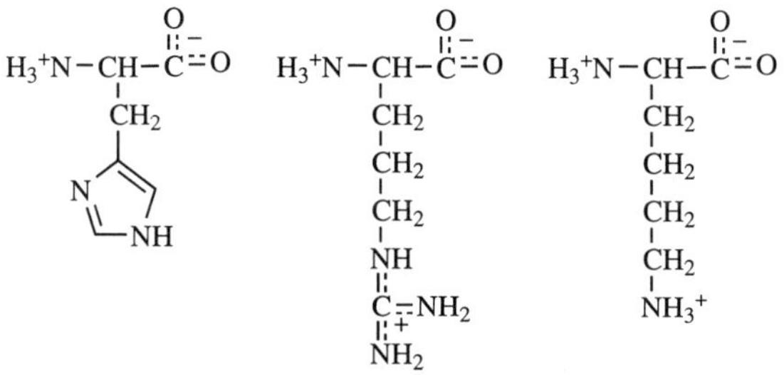

Figure 2.4 Zwitterions of basic amino acids at physiological pH.

labeling becomes important in identifying the ligand atom, the possible metal binding site, in metalloproteins. In histidine, metals bind at either the δ or ε positions. In superoxide dismutase, his bridges the metal ions with Zn(II) coordinated at the δ nitrogen and Cu(II) at the ε nitrogen. In aspartic acid, metals may bind in monodentate mode (Asp-M), in bidentate chelate mode as shown in Figure 2.5, or in bidentate bridging mode (Asp-M_2).

2.2.2 Protein Structure

All proteins have at least three levels of structure: primary, secondary, and tertiary. Proteins with more than one polypeptide chain—hemoglobin and nitrogenase are examples—also possess quaternary structure. The primary, secondary, tertiary, and quaternary structure of proteins control their three-dimensional shape, which, in turn, affects their activity. The primary structure of proteins is formed by a condensation reaction forming a peptide bond as shown in Figure 2.6. In this example the zwitterions of cysteine and isoleucine combine to form the peptide bond via a condensation reaction between the α-carboxyl group of one amino acid and the α-amino group of another amino acid. A water molecule is eliminated as each successive peptide bond is formed. The peptide bond is rigidly planar and has a bond length intermediate between a single and a double bond. This rigid bond

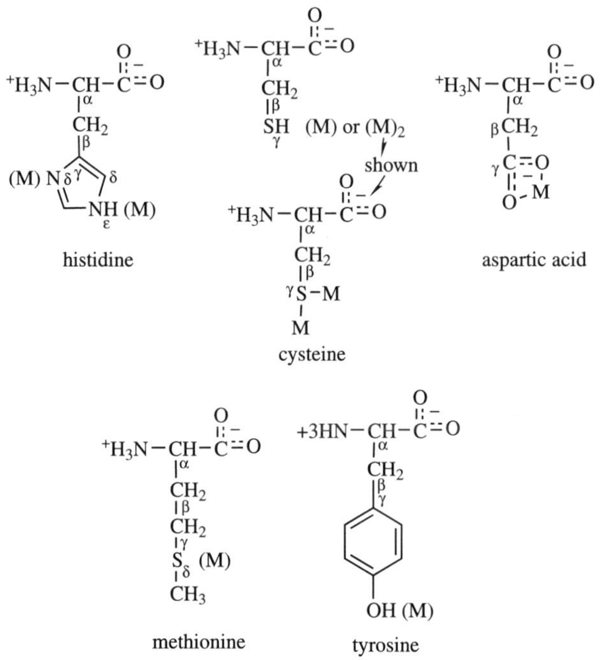

Figure 2.5 Common metal ion bonding modes to amino acid residues in proteins.

forms the backbone of the protein. Other single bonds in a polypeptide chain are flexible and can and do rotate.

The N-terminal end of a polypeptide chain, with its free amino group, is conventionally known as beginning of the chain, while the last amino acid, with free carboxyl group, is the C-terminal end of the chain (see Figure 2.7).

A protein's secondary structure arises from the formation of intra- and intermolecular hydrogen bonds. All carboxyl group oxygens and amine hydrogens of a polypeptide participate in H-bonding. Protein secondary structure also derives from the fact that although all C–N bonds in peptides have some double bond character and cannot rotate, rotation about the C_α–N and C_α–C bonds is possible and is

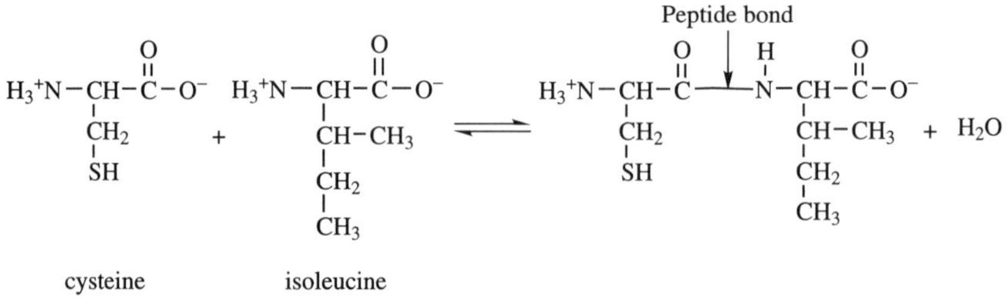

Figure 2.6 Formation of a peptide bond.

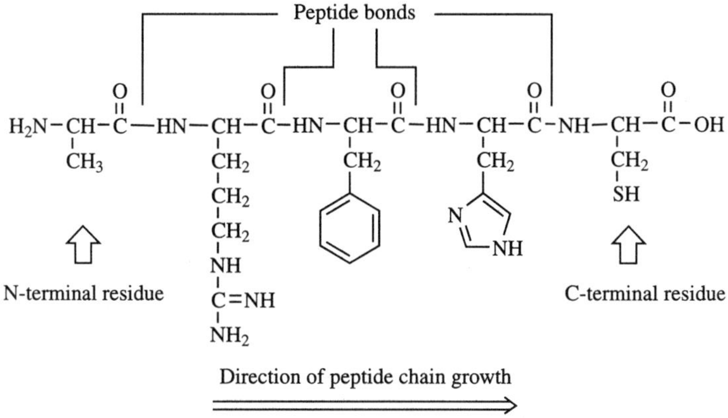

Peptide Nomenclature
alanylarginylphenylalanylhistidylcysteine

Figure 2.7 A peptide of four amino acids.

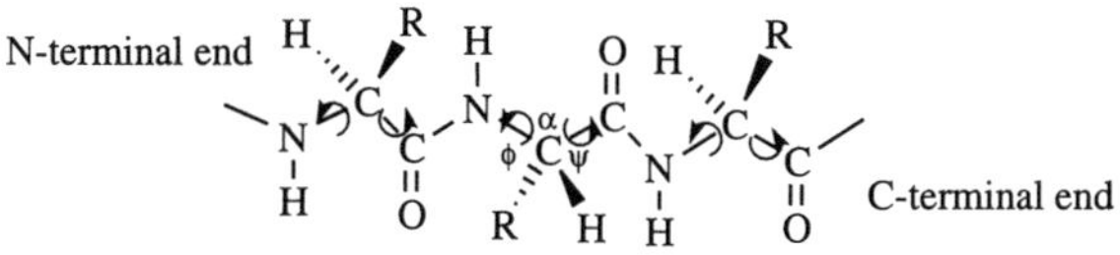

Figure 2.8 Illustration of C_α–N bond angle ϕ (phi) and C_α–C bond angle ψ (psi) in a peptide.

affected by peptide R groups that have differing space and charge constraints. Figure 2.8 indicates the C_α–N bond angle ϕ (phi) and C_α–C bond angle Ψ (psi) about which rotation is possible.

Two types of protein secondary structure are the α-helix and β-pleated sheet, and their representations can be seen in Figure 2.9. The right-handed α-helix occurs in globular proteins and is formed by intramolecular hydrogen bonds between the carboxyl group oxygen of one amino acid and the amine hydrogen of the fourth amino acid away from it ($i + 4$ in Table 2.1). This helix completes one turn every 3.6 residues and rises approximately 5.4 Å with each turn. Other known types of protein helices are the π-helix and the 3–10 helix. The left-handed α-helix is known but not found in protein structures. As can be seen in Table 2.1, these helical structures have differing Ψ and ϕ angles and different H-bonding patterns.[3]

The β-pleated sheet structure occurs in fibrous as well as globular proteins and is formed by intermolecular hydrogen bonds between a carboxyl group oxygen of one amino acid and an amine hydrogen of an adjacent polypeptide chain. Parallel β-pleated sheets form when the adjacent polypeptide chains are oriented in one direction (from N-terminal to C-terminal end or vice versa). Antiparallel β-pleated

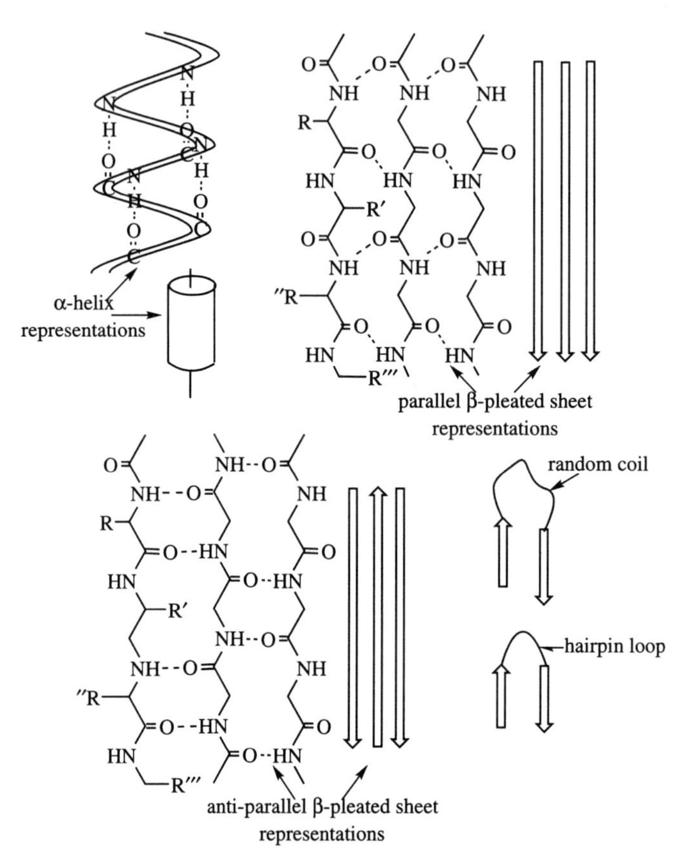

Figure 2.9 Protein secondary structures (many R groups omitted for clarity). (Adapted with permission from Figure 1.19 of Cowan, J. A. *Inorganic Biochemistry, An Introduction*, 2nd ed., Wiley-VCH, New York, 1997. Copyright 1997, Wiley-VCH.)

sheets form when the adjacent polypeptide chains are traveling in opposing directions (one chain N-terminal to C-terminal and the other C-terminal to N-terminal). Representations are shown in Figure 2.9. The secondary structure motifs of α-helix and β-sheet are joined by unstructured areas called loops or coils, also shown in Figure 2.9.

Table 2.1 Characteristics of Protein Helical Structures

Helix Type	Phi, ϕ (deg)	Psi, ψ (deg)	H-Bond Pattern
Right-handed α-helix	−57	−47	i + 4
π-helix	−57	−70	i + 5
3–10 helix	−49	−26	i + 3

The tertiary structure of proteins arises from the interactions of the various R groups along the polypeptide chain. Some of the forces responsible for the tertiary structures include van der Waals forces, ionic bonds, hydrophobic bonds, and hydrogen bonds. Usually, hydrophilic R groups arrange themselves on the exterior of the tertiary structure so as to interact with the aqueous environment, whereas hydrophobic R groups usually orient themselves on the interior of the protein's tertiary structure so as to exclude water. One ionic bond often found in establishing tertiary structure is a disulfide bond between two cysteine side-chain groups—for instance, in the copper enzyme azurin. The image in Figure 2.10 illustrates tertiary protein structure. This visualization, a Wavefunction, Inc. Spartan '02 for Windows™ ribbon diagram, represents X-ray crystallographic structural data taken from the protein data bank (PDB) for the copper enzyme azurin, a plant electron transport protein. The PDB designation for this molecule is 1JOI.[4] The PDB offers structural data on X-ray crystallographic and NMR solution structures of proteins. (See Sections 3.13.1 and 3.13.2 for further information on the PDB and visualization software, and Section 5.2.1 for more information on azurin). In Figure 2.10 one can see an α-helical section of protein tertiary structure near the top center. Representations of β-pleated sheet are strip-like areas. The β-pleated sheet regions define a barrel-type tertiary structure for azurin. Azurin and other similar copper-containing enzymes described in Chapter 5 adopt this so-called β-barrel tertiary structure. The copper(II) ion present in azurin is shown in space-fill

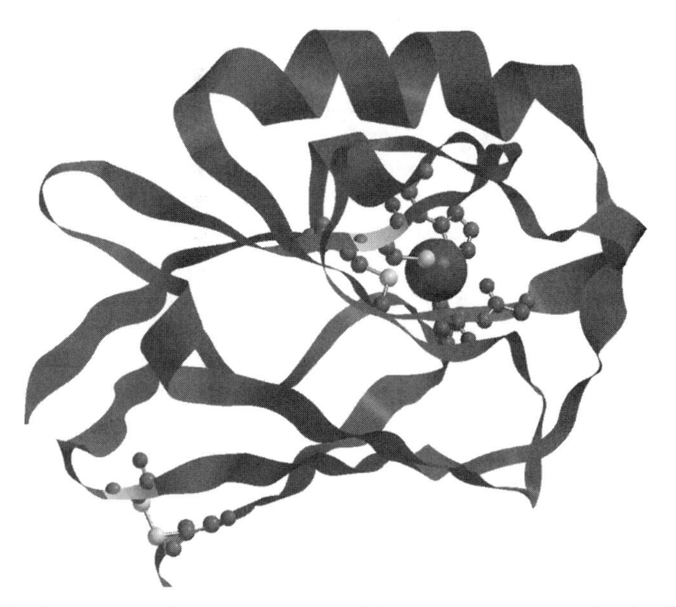

Figure 2.10 Secondary and tertiary structure of the copper enzyme azurin visualized using Wavefunction, Inc. Spartan '02 for Windows™ from PDB data deposited as 1JOI. See text for visualization details. Printed with permission of Wavefunction, Inc., Irvine, CA. (See color plate.)

form, while the copper ligand aa residues are shown in ball-and-spoke form. The copper ligands include two histidine residues: (a) his46 (partially hidden) at the lower right in coordination shpere, and (b) his117 at the upper right. Both his ligands attach to the copper(II) ion through the δ nitrogen with Cu–N bond lengths of approximately 2.0 Å. The cys112 ligand at the upper left forms a Cu–S bond of 2.2 Å in length. These three ligands define a slightly distorted CuN_2S trigonal planar coordination sphere.[5] Additionally met121 to the left of the copper ion, at a distance of >3.0 Å, and gly45 (between his46 and his117), form weak axial interactions with the copper ion so that the overall metal-ligand geometry approximates a trigonal bipyramid. Also note the disulfide bridge (approximately 3.3 Å long) between the ball-and-spoke representations of cys3 and cys26 at the lower left of the barrel-shaped protein tertiary structure. By convention the amino acid residues are numbered from one beginning at the N-terminal end of the protein so that the N-terminal end is beyond cys3 at the lower left edge of Figure 2.10.

Most proteins contain more than one polypeptide chain. The manner in which these chains associate determines quaternary structure. Binding involves the same types of noncovalent forces mentioned for tertiary structure: van der Waals forces, hydrophobic and hydrophilic attractions, and hydrogen bonding. However, the interactions are now interchain rather than intrachain (tertiary structure determination). The quaternary structure of hemoglobin (four almost identical subunits) will be discussed in Chapter 4, that of superoxide dismutase (two identical subunits) will be discussed in Chapter 5, and that of nitrogenase (multiple dissimilar subunits) will be discussed in Chapter 6.

Secondary, as well as tertiary and quaternary, structure of proteins is intimately dependent on the primary sequence of amino acids in the chain. In fact, the manner in which proteins fold into their ultimate structures in biological species is a subject of much research and continued uncertainty even though much is known. Predicting protein folding is an enormous challenge. Most proteins contain dozens or hundreds of amino acids, so there is an astronomical number of ways in which these might be arranged into a compact, folded structure. While only a tiny fraction of the possible folds—perhaps 1000 to 10,000—are found in natural proteins, the challenge is to deduce the best fit of a particular protein sequence to one of these folds. This is called the protein-threading problem. Traditionally, the problem is tackled by assuming that each amino acid prefers to be surrounded by others of a specific kind, and then to look for the best compromise between the needs of all the amino acids. Success using this approach depends on how well we know what the amino acids prefer. Instead of trying to deduce this from physical and chemical principles, Jayanth Banavar of Pennsylvania State University and colleagues use a set of known protein structures to train a computer program to recognize the preferences of each amino acid. Once trained, the program, a neural network, can then predict unknown structures. These researchers have shown that the learning-based method is more successful than one based on a priori assumptions about amino acid preferences. The neural network correctly predicted the structures of 190 out of 213 test proteins, whereas the conventional approach identified only 137 structures correctly.[6]

The amino acid composition of a protein can be determined by cleaving all peptide bonds and identifying the constituent amino acids. The sequence of a given protein is determined by using various methods to cleave only selected peptide bonds and then assembling the information to deduce the amino acid sequence. Proteins with multiple subunits are usually broken down into individual subunits (denatured with heat or chemical reagents) before composition and sequencing analyses are carried out. Characterization of proteins on the basis of size and/or charge can be accomplished by a number of methods. These methods are described in Section 2.2.3.

2.2.3 Protein Sequencing and Proteomics

Protein chemists have followed a basic strategy in sequencing proteins as will be described in this section. Denaturation of multiple-subunit proteins through heat, changes in pH, or chemical reagents (urea, organic solvents, acids, or bases) first produces the constituent subunits. Disulfide bonds are broken by selective reduction or oxidation. Amino acid composition is determined by breaking the peptide bond through exhaustive enzymatic degradation or by hydrolysis with strong acid (6 N HCl) or strong bases. Separation by chromatography is followed by identification and quantification of the individual amino acids by producing colored or fluorescent products, the measured intensity of which are proportional to the concentration of the amino acids. Initial N-terminal and C-terminal sequence determination may be made through the use of end-group analysis. Addition of the reagent fluorodinitrobenzene (FDNB), for instance, reacts with free amine to form a dinitrophenyl derivative used to identify the N-terminal amino acid. Carboxypeptidase will remove the C-terminal amino acid. Closed end (circular) peptides or those with modified N-terminal or C-terminal amino or carboxyl groups will not be detected. The Edman degradation is a method for removing the N-terminal amino acid; this may be repeated successively to sequence an entire peptide. Polypeptides may be broken down into fragments by cleaving at specific acids. Trypsin, for instance, will cleave residues with positively charged R groups (lys and arg) on their carboxyl side, and chymotrypsin will cleave residues with aromatic (phe, tyr, trp) or bulky aliphatic (ile, val) R groups. CNBr cleaves at the carboxyl side of the methionine peptide bond. End-group analysis, Edman degradations, and enzyme treatments may be repeated on different protein fragments to determine sub-sequences and create "overlappings." Large peptide fragments are positioned relative to one another after a second or third treatment creates other fragments whose sequences extend across the initial cleavage points. Eventually the original polypeptide sequence is reconstructed.

Many other methods are used in the analysis of protein mixtures or purified proteins. Ultracentrifugation separates proteins according to size and can determine their molecular weight. Gel filtration (or size exclusion) chromatography separates proteins according to size. Elution time of the protein molecules through a column is related to molecular size, with the largest proteins eluting first and the smallest last. Ion exchange chromatography separates proteins based on their charge, whereas affinity chromatography separates proteins that bind specifically to certain chemical groups. Electrophoresis separates different proteins on the basis of their net charge and on the basis of mass if they have the same charge. Isoelectric focusing separates proteins on the basis of their isoelectric points—the pH at which the individual amino acids exist as zwitterions. More information on these methods can be found in reference 7. The three-dimensional structure of a protein—its secondary, tertiary, and quaternary structure—has been determined for hundreds of proteins (and thousands of protein variations) using X-ray crystallography in the single-crystal solid-state and nuclear magnetic resonance (NMR) spectroscopy in solution. These techniques are discussed in detail in Chapter 3. The amount and type of secondary structure in a protein may be determined by circular dichroism (CD) and magnetic circular dichroism (MCD) spectroscopy. A detailed description of CD and MCD spectroscopy applied to bioinorganic molecules is found in Chapter 5 of reference 8. Mass spectrometry (MS) can determine the structure of very small quantities of protein and is used as an analytical technique in proteomics as discussed below.

The procedures above describe treatment of an individual protein or mixtures of protein subunits obtained by breaking down quaternary structures of individual proteins. More recently, researchers have been interested in numbers of proteins having separate and distinct functions but that are grouped together in a cellular matrix. Proteomics, a term coined around 1995, attempts to study the interactions of the proteins as a group, and in their combined molecular behavior. Proteomics can be defined as the analysis of *all* the proteins expressed at a given time or in a given location rather than the study of one isolated protein as has been the scientific practice until recently. In describing the behavior of numbers of proteins functioning together in an organelle, the proteome becomes defined as the protein complement of the organism's genome. One can think of the human proteome as the universe of proteins encoded by gene sequences in the human genome. The genome, discussed further in Section 2.3.6, is defined as all of the genes contained in a given organism.

Scientists believe that the molecular function of an isolated protein may be very different from the function of that protein in its complex cellular environment—that is, in the proteome. It is also thought that an understanding not only of the structure and function of the individual proteins, but also of their interactions with other proteins and other molecules, may lead to better disease diagnosis and treatment. Proteomics is a more complex topic than genomics because organisms may have well over an order of magnitude more proteins than genes. Because proteomics studies large, complex groups of proteins together at the same time, new techniques and technologies are being developed to assist in this effort. A few of these techniques and technologies are mentioned here. The reader is referred to the literature for further information.[9]

Much of the information about proteomics techniques and technologies included here comes from the Chembytes ezine article written by Michael Dunn and available in 2001 on the website http://www.chemsoc.org/chembytes/ezine/1998/dunn.htm. The first requirement for proteome analysis, also known as protein profiling, is separating the complex protein mixture into its various protein components. Two-dimensional polyacrylamide gel electrophoresis (2-DE or 2-D gels) was the best method current to 2001 for separating sample proteins according to their different properties. The first dimension separation, called isoelectric focusing (IEF), separates proteins according to charge as the different proteins are focused at their respective isoelectric points (pI, the pH at which the protein's net charge equals zero). The second dimension separation, by sodium dodecyl sulfate polyacrylamide gel electrophoresis (SDS-PAGE), separates the proteins by size—molecular weight, M_r. The orthogonal combination of the two separations carried out at right angles results in the proteins being distributed across the 2-D gel profile. Automated computer analysis algorithms, available commercially, are needed at this point for rigorous qualitative and quantitative analysis of the complex and partially overlapping patterns of proteins visualized by the 2-D separation process.

Following the separation process, the individual proteins must be identified and characterized. The 2-D separation has provided M_r, pI, and relative abundance data but no information on protein identities or functions. Computer algorithms are available for matching the M_r, pI, and relative abundance data to known proteins in mixtures, but these are unlikely to provide unequivocal protein identification. One method used to identify individual proteins has been to excise a spot containing one protein from the 2-D gel and then digest the protein with a proteolytic enzyme to form peptide fragments which are then analyzed by mass spectrometry (MS). Peptide mass profiles obtained by MS analysis of given proteins can yield unique profiles leading to protein identification when compared directly to known protein MS peptide mass profile databases. One may also use Edman degradations—the same chemical method described above for successively releasing individual N-terminal residues of a protein—followed by MS of the degradation products to obtain peptide mass profiles of the protein. These methods are slow if one wishes to identify the hundreds of different proteins that may be separated on the 2-D gel. The techniques needed must be capable of high-throughput sensitive screening of proteins separated on 2-D gels so that only those proteins that cannot be identified unequivocally or appear to be novel need further characterization by protein sequencing. High-pressure liquid chromatographic (HPLC) analyses assist here because some proteins have unique amino acid compositions that can be used for their identification when compared to databases of protein amino acid compositions. Gel electrophoresis, MS, and HPLC techniques are continually being refined and extended for more efficient proteome analyses. One final requirement for proteomic technology is that the generated data must be stored in databases that can be interrogated effectively in the laboratory and made available to other scientists worldwide through the internet or other mechanisms. A list of protein 2-D databases already existed in 2001 at www.expasy.ch/ch2d/2d-index.html.

The proteomics research of a number of scientists was described in a *C&E News* report of the 2001 Pittcon meeting.[10] One group, that of Catherine Fenselau at the University of Maryland, has studied a new method for proteolytic stable isotope labeling to provide quantitative and concurrent comparisons between individual proteins from two entire proteome pools.[11] Two ^{18}O atoms are incorporated into the carboxyl termini of all tryptic peptides during the proteolytic cleavage of all proteins in the first pool. Proteins in the second pool are cleaved analogously with the carboxyl termini of the resulting peptides containing two ^{16}O atoms (no labeling). The two peptide mixtures are pooled for fractionation and separation, and the masses and isotope ratios of each peptide pair (differing by 4 Da) are then measured by high-resolution mass spectrometry. Short sequences and/or accurate mass measurements combined with proteomics software tools allow the peptides to be related to the precursor proteins from which they were derived. In general, proteolytic ^{18}O labeling enables a shotgun approach for proteomic studies with quantitation capability. The authors propose this method as a useful tool for comparative proteomic studies of very complex protein mixtures. Research in the same laboratory proposes using proteomics to study groups of cellular proteins that may be responsible for acquired drug resistance rather than looking at one protein at a time, the current practice. Acquired drug resistance (ADR), which causes chemotherapeutic agents to lose effectiveness over time, is discussed further in Section 7.2.3.3.

2.2.4 Protein Function, Enzymes, and Enzyme Kinetics

Protein functions are infinitely varied in biological species. Proteins may transport substances—myoglobin and hemoglobin (discussed in Chapter 4), transport oxygen, and carbon dioxide, in mammalian blood. Proteins catalyze necessary biochemical reactions most of these catalysts are called *enzymes*. The active site of an enzyme contains those amino acids that come in direct contact with the substrate—substance acted upon—and bind it. Some enzymes require nonprotein groups, called *cofactors*, for their activity. Cofactors may be metal ions such as $Fe^{2+,\ 3+}$, $Cu^{1+,\ 2+}$, $Ni^{1+,\ 2+,\ 3+}$, $Co^{1+,\ 2+,\ 3+}$, and so on. Cofactors may also be organic molecules such as nicotinamide adenine dinucleotide (NAD^+), shown in Figure 2.11. These organic molecules are also called *coenzymes*. An enzyme without its cofactor, called an *apoenzyme*, usually will be inactive in its catalytic role. An active enzyme with its cofactor is called a *holoenzyme*. Enzymes requiring metal

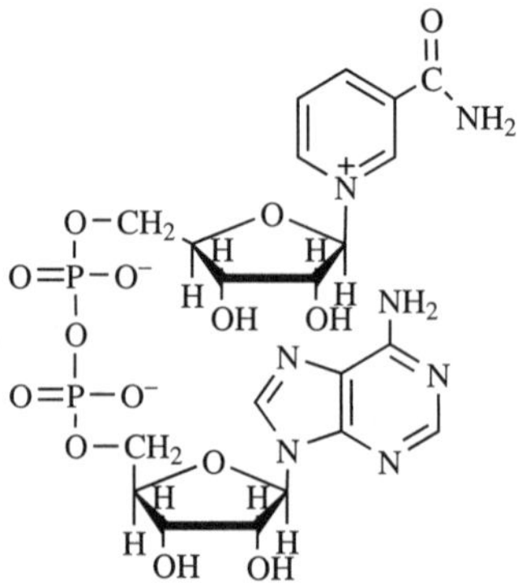

Figure 2.11 Cofactor nicotinamide adenine dinucleotide (NAD^+).

ions as cofactors are called *metalloenzymes*. Copper-containing enzymes with varied functions necessary for plant life (azurin, plastocyanin) and mammalian existence (superoxide dismutase) are described in Chapter 5. The enzyme nitrogenase catalyzes the formation of ammonia from elemental dinitrogen and dihydrogen; the description of this complex enzyme as well as the attempts to create models for its catalytic behavior are discussed in Chapter 6. Essential cofactors and coenzymes that must be ingested in the diet include water-soluble vitamins such as the cobalt-containing vitamin B_{12}. *Isozymes* may be structurally different forms of the same enzyme or oligomeric proteins with differing sets of subunits. Isozymes may also be defined as multiple forms of an enzyme whose synthesis is controlled by more than one gene. Usually, isozymes catalyze similar types of reactions. *Allosteric* enzymes have their activity modulated by the binding of a second molecule, whereas nonallosteric enzymes do not. Proteins maintain biological structures; that is, the protein collagen constitutes part of fibrous connective tissues in skin, bone, tendon, cartilage, blood vessels, and teeth. Proteins may facilitate movement; that is, the proteins actin and myosin mediate muscle contraction. Proteins serve as storage compounds; that is, the protein ferritin stores iron in the human body. Immunoglobins are proteins that react with and neutralize foreign compounds (antigens) in the body. The protein hormone insulin regulates glucose levels in the blood.

Many enzymes are named in classes by attaching "ase" to the type of reaction they catalyze. Oxidoreductases catalyze oxidation–reduction reactions—alcohol dehydrogenase catalyzes the oxidation of an alcohol to an aldehyde. Transferases catalyze the transfer of groups from one molecule to another. Hydrolases catalyze the breaking of covalent bonds using water; for example, peptidase hydrolyzes a peptide bond. Lyases either remove a group by splitting a bond and forming a double bond or add a group to a double bond to form a single bond; for example, decarboxylase removes a carboxyl group to form carbon dioxide. Isomerases catalyze internal atom rearrangements in a molecule. Ligases catalyze the formation of covalent bonds.

The general theory of enzyme kinetics is based on the work of L. Michaelis and M. L. Menten, later extended by G. E. Briggs and J. B. S. Haldane.[1a] The basic reactions (E = enzyme, S = substrate, P = product) are shown in equation 2.1:

$$\begin{aligned} E + S &\Leftrightarrow ES \\ ES &\Leftrightarrow E + P \end{aligned} \tag{2.1}$$

Assuming that the reactions are reversible and that a one-substrate enzyme-catalyzed reaction is being studied, one can derive the Michaelis–Menten rate:

$$V = \frac{V_{max}[S]}{[S] + K_m} \tag{2.2}$$

where V is the initial rate for first-order breakdown of the enzyme–substrate ([ES]) complex into enzyme (E) and product (P); V_{max} is the maximum reaction rate for a given concentration of enzyme in the presence of saturating levels of substrate; [S] is substrate concentration; and K_m is the Michaelis constant, the concentration of substrate required to achieve one-half the enzyme's maximal velocity. Equation 2.2 applies to single-substrate reactions at a constant enzyme concentration. Only nonallosteric enzymes—those not dependent on binding of a second molecule in addition to the substrate—are treated by the Michaelis–Menten rate equation. When the substrate concentration is low, the rate of reaction (V) increases in direct proportion to the substrate concentration (first-order kinetics). When the substrate concentration becomes high enough to saturate all available enzyme active sites, the reaction rate becomes constant (zero-order kinetics). The graphical representation of this behavior is shown in Figure 2.12.

The Lineweaver–Burk plot uses the reciprocal of the Michaelis–Menten equation in the form of the equation of a straight line, $y = mx + b$, having the form shown in equation 2.3:

$$\frac{1}{V} = \left(\frac{K_m}{V_{\max}}\right)\frac{1}{[\mathrm{S}]} + \frac{1}{V_{\max}} \tag{2.3}$$

Plotting $1/V$ versus $1/[\mathrm{S}]$, one obtains a straight line having a slope of $K_m/V_{\max}$ with a y-axis intercept of $1/V_{\max}$ and an x-intercept of $-1/K_m$ as shown in Figure 2.13. Lineweaver–Burk plots of enzyme activity in the presence of an inhibitor can distinguish the type of inhibitor. Competitive inhibitors have a molecular structure similar to that of the substrate and will alter K_m but not $V_{\max}$ because they compete with the substrate for binding at the enzyme's active site but do not change the enzyme's affinity for substrate. Noncompetitive inhibitors bear no structural similarity to the substrate but bind the free enzyme or enzyme–substrate

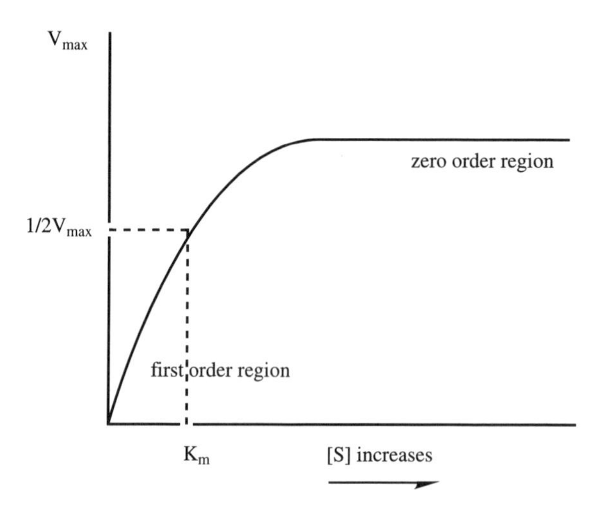

Figure 2.12 Graphical representation of the Michaelis–Menten equation for nonallosteric enzymes.

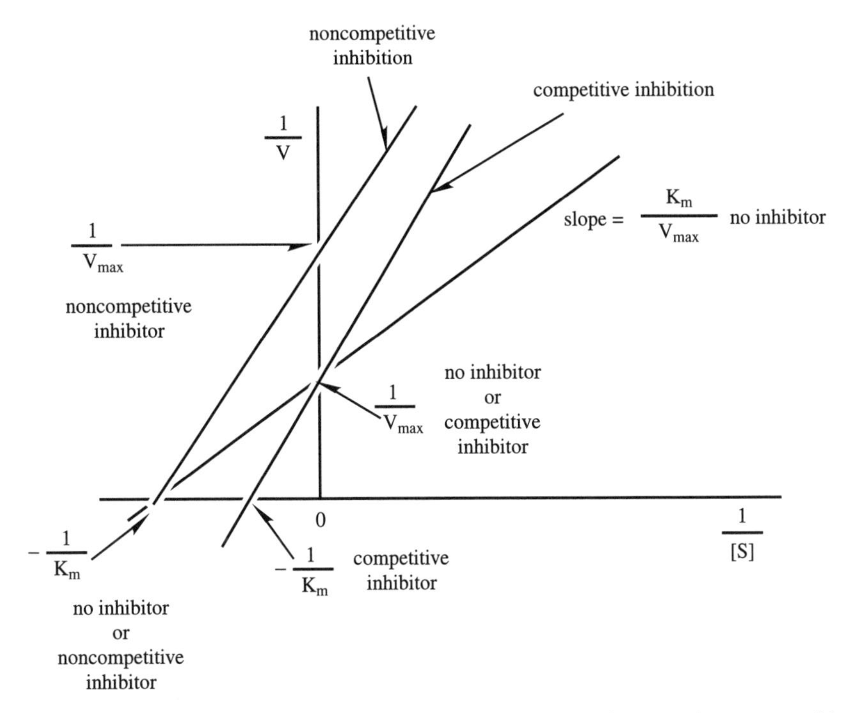

Figure 2.13 Lineweaver–Burk plot for no inhibitor, competitive, and noncompetitive inhibition.

complex at a site other than the active site. They reduce the enzyme's affinity for substrate and decrease V_{max} while not changing K_m. The lines indicating competitive and noncompetitive inhibition are included in Figure 2.13.

2.3 NUCLEIC ACIDS

2.3.1 DNA and RNA Building Blocks

The sequence of amino acids in a given protein determines its structure. The amino acid sequence is controlled ultimately by sequences of genes that are made up of deoxyribonucleic acid (DNA). DNA is composed of repeating units of the nitrogenous bases adenine and guanine (purines) and thymine and cytosine (pyrimidines) linked to the five-carbon cyclic sugar β-D-2-deoxyribose. In turn, the sugars are connected by 3′–5′ phosphodiester bonds forming the linkages in the DNA polymeric chain. DNA chains form the well-recognized double helix through hydrogen bonding of adenine to thymine (A–T pairs) and guanine to cytosine (G–C pairs). More about the forms of the double helix is found in the next section. The complete nitrogenous base, sugar, and phosphate unit is called a *nucleotide*.

Nitrogenous base plus sugar moiety are called *nucleosides*. Ribonucleic acids (RNA) resemble DNA in that nucleoside monophosphates are joined through phosphodiester bonds. RNAs differ in that the sugars are β-D-ribose units and the pyrimidine uracil is found in place of thymine. Molecular structures and nomenclature for nitrogenous bases, nucleosides, and nucleotides are delineated in Table 2.2.

Common metal binding sites on nucleobases are indicated in Table 2.2. Many of these are endocyclic (in the ring) nitrogen atoms that have more available lone-pair characteristics than exocyclic (exterior to the ring) amino groups whose lone pairs are delocalized into the ring by resonance. Experimental and theoretical studies

Table 2.2 Nitrogenous Bases, Nucleosides, Nucleotides, and Sugars Found in DNA and RNA

Base[a]	Nucleoside	Nucleotide Example	Sugar
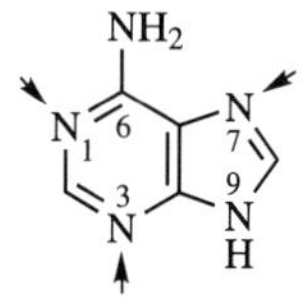 Adenine, A	Adenosine (RNA) or deoxyadenosine (DNA)	Adenosine triphosphate (ATP) 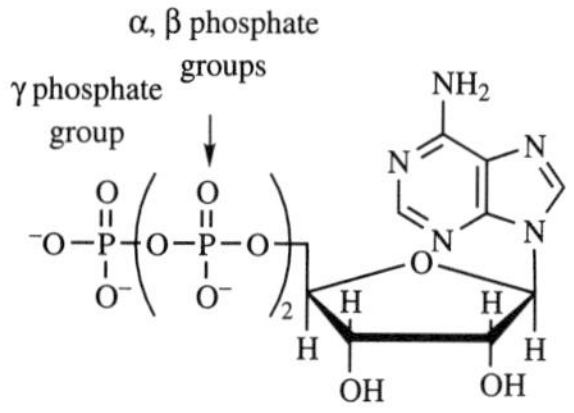	β-D-2-Deoxyribose[b] in DNA or β-D-Ribose[c] in RNA 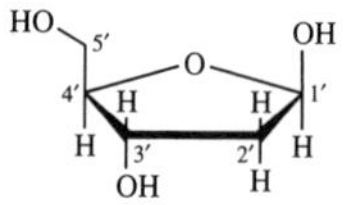
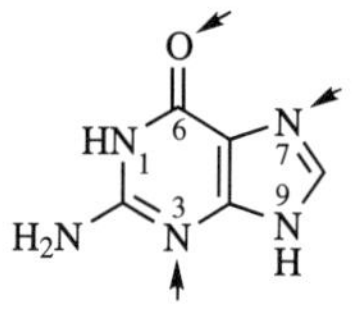 Guanine, G	Guanosine (RNA) or deoxyguanosine (DNA)	Deoxyguanosine 5′-monophosphate (5′-dGMP) 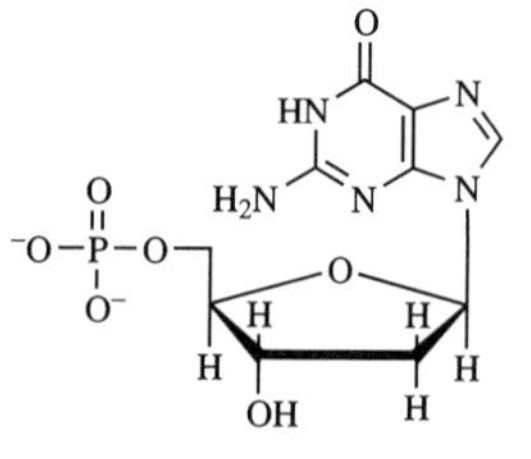	β-D-2-Deoxyribose[b] in DNA or β-D-Ribose[c] in RNA 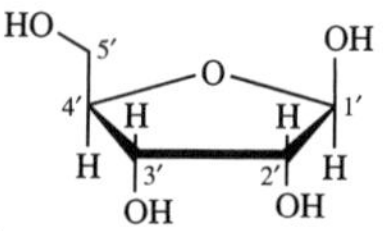

γ phosphate group

α, β phosphate groups

Table 2.2 (*Continued*)

Base[a]	Nucleoside	Nucleotide Example	Sugar
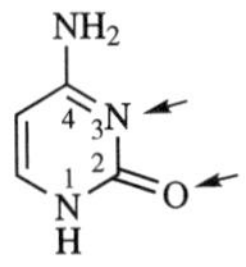 Cytosine, C	Cytidine (RNA) or deoxycytidine (DNA)	Deoxycytidine 3′-monophosphate (3′-dCMP) 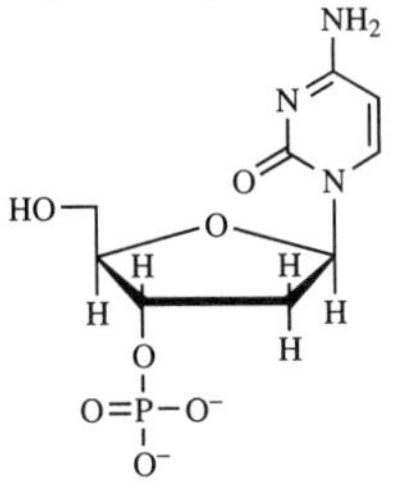	β-D-2-Deoxyribose[b] in DNA or β-D-Ribose[c] in RNA
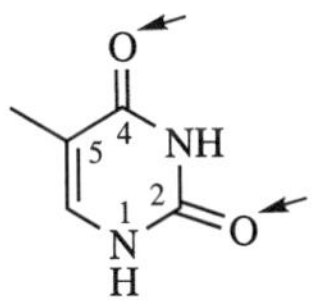 Thymine, T 5-Methyl uracil	Deoxythymidine (DNA)	Deoxythymidine 5′-diphosphate (5′-dTDP) 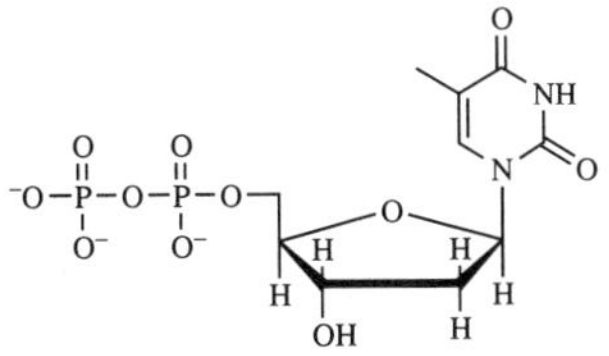	β-D-2-Deoxyribose[b] in DNA
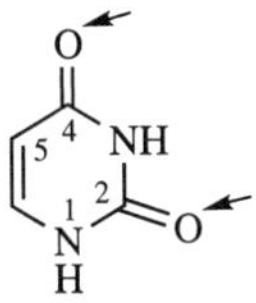 Uracil, U	Uridine	Uridine 3′-monophosphate (3′-UMP) 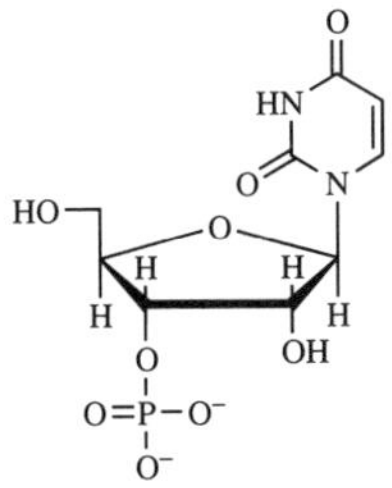	β-D-Ribose[c] in RNA

[a] Arrows indicate common metal-binding sites.
[b] H at 2′ position.
[c] OH at 2′ position.

indicate that the purine N7 sites are the best nucleophiles among all the possible metal binding sites.[2] They are exposed to solvent in the B DNA major groove and are not involved in Watson–Crick base pairing. The N7 sites of adenine and especially guanine are the preferred platinum ion coordination location for the platinum-containing anticancer drugs that will be discussed in Chapter 7. Hard metal ions such as Na^+, K^+, and Mg^{2+} prefer the hard negatively charged oxygen atoms of the phosphodiester groups, with the result that enzymes requiring ATP often require Mg^{2+} as well.[2] Various metal chelates of the α, β, and γ phosphate groups have been identified by X-ray crystallography and ^{31}P NMR.

2.3.2 DNA and RNA Molecular Structures

The nitrogenous base pairs in DNA (A–T and C–G) link through the formation of hydrogen bonds within the double-stranded (ds) structure for DNA. One type of hydrogen bonding is shown in Figure 2.14. Watson–Crick pairing is shown; however, other hydrogen bonding conformations are found. Some of these are known as Hoogsteen pairs, reversed Hoogsteen pairs, and reversed Watson–Crick pairs. Note that in Watson–Crick base pairing, A–T pairs form two hydrogen bonds while C–G pairs form three. RNA usually exists in the single-stranded (ss) form but may fold into secondary and tertiary structures through the formation of specific base pairings (A-U and C-G). Double-stranded RNA is also known.

Double-stranded DNA exhibits complementarity in forming the double helix. The complementary sequences have opposite polarity; that is, the two chains run in opposite directions as in the following illustration:

$$5' \ldots \text{ATCCGAGTG} \ldots 3'$$
$$3' \ldots \text{TAGGCTCAC} \ldots 5'$$

This is described as an antiparallel arrangement. This arrangement allows the two chains to fit together better than if they ran in the same direction (parallel arrangement). A sequence illustrating complementarity is shown in Figure 2.15.

In forming the double-helix polymeric DNA structure, the two sugar–phosphate backbones twist around the central stack of base pairs, generating a major and minor groove. Several conformations, known as DNA polymorphs, are possible.

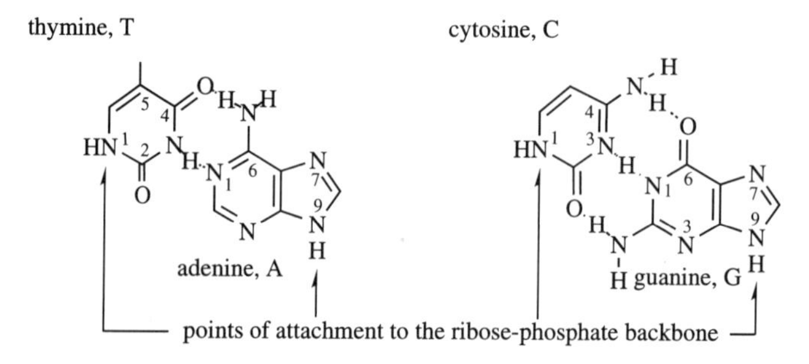

Figure 2.14 Watson–Crick base pairing in DNA.

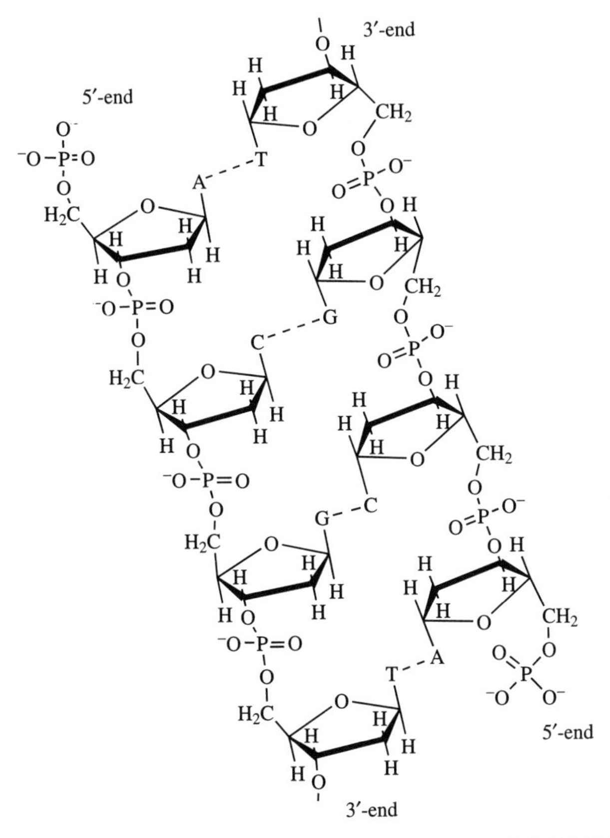

Figure 2.15 Complementary antiparallel double-stranded (ds) DNA.

The classical form is B DNA with well-defined major and minor grooves; $C_{2'}$-*endo* deoxyribose sugar puckers (see Figure 2.16), and parallel stacked base pairs perpendicular to the helix axis. A and B DNA form right-handed helices, and Z DNA forms a left-handed helix. Usually, DNA adopts the B conformation; but in high salt or organic solvent conditions, the Z conformation may be found. Sequences containing alternating GC nucleotides favor the Z DNA conformation. Several different base-sequence-dependent DNA polymorphs may exist within a given genome. Double-stranded RNA or RNA–DNA hybrids normally form the A DNA conformer. Visualizations of all forms are found on the Jena image library of biological molecules site at http://www.imb-jena.de/IMAGE_DNA_MODELS.html. DNA conformations are dependent on a number of parameters, including the type of pucker found in the nonplanar ribose sugar ring, the syn or anti conformation of the nucleobase relative to the sugar moiety, and the orientation about the C4′–C5′ bond. Figure 2.16, as adapted from references 12 and 13, illustrates some of the possibilities.[12]

anti-deoxyguanosine
anti-dG

syn-deoxyguanosine
syn-dG

anti-cytidine, thymidine, uridine

syn-cytidine, thymidine, uridine

O_2 = O at position 2 of pyrimidine

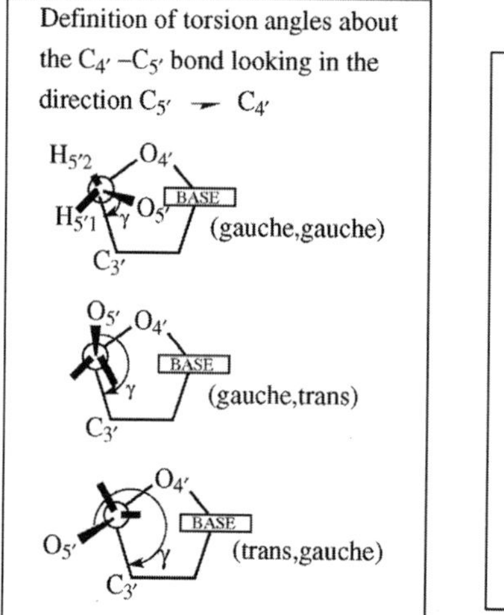

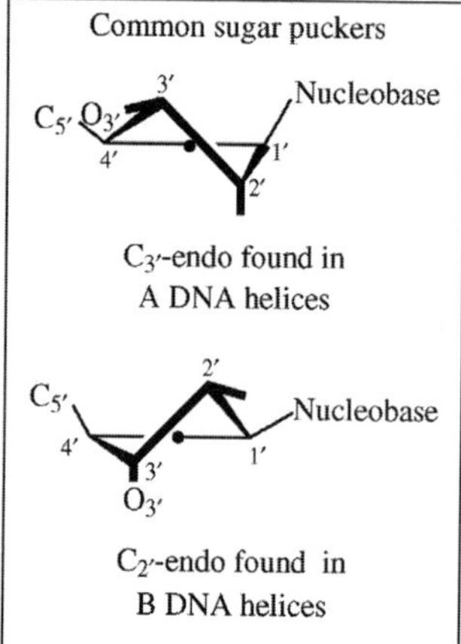

Figure 2.16 Orientations found in DNA helices. (Adapted with permission from: Figure 2.11 of Saenger, W. *Principles of Nucleic Acid Structure*, Springer-Verlag, New York, 1984; copyright 1984, Springer-Verlag, New York; and Figure 1.22 A, B, C of Cowan, J. A. *Inorganic Biochemistry, An Introduction*, 2nd ed., Wiley-VCH, New York, 1997. Copyright 1997, Wiley-VCH.)

The differing conformations of the ribose have been named with respect to that ring atom which puckers out of the plane given by the other ring atoms. The most prominent conformations are $C_{2'}$-*endo* in B-helices and $C_{3'}$-*endo* in A-helices. At room temperature both conformers are in a dynamic equilibrium. Intermediates between $C_{2'}$-*endo* and $C_{3'}$-*endo* are found in several (time-averaged) structures obtained from X-ray crystallography or NMR. When changing the pucker phase

Table 2.3 Some Properties of A DNA, B DNA, and Z DNA

Property	A DNA	B DNA	Z DNA (the repeat unit is the dimer —G C—)
Helix handedness	Right	Right	Left
Sugar pucker	$C_{3'}$-*endo*	$C_{2'}$-*endo*	$C_{3'}$-*endo*
Number of nucleotides per pitch or base pairs per turn	11	10.4	12
Turn angle per nucleotide or twist angle (t, °)	32.7	34.6	30
Helical rise (h, Å)	2.56	3.38	3.7
Pitch (P, Å)	24.6	34.0	45.6
Diameter (Å)	25.5	23.7	18.4
Conformation of gylcosidic bond	Anti	Anti	Anti at C Syn at G
Major (minor) groove width (Å)	2.7 (11.0)	11.7 (5.7)	
Major (minor) groove depth (Å)	13.5 (2.8)	8.5 (7.5)	

angle from 0° to 360°, one steps through all possible conformations of the ribose ring. This pseudorotation cycle can be depicted as a conformational wheel such as that found in Figure 2.8 of reference 12. Figure 9.3 of reference 12 shows a view of $C_{2'}$-*endo* and $C_{3'}$-*endo* conformations observed in A- and B-type polynucleotides.

The distance between two subsequent base pairs along the helical axis is called helical rise (h). The pitch (P) is the length of the helix axis for one complete helix turn. The turn angle per nucleotide or twist angle (t) is given by 360°/number of nucleotides per turn. Data describing some properties of A, B, and Z DNA structure are found in Table 2.3 as adapted from Table 1.10 of reference 13 and the Jena image library website address above.

In the nucleus of eukaryotic cells, a compacted DNA is formed by winding in a shallow, left-handed superhelix around a group of eight histone proteins to form nucleosome core particles. The core particles are further organized into chromatin through interaction with additional histone proteins. Chromatin is anchored to a scaffold of proteins within the chromosomes. In order for replication, transcription, or translation to take place, supercoiled DNA must unravel and the DNA strands separate, at least temporarily. It is thought that the unraveling process, or some version of it, may be necessary before platinum antitumor agents such as *cis*-dichlorodiammineplatinum(II), cisDDP, can bind to DNA, their major target in vitro and in vivo. Much more about the platinum-containing drugs will be found in Chapter 7.

2.3.3 Transmission of Genetic Information

Three major components in the transmission of genetic information are deoxyribonucleic acids (DNAs), ribonucleic acids (RNAs), and proteins. The genetic code expressed through DNA ultimately determines which proteins a cell will produce. Coiled and supercoiled DNA molecules contain numerous sequences of nucleotides that may be transcribed as RNAs and translated to many different proteins. DNA molecules also contain long sequences of nucleotides not coding for protein and whose purpose is not completely understood. A gene is a specific sequence of DNA that encodes a sequence of messenger RNA (mRNA) codons needed to synthesize a complete protein. Genes may be collected together in a chromosome, the DNA packaging unit in eukaryotes. The human genome is an ordered sequence of almost 3 billion adenine, guanine, cytosine, and thymine bases found on 46 chromosomes. Most cells transfer genetic information DNA → RNA, but retroviruses (such as the human immunodeficiency virus, HIV) transmit their information RNA→ DNA. Cells have three major types of RNA. Messenger RNA (mRNA) is produced directly from a DNA template (the sequence of nitrogenous bases along the DNA chain) in a process called transcription. Transfer RNA (tRNA), found in the cytosol, transfers an amino acid to the ribosome for protein synthesis in the translation process. Ribosomal RNA (rRNA) is an essential component of the ribosome, the organelle in which particles of RNA and protein come together to synthesize proteins. The processes of replication, transcription, and translation are illustrated schematically in Figure 2.17 and are further explained in the following paragraphs.

When a cell divides, DNA duplicates itself in a multistep process called replication. DNA replication takes place in a semiconservative manner, meaning that the DNA double strand unwinds and each single strand acts as a template for a new complementary strand. Thus each daughter DNA strand is half comprised of molecules from the old strand. DNA replication is neither a passive nor spontaneous process. Many enzymes are required to unwind the double helix and synthesize a new DNA strand. The enzyme topoisomerase, for instance, is responsible for the initiation of DNA unwinding. Nicking a single DNA strand releases the tension holding the helix in its coiled and supercoiled structure. Another enzyme, helicase, requires energy in the form of ATP to unwind the original double strand, held together by A–T and G–C hydrogen bonds, finally forming the single-stranded DNA (ssDNA) to be replicated. DNA polymerase holoenzyme, a complex aggregate of several different protein subunits, then proceeds along the single DNA strand, bringing free deoxynucleotide triphosphates (dNTPs) to hydrogen bond with

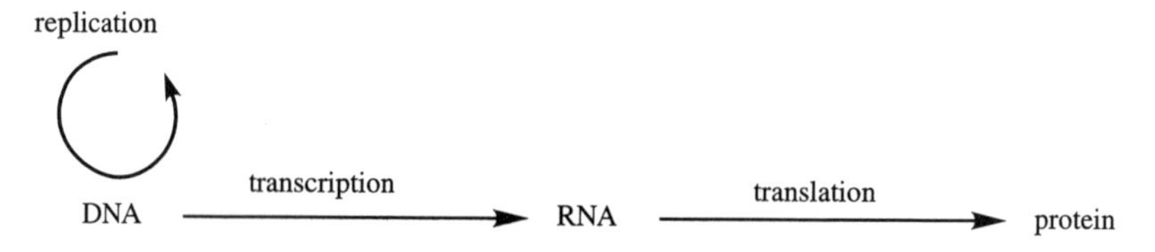

Figure 2.17 Schematic representation of replication, transcription, and translation.

their complementary nucleotide on the single strand. Energy stored in the incoming dNTP is used to covalently bind each deoxynucleotide to the growing second strand assisted by DNA polymerase III. Primase is an enzyme component of the protein aggregate called the primeosome. It attaches a small RNA primer to the single-stranded DNA to act as a substitute 3′-OH for DNA polymerase to begin synthesis of the second strand. The RNA primer is eventually removed by RNase H and the gap filled in by DNA polymerase I. The enzyme ligase can catalyze the formation of a phosphodiester bond given adjacent but unattached 3′-OH and 5′-phosphate groups. As will be discussed in much more detail in Chapter 7, platinum-containing anticancer agents attach themselves preferentially to adjacent guanines in double-stranded DNA. The DNA platination causes kinking in the dsDNA, and this is believed to prevent DNA replication by the mechanism just discussed. There are many repair mechanisms that can remove the DNA platination lesion and other mechanisms for skipping over the platinated portion of dsDNA. These repair mechanisms are believed responsible for some of the acquired resistance that cancerous cells develop, eventually rendering the platinum drugs ineffective.

Transcription is the process that transfers sequence information from the gene regions of DNA to messenger RNA (mRNA) so that it can be carried to the ribosomes in the cytoplasm. The primary protein aggregate responsible for effecting this process is RNA polymerase holoenzyme. This enzyme aggregate directs the synthesis of mRNA on a DNA template. To begin the process, RNA polymerase must be able to recognize a particular DNA sequence at the beginning of genes called the promoter. The promoter is a unidirectional sequence on one strand of the DNA that tells RNA polymerase both where to start and in which direction (on which strand) to continue synthesis. The DNA strand from which RNA polymerase copies is called the antisense or template strand, and the other DNA strand is called the sense or coding strand. The RNA polymerase gathers ribonucleic nucleotide triphosphates (NTPs or rNTPs) proceeding to synthesize the single RNA strand in the 5′ to 3′ direction. Transcription terminates when the RNA polymerase reaches a stop signal on the gene.

In the translation process, proteins are synthesized in the cellular factory called the ribosome. The ribosome consists of structural RNA and about 80 different proteins. When inactive, the ribosome contains a large subunit and a small subunit. When the small subunit encounters mRNA, the process of translation to protein begins. Amino acids necessary for protein synthesis bind at two sites on the large subunit, these sites being close enough together to facilitate formation of a peptide bond. One site (A) accepts a new tRNA carrying an amino acid, while the second site (P) bears the tRNA attached to the growing protein chain. Each tRNA has as specific anticodon and acceptor site. The genetic code is composed of 64 triplet codons (4^3, 4 being the nucleobases A, U/T, C, G with the triplet code raising 4 to the third power) so that each of the 20 amino acids has multiple codons. Some examples are as follows: CAU/T and CAC code for histidine (his); GGU/T, GGC, GGA, and GGG code for glycine (gly); U/TAA, U/TAG, and U/TGA are STOP codons that end the protein sequence. Each tRNA also has a specific charger protein, called an aminoacyl tRNA synthetase, that binds to a particular tRNA and attaches the correct amino acid to the acceptor site. The start signal for translation is the AU/TG codon, coding for methionine (met). Not every protein starts with met, so that often this amino acid must be removed during post-translational protein processing. The ribosomal large subunit binds to the mRNA and the ribosomal small subunit and elongation begins. Entering tRNAs with attached amino acids attach to the A site matching their codon to that on the mRNA. A peptide bond is formed with the amino acid attached to the tRNA at the P site, followed by a shift of the new peptide to the P site freeing the A site for the next tRNA. When the ribosome reaches a stop codon, no aminoacyl tRNA binds to the empty A site, the ribosome breaks into its large and small subunits, and the newly synthesized protein and mRNA are released. A newly synthesized protein is often modified in post-translational processing. Some amino acids may be altered; for example, tyrosine residues may be phosphorylated or asparagine residues glycosylated.

2.3.4 Genetic Mutations and Site-Directed Mutagenesis

Two major types of genetic mutation are possible: chromosomal mutations and point mutations. Chromosomal mutations can alter or shift large sections of chromosomes (collections of genes), leading to changes in the way their genes are expressed. In translocations, large DNA segments are interchanged between two different chromosomes, causing possible changes in gene expression. Inversions occur when a region of DNA flips its orientation with respect to the rest of the chromosome, also leading to changes in gene expression. Sometimes large sections of a chromosome may be deleted, leading to loss of important genes. Chromosome nondisjunction arises when one daughter cell receives more or less DNA than it should during cell division. Point mutations arise from single base-pair changes. A nonsense mutation creates a stop codon where none previously existed, possibly removing essential regions in the resulting protein—that is, U/TAA (stop) instead of U/TAC (tyr). A missense mutation changes the mRNA code, placing a different amino acid into the primary sequence and possibly changing the shape or the metal coordination sphere of the resulting protein. For example, CAA (arg) could be substituted for CAC (his) by a missense point mutation. A silent mutation codes for the same amino acid in the protein primary sequence; that is, CAC and CAU/T both code for histidine. Within a gene, insertion of a small number of bases not divisible by three will result in a frame shift mutation, resulting in an incorrect series of amino acids following the frame shift. An example is shown in Figure 2.18.

Organisms have elaborate proofreading and repair mechanisms that recognize false base-pairings, and other types of DNA damage, and repair these. Genetic mutations that occur in spite of proofreading and repair mechanisms cause genetic diseases. Recent advances in genetic manipulation have allowed the source of genetic diseases to be discovered and in some cases treated. Publication of the human genome sequence in 2001, as discussed below, constitutes a large step forward in the information available for scientists studying genetic disease and treatment.

CUU CCU CAC CGC AGU
leucine proline histidine arginine serine

insert G

↓

CUU CCU GCA CCG CAG U
leucine proline alanine proline glutamine

Figure 2.18 Frame shift mutation caused by insertion of one base, G.

Scientists use mutations to study proteins by introducing amino acid substitutions in crucial locations for determining protein structure or catalytic, binding, and regulatory functions. In order to carry out these mutations, a process called site-directed mutagenesis is invoked, during which the coding gene must be isolated, cloned, sequenced, and then changed in a specific manner. These processes of DNA isolation and cloning, called recombinant DNA techniques, will be discussed further in the next section. Site-directed mutagenesis enables the scientist to change any residue in any protein to any other residue by changing the DNA code and then transcribing and translating the mutated DNA. The essential process involves extracting the plasmid containing the gene of interest from bacterial cells (usually *Escherichia coli* cells) and then using a series of enzymatic reactions to switch codons. The plasmid is then reintroduced to the *E. coli* cells for expression—transcription and translation—into the mutated protein. The techniques are described in more detail in the following section. The three-dimensional structure of the mutated protein is studied by many methods, one of the most popular being X-ray crystallography. One such study has been carried out on the electron transfer protein azurin, the wild type (wt) of which has been discussed and illustrated above in Figure 2.10. In studies of the electron transfer properties of this protein, it was found that the his35 residue was part of a hydrophobic region or patch possibly involved in electron transfer between azurin and itself—electron self-exchange (ese)—or a redox partner such as cytochrome c_{551}—electron transfer (et).[14] The authors created two mutant azurins, one in which a glutamine residue was substituted for his35 (his35gln or H35Q mutant) and the other in which a leucine residue was substituted for his35 (his35leu or H35L mutant). An important finding resulting from X-ray crystallographic structure of the H35Q mutant compared to the wt azurin structure was that the β barrel strand containing the mutation and the unmutated strand to its left were shifted with respect to the configuration found in the wt azurin. The cleft created in the changed H35Q mutant exposes the Q35 residue to solvent, a feature not found for the wt azurin in which his35 is shielded from solvent and buried within the protein structure. Figure 6 of reference 14 shows the differences in tertiary structure in a stereo diagram. (Stereo diagrams permit 3-D visualization of a molecule on paper or a computer screen. Helpful hints for stereoviewing are found in Section 3.13.1.) The technique for introducing point mutations using site-directed mutagenesis is discussed in the following section.

2.3.5 Genes and Cloning

Much of the information contained in this section and the one following has been obtained from the websites http://www.bis.med.jhmi.edu/Dan/DOE/intro.html and http://seqcore.brcf.med.umich.edu/doc/educ/dnap/mbglossary/mbgloss.html and from references 1a, 2, 7, and 13. In eukaryotes, species whose cells contain a nucleus and other intracellular structure, gene expression is more complex than in prokaryotes, simpler life forms whose cells have no intracellular compartments. In eukaryotes, DNA is segmented into exons (regions that encode for protein) and introns (regions that do not). When messenger RNA (mRNA) is initially transcribed, both exon and intron regions are included. Subsequently, the mRNA is spliced together at points so that only the exon regions will be translated into protein in the ribosome. DNA molecules are large and complex to work with in their native state. Usually, scientists wish to work with only the DNA exons, and particularly the exons that code for the specific protein(s) of interest. Recombinant DNA technology (also called cloning and genetic engineering) combines a number of different techniques that allow scientists to manipulate and replicate DNA as well as study its structure and function. Recombinant DNA is DNA that has been created artificially by techniques to be described in the following sections. The recombinant DNA molecule must be replicated many times to become useful for laboratory research; this process is called *cloning*.

The first step in the cloning process is to break the large DNA molecule down into fragments. Sanger sequencing (also called the chain termination or dideoxy method) uses an enzymatic procedure to produce DNA chains of varying length in four different reactions, stopping DNA replication at positions occupied by one of the four bases and then determining the resulting fragment lengths. This method uses naturally occurring enzymes called *restriction endonucleases*. Restriction endonucleases are a class of enzymes, generally isolated from bacteria, that recognize and cut specific sequences—restriction sites—in DNA. One restriction enzyme, BamHI, locates and cuts any occurrence of

$$\begin{array}{c} 5'\text{-GGATCC-}3' \\ 3'\text{-CCTAGG-}5' \end{array}$$

The enzyme clips after the first G in each strand. Both strands contain the sequence GGATCC but in antiparallel orientation. This type of recognition site is called *palindromic*. Another sequencing method, Maxam–Gilbert (also called the chemical degradation method), uses chemicals to cleave DNA at specific bases, resulting in fragments of different lengths. The resulting DNA restriction fragments are separated by electrophoresis on an agarose gel in a method similar to that described for proteins in Section 2.2.3. Even fragments that differ in size by a single nucleotide can be resolved. The gel-separated fragments are converted to single-stranded DNA (ssDNA) by treatment with strong base. In the Southern blot technique, the gel is then covered with nitrocellulose paper and compressed with a heavy plate. The nitrocellulose paper preferentially takes up the ssDNA, producing a pattern of DNA identical to that on the gel. A DNA probe (either an ssDNA or RNA fragment with a complementary base sequence to the DNA of interest) then identifies the desired single strands of DNA through use of a ^{32}P radioactively tagged nucleotide probe. For eukaryotic DNA, one usually begins with an mRNA template that will synthesize the desired DNA by treatment with reverse transcriptase, an enzyme that transcribes RNA to DNA, the reverse of the normal transcription process. The synthesized DNA is called cDNA because it carries a complementary base sequence to the mRNA template. In this cloning method, only the DNA exon portion that codes for the desired protein is transcribed; the intron portion that does not code for protein is not transcribed.

Once the desired cDNA or gene fragment has been obtained, it is produced (expressed) in large quantity through the cloning process. DNA cloning in vivo (in a living cell) can be carried out in a unicellular prokaryote such as the bacterium *Escherichia coli*, in a unicellular eukaryote such as yeast, or in mammalian cells grown in tissue culture. In any case the recombinant DNA must be taken up by the cell in a form that can be replicated and expressed. This is accomplished by incorporating the cDNA into a vector, often a plasmid. Plasmids are small (a few thousand base pairs) circular DNA molecules that are found in bacteria separate from the bacterial chromosome. Usually they carry only one or a few genes. The same restriction enzyme that yielded the cDNA of interest can be used to cleave the plasmid, and then the enzyme DNA ligase is used to splice the cDNA of interest into the plasmid. The result is an edited or recombinant DNA molecule. When the recombinant DNA plasmid is inserted into the host, say *E. coli*, the *E. coli* cells will produce many copies of the DNA, which will in turn be transcribed and translated into the desired protein. The process, much simplified, is illustrated in Figure 2.19.

One of the in vitro (in the test tube) processes used to clone DNA is called the *polymerase chain reaction* (PCR). A vial in which PCR is to be carried out contains all the necessary components for DNA duplication: the piece of DNA to be cloned; large quantities of the four nucleotides, A, T, C, G; large quantities of a primer sequence, a short sequence of about 20 nucleotides synthesized by the primase enzyme; and DNA polymerase. To conduct the process, the vial is first heated to 90–95°C for 30 seconds to separate the two DNA chains in the double helix. Next the vial is cooled to 55°C so that the primers will bind (anneal) to the DNA strands. This process takes about 20 seconds. In the final step, the DNA polymerase (usually the Taq polymerase from the hot springs bacterium *Thermophilis aquaticus*, which works best at around 75°C) starts making complete copies of the DNA template. The complete sequence of steps for the PCR process takes about two minutes, and a million or so copies of the desired DNA can be produced in one or two hours. The PCR method has found applications in clinical medicine, genetic disease diagnostics, forensic science, and evolutionary biology. For example, very small samples of DNA, perhaps from a crime scene or from an ancient mummy, can produce sufficient copies to carry out forensic tests such as DNA profiling.

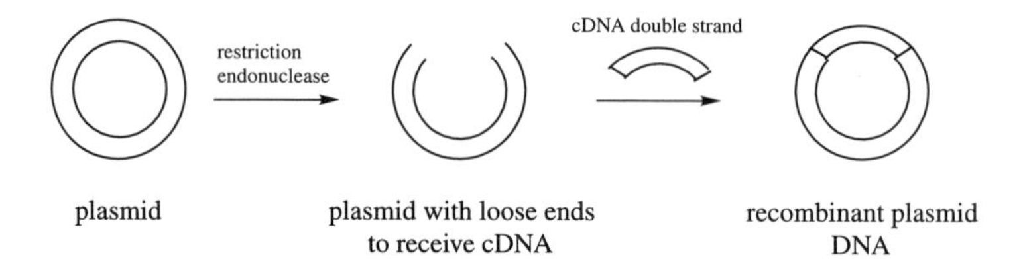

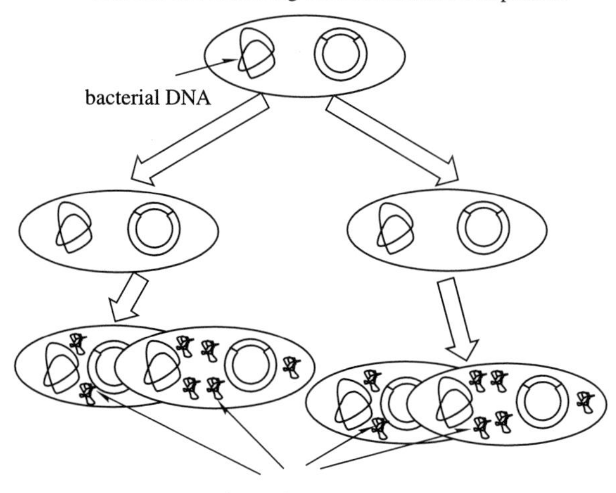

Figure 2.19 Production of protein via a plasmid containing a cloned cDNA fragment. (Adapted with permission from Figure 3.8 of Lippard, Stephen J.; Berg, Jeremy M. *Principles of Bioinorganic Chemistry*, University Science Books, Mill Valley, CA, 1994. Copyright 1994, University Science Books, Sausalito, CA.)

2.3.6 Genomics and the Human Genome

Each cell in an organism contains its complete genome; but depending on the cell type, only the genes necessary for conducting the work of that cell type are expressed. The human genome sequence consists of an ordered listing of the adenine, cytosine, guanine, and thymine bases found on the 46 human chromosomes. Only about 1% of the genome sequence codes for proteins necessary for human life. Most of the rest of the genome consists of large repetitive noncoding regions whose function is not well understood. It is known, however, that critical clues to diseases such as cancer, diabetes, and osteoporosis lie in areas of the genome that do not code for protein. About one-fourth of the genome contains long, gene-free segments, whereas other regions contain much higher gene concentrations. The number of protein coding regions (genes) in the human genome, estimated at 30,000–40,000, is surprisingly small, given that the fruit fly has 13,000 genes and the thale cress plant has 26,000.[15]

The National Human Genome Research Institute (led by Francis Collins) and the Department of Energy's Human Genome Program (headed by Ari Patrinos) managed and coordinated the sequencing of the human genome program, initiated in 1990. A substantially complete version of the 2.9 billion base-pair human genome sequence was published in 2001.[16] Reference 16 reports the results of an international collaboration to produce and make freely available a draft sequence of the human genome. The article presents an initial analysis of the data and describes some of the insights that can be gleaned from the sequence.

The Human Genome Project uses the shotgun sequencing method in which enzymes cut DNA into hundreds or thousands of random bits that are then sent to automated sequencing machines capable of handling DNA fragments up to 500 bases long. After sequencing, the fragments are pieced back together to become part of the sequenced genome. The shotgun approach is applied to cloned DNA fragments that have already been mapped; that is, the fragment's location on the genome is already known. By 2003, the Human Genome Project hopes to deliver a complete human genome sequence available to scientists in a freely accessible database. The National Institutes of Health website address for current human genome sequence data is http://www.ncbi.nlm.nih.gov/genome/guide/H_sapiens.html and the National Human Genome Research Institute's researcher resources website is at http://www.nhgri.nih.gov/Data/.

The Human Genome Project's goal is to produce high-quality, accurate, finished DNA sequences according to the following standards: (1) The DNA sequence is 99.99% accurate. (2) The sequence must be assembled; that is, the smaller lengths of sequenced DNA have been incorporated into much longer regions reflecting the original piece of genomic DNA. (3) The task must be affordable (the project funds technology development to reduce costs as much as possible). (4) The data must be accessible. To this end, verified DNA sequencing data are deposited in public databases on a daily basis. In the previous section, the Sanger and Maxam–Gilbert methods for DNA cleavage followed by gel electrophoresis for DNA sequencing was described. These so-called first-generation gel-based sequencing technologies can be used to sequence small regions of interest in the human genome, but these methods are too slow and too expensive for individual chromosomes let alone a complete genome. The Human Genome Project, in carrying out its goal of affordability, has focused on the development of automated sequencing technology that can accurately sequence 100,000 or more bases per day at a cost of less than $0.50 per base. Second-generation (interim) sequencing technologies focusing on important disease genes, for instance, use technologies such as (a) high-voltage capillary and ultrathin electrophoresis to increase fragment separation rate and (b) resonance ionization spectroscopy to detect stable isotope labels. Third-generation gel-less sequencing technologies aim to increase efficiency by several orders of magnitude. These developing technologies include (1) enhanced fluorescence detection of individual labeled bases in flow cytometry, (2) direct reading of the base sequence on a DNA strand using scanning tunneling or atomic force microscopies (described in Section 3.7.1, (3) enhanced mass spectrometric analysis of DNA sequences, and (4) sequencing by hybridization to short panels of nucleotides of known sequence.

Concurrently with work published by the Human Genome Project, a complete human genome sequence was reported by a consortium of 14 academic, nonprofit, and industrial research groups with the work coordinated by Celera Genomics.[17] The following text is excerpted from the abstract of reference 17. In this work a 2.91-billion base-pair (bp) consensus sequence of the euchromatic portion (the portion containing genes) of the human genome was generated by the whole-genome shotgun sequencing method. Two assembly strategies—a whole-genome assembly and a regional chromosome assembly—were used, each combining sequence data from Celera and the publicly funded genome effort. Analysis of the genome sequence revealed 26,588 protein-encoding transcripts for which there was strong corroborating evidence and an additional approximately 12,000 computationally derived genes with mouse matches or other weak supporting evidence. Although gene-dense clusters are obvious, almost half the genes are dispersed in low G + C sequence separated by large tracts of apparently noncoding sequence. Only 1.1% of the genome is spanned by exons, whereas 24% is in introns, with 75% of the genome being intergenic DNA. DNA sequence comparisons between the consensus sequence and publicly funded genome data provided locations of 2.1 million single-nucleotide polymorphisms (SNPs). A random pair of human haploid genomes differed at a rate of 1 bp per 1250 on average, but there was marked heterogeneity in the level of polymorphism across the genome. Less than 1% of all SNPs resulted in variation in proteins, but the task of determining which SNPs have functional consequences remains an open challenge.

Scientists will continue to use the information generated by the human genome sequencing publications to understand how genes function, how genetic variations predispose the organism to disease, and how gene function can be used in disease detection, prevention, and treatment regimens.

2.4 ZINC-FINGER PROTEINS

Zinc-finger proteins, discussed briefly here, provide examples of a bioinorganic topic intimately associated with biochemical knowledge of both proteins and nucleic acids. Sporting a well-recognized finger-like motif, these proteins are known to participate in one of the many and varied protein–DNA interactions that command the attention of bioinorganic researchers. It was known in the 1970s that zinc was crucial to DNA and RNA synthesis and to cell division. In the 1980s it was discovered that the African clawed toad *Xenopus*' transcription factor IIIA (TFIIIA) contained 2–3 mol zinc/mol of protein.[18] TFIIIA is a site-specific DNA-binding regulatory protein that activates the transcription of the 5S RNA gene into DNA. It was found that protein isolated from the 5S RNA complex, containing zinc, was bound to a 45-base-pair DNA sequence and that the protein protects the DNA from nuclease digestion.[2] It was soon discovered that the two cysteine (cys) and two histidine (his) residues per a 30-amino-acid unit of TFIIIA form a tetrahedral coordination complex with each of 7–11 zinc ions. These generate peptide domains, now called zinc fingers, that interact with DNA.[19]

Further research has led to a classification of zinc-finger proteins to include various criteria: (1) one or multiple repeat units of about 30 amino acids per finger; (2) both two cys and two his (cys_2his_2) with their spacing conserved (the same amino acids occurring in the same locations in many different biological species); (3) two aromatic residues, usually phenylalanine (F), and tyrosine (Y), and the hydrophobic aa leucine (L) conserved in the TFIIIA repeat units; and (4) a three-dimensional structure of the sequence that resembles a finger (see Figure 2.20).[20] Comparisons with metalloproteins of known structure have allowed the development of a detailed three-dimensional model for these domains consisting of two antiparallel β-sheets followed by an α-helix (see Figure 2.21). The proposed structure provides a basis for understanding the detailed roles of the conserved residues and allows construction of a model for the interaction of these proteins with nucleic acids in which the proteins wrap around the nucleic acids in the major groove.[21] Usually, zinc-finger proteins play a regulatory function and are found in nucleic acid polymerases and transcription factors. The role of zinc ion is not catalytic but structural, maintaining the structure of proteins that bind to DNA to activate and deactivate genes.

The consensus aa sequence for zinc-finger repeats, commonly called cys_2his_2-type zinc fingers, is often (phe, tyr)-X-**cys**-$(X)_{2\text{-}5}$-**cys**-$(X)_3$-phe-$(X)_5$-leu or other hydrophobic residue-$(X)_2$-**his**-$(X)_{3\text{-}5}$-**his**-$(X)_5$, where X is any amino acid.[22] The residues in bold type indicate amino acid ligands involved in zinc ion binding.

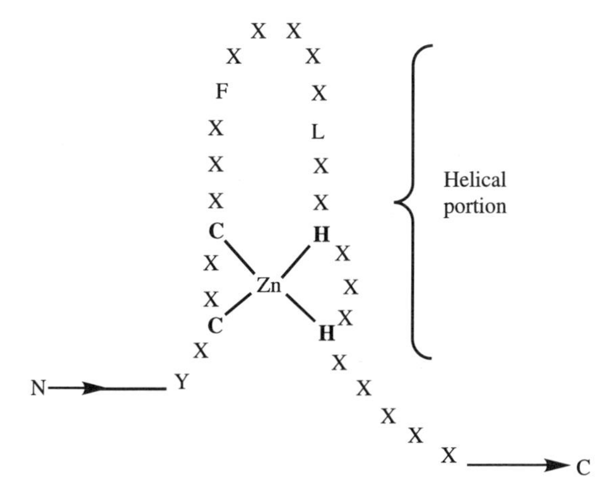

Figure 2.20 Schematic representation of a zinc-finger domain from TFIIIA and related proteins. X represents any amino acid; conserved amino acids are histidine (H), cysteine (C), tyrosine (Y), phenylalanine (F), and leucine (L).

TFIIIA zinc-finger units and others with very similar aa sequences and DNA binding motifs—Zif268, TTK-2, YY1-3, 1MEY, and others—have come to be known as "canonical fingers." Many other known zinc fingers adopt different aa sequences and DNA-binding contacts and have become known as "nonstandard fingers." Figure 3 of reference 22 summarizes both types. Zinc-finger aa sequences fold in the presence of zinc to form a compact ββα domain—two antiparallel β-sheets followed by an α-helix. (Figure 2.21 illustrates the three-dimensional structure.) The ending X_5 sequence often has the amino acid sequence TGEKP forming a flexible linker between the multiple zinc fingers in a specific protein. The zinc ion coordinates tetrahedrally to two cysteines lying at one end of the β-sheet and to two histidines in the C-terminal end of the α-helical portion of the protein sequence. Zinc coordination establishes the protein folding pattern, and substituting a residue other than cys or his at one of the ligand positions usually results in a loss of function. In binding to DNA, zinc fingers have several kinds of contacts, the most important of these being hydrogen bonding contacts between aa residues on the α-helix and bases primarily on one strand of dsDNA. This strand is called the *primary strand.* DNA conformational changes that occur on zinc-finger binding include an enlarged major groove resulting from a combination of negative base-pair displacement and unwinding of the DNA. Other kinds of interactions between zinc fingers and DNA include hydrophobic and phosphate contacts. Most phosphate contacts are made to the primary DNA strand; the most conserved of these is made through the Nδ of the histidine at position 7 of the protein α-helix. This histidine is also a zinc ligand through Nε, bringing the phosphate–zinc finger interaction into close proximity to the DNA strand recognized by the finger.

The criteria described have broadened as more zinc-finger proteins have become known, so that the "zinc-finger" designation now serves to describe any relatively short protein sequence that contains four or more cys and/or his residues and which is believed to interact with a nucleic acid binding domain.[23] Small differences in the primary, secondary, and tertiary structure of the proteins are common; for instance, the α-helical region of the peptide for TFIIIA extends through both histidine ligands; however, the α-helical region for the ADR1 peptide extends only to the first histidine ligand. In individual zinc fingers the helical region varies from 5 to 11 amino acids long. Other zinc fingers exhibit 3–10 helices rather than the more common α-helix. In all cases the helical region is intimately involved in DNA interaction.

In summary, the zinc ions in zinc-finger proteins provide a structural center to direct folding of the protein. The metal ions influence the protein's three-dimensional shape and define the shape or folding pattern of the peptide domain that interacts specifically with DNA. The proteins target specific sites on DNA, meaning that at least one element of the site recognition by DNA regulatory proteins appears to be recognition of complementary shapes. Zinc fingers change the three-dimensional structure of the B DNA to which they are bound, opening up the major groove. It is now believed that the zinc-finger domain represents a ubiquitous structural motif for eukaryotic DNA-binding proteins.[24] Zinc may be chosen for this purpose for at least two reasons: (1) its natural abundance and (2) the absence of redox activity associated with Zn(II) ions avoiding DNA damage, as might be the case with redox active metal centers such as Fe(II)/(III) or Cu(I)/(II). Other zinc-containing structural motifs are known, including the protein GAL4, a transcription factor required for galactose utilization in *Saccharomyces cerevisiae*.[25] The crystal structure of this protein bound to an oligonucleotide indicates that GAL4 is a protein dimer. Each monomer unit contains a binuclear zinc cluster with two zinc ions tetrahedrally coordinated by six cysteines. Two of the cysteines are bridging.

Zinc-finger proteins have been found in many species including *Xenopus* TFIIIA as mentioned above, the yeast alcohol dehydrogenase regulatory gene ADR1,[26] the mouse protein ZIF268 extensively studied by many techniques including X-ray crystallography,[27] and the human oncogene GLI protein,[28] one of perhaps 1000 zinc-finger proteins encoded within the human genome. The number of zinc-finger domains in a single protein ranges from one to as many as 37. A minimum of three fingers seems to be needed for optimal DNA recognition and binding. Several representative zinc-finger domains are shown in Figure 7.4 of reference 2. Some discussion of structural characterization by NMR and X-ray crystallography of individual zinc-finger proteins and attempts to design zinc-finger proteins for binding to specific DNA sequences are discussed in the following sections.

2.4.1 Descriptive Examples

The NMR solution structure of the first zinc-finger domain of the yeast transcription factor SW15 shows a zinc-finger protein with two typical β-sheets and one α-helical region—the ββα motif—but also additional structural elements in the N-terminal region.[29] Figure 2.21 is a Wavefunction, Inc. Spartan '02 for Windows™ visualization of the SW15 zinc-finger domain rendered from data deposited in the protein data bank (PDB code 1NCS). The novel structural elements are a β-strand and a short α-helix not previously observed in other zinc–finger structures. These appear at the bottom right-hand side of Figure 2.21, below the right-hand his56 ligand. The zinc ion is shown in green space-fill form and the cys and his ligands are shown in ball-and-spoke form. The ligands cys34 and cys39 are positioned in the loop region to the left of the zinc ion in Figure 2.21, conforming to the zinc-finger pattern CX_4C. The ligands his52 (above) and his56 (to the right) are positioned in the α-helical region, and conform to the zinc-finger pattern HX_3H. Sequence analysis and comparison to other known proteins suggests that other zinc-finger proteins may also have this structure. Biochemical studies show that the additional N-terminal region structure (the β-strand and short α-helix on the botton right-hand side of Figure 2.21) increases DNA-binding affinity. The authors postulate that the structure has implications for DNA recognition by extending the potential DNA-binding surface of a single zinc-finger domain, and they believe that this additional structure may enhance stability of the DNA–zinc-finger adduct.

The X-ray crystallographic structure from the mouse protein Zif268 and a consensus DNA-binding site has been determined at 2.1-Å resolution, as reported by the authors of reference 27. In this complex, the zinc fingers bind in the major groove of B-DNA and wrap part way around the double helix. Each zinc-finger

Figure 2.21 Zinc-finger protein from the yeast transcription factors SWI as visualized using Wavefunction, Inc. Spartan '02 for Windows™ from PDB data deposited as 1NCS. See text for visualization details. Printed with permission of Wavefunction, Inc., Irvine, CA. (See color plate.)

domain consists of two antiparallel β-sheets containing two cys Zn ligands and an α-helix containing two his Zn ligands. The interatomic distance between Zn(1) and Zn(2) is 26.6 Å, and that between Zn(2) and Zn(3) is 27.4 Å. Each finger has a similar relation to the DNA and makes its primary contacts in a three-base-pair subsite. Each of the zinc fingers uses arg, his, and asp amino acid residues from the N-terminal portion of an α-helix to make contact with the bases in the major DNA groove. Most of the zinc-finger–DNA contacts are made with the guanine-rich strand of the DNA. The N^{ε} of his coordinates the zinc ion, while the N^{δ} hydrogen bonds to phosphodiester oxygens of DNA.

Earlier experiments had already shown that zinc-finger protein design could test DNA binding specificity. For instance, Desjarlais and Berg in 1993 had designed three zinc-finger proteins with different DNA binding specificities.[30] The design strategy combined a consensus zinc-finger framework sequence with previously characterized recognition regions such that the specificity of each protein was predictable. The first protein consisted of three identical zinc fingers, each of which was expected to recognize the subsite GCG. The designed protein binds specifically to the sequence 5′-GCG-GCG-GCG-3′ with a dissociation constant of approximately 11 μM. The second protein had three zinc fingers with different predicted preferred subsites and would bind to the predicted recognition site 5′-GGG-GCG-GCT-3′ with a dissociation constant of 2 nM. A permuted version of the second protein was also constructed and shown to preferentially recognize the corresponding permuted site 5′-GGG-GCT-GCG-3′ over the nonpermuted site. The results indicated that observations on the specificity of zinc fingers can be extended to generalized zinc-finger structures and realized the use of zinc fingers for the design of site-specific DNA-binding proteins.[30]

The Pabo group has described a method for selecting DNA-binding proteins that recognize desired DNA sequences.[31] The research began with amino acid sequences and secondary structure of the three Zif268 zinc fingers: Zif1, Zif2, and Zif3. These are shown schematically in Figure 2.22A as adapted from reference 31. Figure 2.22B shows how the residues in the α-helical regions of fingers Zif1, Zif2, and Zif3 interact with 3-base-pair subsites in the bound DNA. The bold numbers above the helical residue positions denote the residue within the helix, with −1 denoting a residue outside the N-terminal end of the helix. Note that many

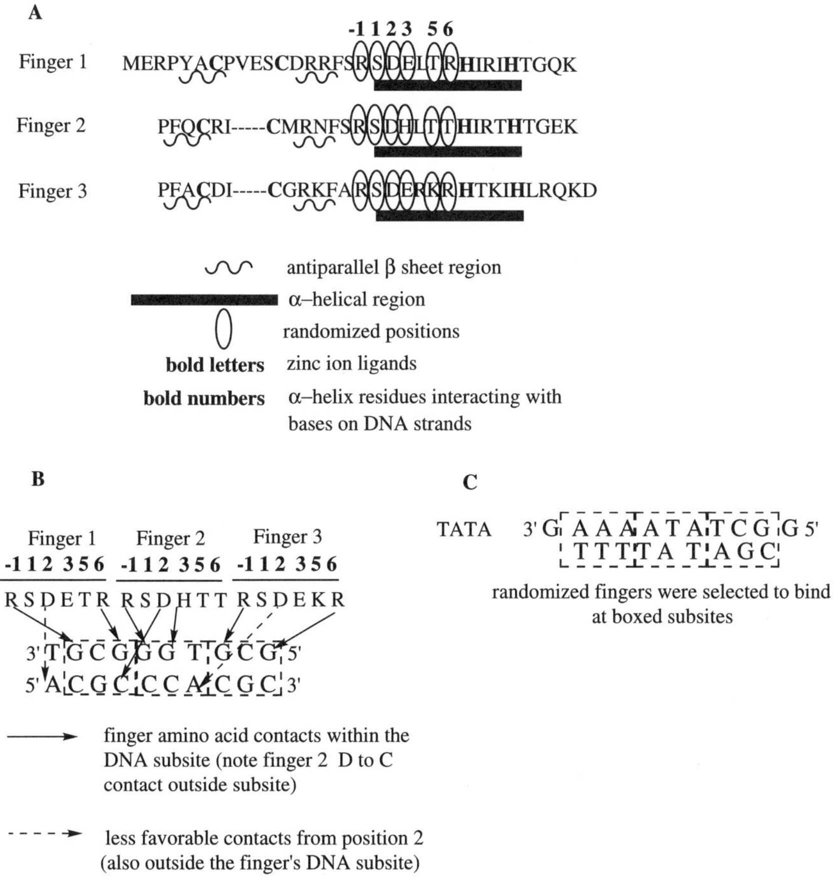

Figure 2.22 (A) Amino acid sequence and secondary structure of Zif1, Zif2, and Zif3 zinc fingers of Zif268. (B) Base contacts of the −1, 1, 2, 3, 5, 6 α-helical amino acid residues of Zif1, Zif2, and Zif3. (C) DNA sequence of the TATA box strand for which zinc fingers were designed. (Figures adapted with permission from Figure 1 of Greisman, H. A.; Pabo, C. O. *Science*, 1997, **275**, 657–660. Copyright 1997, American Association for the Advancement of Science.)

contacts are from arginine (R) residues to guanines on the same (primary) DNA strand. Other contacts from position 2 of the α-helical portion often contact with bases on the complementary strand at positions not within the 3-base-pair subsite. In this research the amino acid residues in α-helical portions of Zif268's Zif1, Zif2, and Zif3 zinc fingers known to make contact with DNA were randomized into a library. The proteins of the library then were subjected to multiple cycles of selection and amplification by genetic engineering techniques called phage display. Phage display has proven to be a powerful enabling technology in genomics and drug development. It allows the directed evolution of proteins engineered for specific properties and selectivity and provides an approach for the engineering of human antibodies, as well as protein ligands, and for such diverse applications as arrays, separations, and drug development. The use of phage display in screening for novel high-affinity ligands and their receptors has been useful in functional genomics and proteomics. Display methods promise to have benefit in the development of therapeutics targeting many different disorders, including cancer, AIDS, autoimmune disorders, and other diseases. In the Pabo group's work the phage display-selected proteins were successively substituted for Zif1, Zif2, and Zif3 in binding to different DNA strands such as the so-called TATA box sequence 3′-GAAAATATCGG-5′ shown in Figure 2.22C. The protein sequences were deemed successful if they would bind tightly to the selected DNA sequences—that is, bind with nanomolar dissociation constants and discriminate at greater than 20,000-fold in binding to nonspecific DNA sequences. The authors believe that the protocol described could be adapted to finding zinc fingers capable of binding to many different DNA- and RNA-binding domains and also that their sequential selection strategy could be applicable to the designing of zinc-finger proteins to be used in gene therapy.

Various other research groups have been working since the early 1990s toward rule formation for zinc-finger proteins and their interactions with DNA. Researchers believe that there are key amino acid positions on the zinc-finger protein that interact with base or phosphate positions on DNA in similar ways for different zinc-finger–DNA systems. These positions may form a recognition code for zinc-finger binding to DNA. Wavefunction, Inc. Spartan '02 for Windows™ visualizations of such a system—in this case the so-called Tramtrack (TTK) transcriptional regulator from the *Drosophila* development gene *fushi-tarazu*—are shown in Figures 2.23 and 2.24. The TTK structural data are deposited as 2DRP in the Protein Data Bank (PDB). Two identical zinc-finger-protein–DNA double-strand interactions are found in the 2DRP crystal's unit cell. A description of the protein–DNA interaction follows as adapted from reference 32. The 2DRP X-ray crystallographic structure was solved to 2.8-Å resolution. Two zinc-finger motifs (F1 and F2) form independent DNA-binding modules, each positioned with the N terminus of the protein's α-helix pointing into the major groove and making base-specific contacts. Most of the protein–DNA contacts are made on the primary strand of the 18-base-pair dsDNA, although one contact in F2—namely, asp154 (**2**) to C32—connects to the complementary DNA strand. The bold number shown in parentheses represents the position of the residue on the α-helix. In Figure 2.24, the amino acid-nucleobase

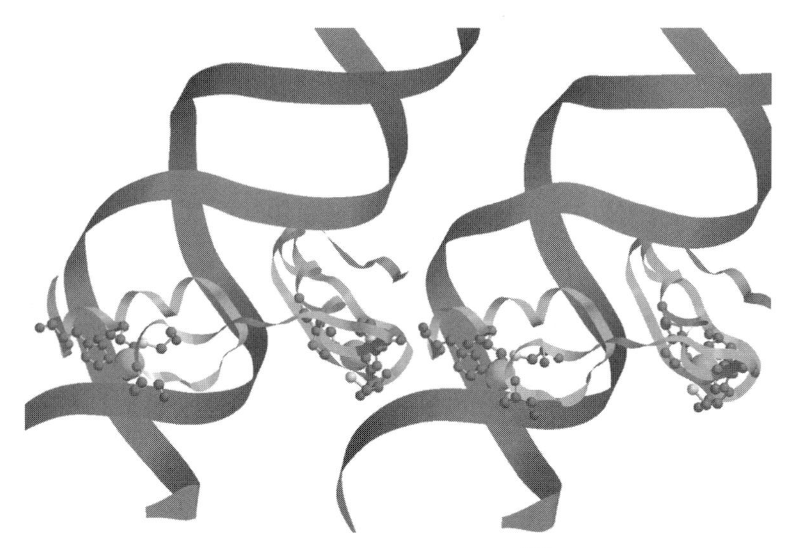

Figure 2.23 Wavefunction, Inc. Spartan '02 for Windows^{TM} visualization (side-on) of the PDB structural data (2DRP) for zinc-finger–dsDNA contacts as described in reference 32. See text for visualization details. Printed with permission of Wavefunction, Inc., Irvine, CA. (See color plate.)

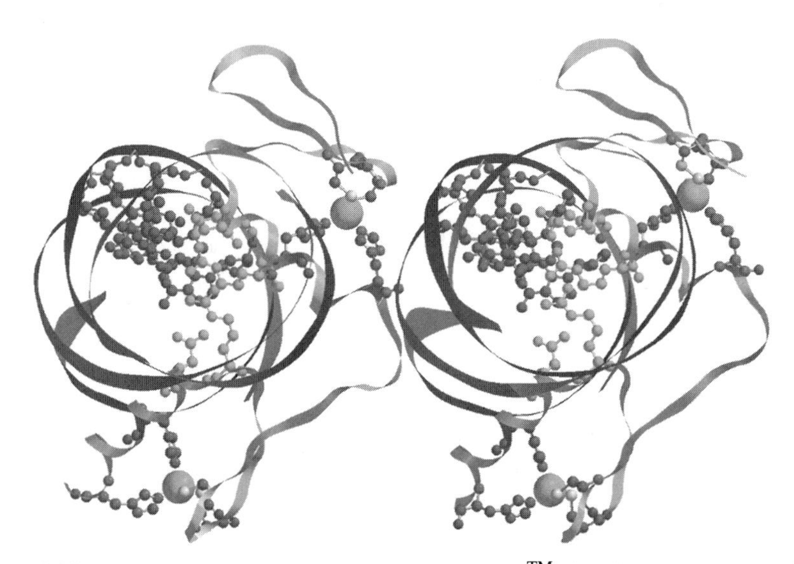

Figure 2.24 Wavefunction, Inc. Spartan '02 for Windows^{TM} visualization (end-on) of the PDB structural data (2DRP) for zinc-finger–dsDNA contacts as described in reference 32. See text for visualization details. Printed with permission of Wavefunction, Inc., Irvine, CA. (See color plate.)

contacts are shown in ball-and-spoke form–amino acids in green and nucleotides in red. In Figures 2.23 and 2.24, the rest of the zinc-finger motifs are shown in green ribbon form, except for the cys and his ligands binding to zinc. These are shown in ball-and-spoke form. In Figures 2.23 and 2.24, the DNA strands are shown as orange or red ribbons. The three amino acids of F1 making contact with the primary strand of DNA are: ser124 (**2**) to T11, asn125 (**3**) to A10, arg128 (**6**) to G9. The bold numbers in parentheses indicate the position of the aa residue along the DNA-contacting α-helix. (See Figure 2.22A.) The two amino acids of F2 making contact with the strand of major DNA contact are: arg152 (−**1**) to G8 and asn155 (**3**) to A7. In F1, the Zn(II) ion (shown in space-fill form) is coordinated by two histidines from an α-helical region—his159 and his164—and two cysteines from a loop region—cys143 and cys146. In F2, the Zn(II) ion (shown in space-fill form) is coordinated by two histidines from an α-helical region—his129 and his134—and two cysteines from a loop region—cys113 and cys116. The string of amino acids in sequence, with beginning and ending residue numbers shown, for F1 is ---Y(111)X**C**X_2**C**X_3YX_5FX_2**H**X_4**H**X_6Y(141)---, with the zinc ion ligand atoms shown in boldface. The amino acid sequence for F2, beginning with the right-hand ending Y from the F1 sequence, is ---Y(141)X**C**X_2**C**FX_2FX_8**H**X_4**H**X_2(166)---. The α-helical region of F1 runs from residues 123 to 134, encompassing all its DNA strand contacts and the histidine zinc ligands. The α-helical region of F2 runs from residues 153 to 164, encompassing all its DNA strand contacts and the histidine zinc ligands. Figure 2.23 shows the side-on representation of the zinc-finger–DNA contacts, while Figure 2.24 shows the same molecules in an end-on representation. In Figure 2.24, the coordination sphere of the lower zinc ions show cys143 in front of the zinc ions, with cys146 to the right. His159 points down toward each of the lower zinc ions, and his164 completes the distorted tetrahedral coordination sphere at each zinc ion's left-hand side. Also in Figure 2.24, the coordination sphere of the upper zinc ions show cys113 in front of the zinc ion, with cys116 partially hidden behind them. His134 points up toward each of the zinc ions and his129 lies at the left-hand side of each.

Shi and Berg have reported on zinc-finger proteins that bind to DNA–RNA hybrids.[33] Previous X-ray crystallographic structures such as that reported in references 27, 31, and 32 have indicated that zinc-finger proteins have more contacts with one strand of dsDNA than with the other. Also, the structures of the strands in complexes with zinc-finger proteins have characteristics of both A- and B-form DNA. These findings led Shi and Berg to examine the binding of two different zinc-finger proteins (Sp1, a human transcription factor that contains three zinc fingers, and ZF-QQR) to dsDNA, to double-stranded RNA, and to DNA–RNA hybrids. RNA, of course, differs from DNA in the presence of the 2′-OH group in RNA, which will in turn affect contacts between the zinc-finger proteins and the nucleic acids. While the binding of Sp1 to its preferred DNA–DNA and DNA–RNA segments was similar, ZF-QQR not only bound more tightly to its preferred DNA–DNA segments than Sp1 but bound DNA–RNA segments five times more strongly than the DNA–DNA segments. Both Sp1 and ZF-QQR proteins bound much less well to RNA–DNA hybrids. The authors found that interactions between the zinc-finger proteins and the DNA–RNA hybrids were dependent on which strand was RNA (DNA–RNA preferred over RNA–DNA) and were also sequence-specific. It is also important to realize that although DNA–RNA hybrids are not well understood, it is known that they do adopt structures that are intermediate between A- and B-form DNA. The authors concluded that interactions with DNA–RNA hybrids should be considered with regard to the biological roles of zinc-finger proteins because these proteins could be designed to target specific DNA–RNA hybrid structures in vivo.

In conclusion, research on zinc-finger–DNA interactions has indicated that secondary structure of the zinc-finger proteins as well as their primary amino acid residue sequences are important in designing proteins to interact with specific DNAs. The points of DNA contact appear to be principally on one strand of dsDNA and, for many but certainly not all interactions, occur at guanine bases. Although early research suggested that the zinc-finger protein–DNA interactions would be readily characterized, copied, and applied to new systems through application of a "recognition code," the situation now appears to be more complex than originally thought. Progress continues on research that will hopefully lead to advanced understanding of zinc-finger–DNA interactions along with the capability to use this knowledge to good effect in gene therapeutic and other medicinal applications. The latest review information on DNA recognition by cys_2his_2 zinc-finger proteins at the time of this writing was contained in the *Annual Review of Biophysics and Biomolecular Structure* article in reference 22.

2.5 CONCLUSIONS

The material introduced in Chapter 2 was chosen to reflect the emphasis of information presented in the rest of this text. In Chapter 3, more information on the X-ray crystallographic and NMR solution structural techniques applied to proteins and nucleic acids will be presented. Analysis of metalloproteins and metalloenzymes discussed in Chapters 4–6 provides important information not only on structures but on functions of these bioinorganic systems. Knowledge about the protein composition and structural analyses of these metalloproteins and metalloenzymes will be assumed on the reader's part from the background presented in this chapter and the next. In Chapter 7, familiarity with protein and nucleic acid structure and function is essential for understanding of the mechanisms of activity of inorganic drugs. The student is invited to refer to Chapter 2 materials often for greater understanding of the bioinorganic systems presented in the following chapters.

REFERENCES

1. (a) Lehninger, A. L.; Nelson, D. L.; Cox, M. M. *Principles of Biochemistry*, 3rd ed., Worth Publishers, New York, 2000. (b) Stryer, L. *Biochemistry*, 4th ed., W. H. Freeman, New York, 1995.

2. Lippard, Stephen J.; Berg, Jeremy M. *Principles of Bioinorganic Chemistry*, University Science Books, Mill Valley, CA, 1994.
3. Creighton, T. E. *Proteins, Structures and Molecular Properties*, 2nd ed., W. H. Freeman, New York, 1992.
4. Zhu, D.; Dahms, T.; Willis, K.; Szabo, A. G.; Lee, X. *Arch. Biochem. Biophys.*, 1994, **308**, 469–470. (PDB: 1JOI)
5. Faham, S.; Mizoguchi, T. J.; Adman, E. R.; Gray, H. B.; Richards, J. H.; Rees, D. C. *J. Biol. Inorg. Chem.*, 1997, **2**, 464–469.
6. Chang, I.; Cieplak, M.; Dima, R. I.; Maritan, A.; Banavar, J. R. *Proc. Natl. Acad. Sci. USA* 2001, **98**(25), 14350–14355. (b) Cieplak, M.; Holter, N. S.; Maritan, A.; Banavar, J. R. *J. Chem. Phys.*, 2001, **114**(3), 1420–1423.
7. Lyman, B. A. *Biochemistry*, Springhouse Corporation, Springhouse, PA, 1994.
8. Johnson, M. K., in Que, L., ed. *Physical Methods in Bioinorganic Chemistry: Spectroscopy and Magnetism*, University Science Books, Sausalito, CA, 2000, pp. 233–285.
9. (a) Görg, A.; Boguth, G.; Obermaier, C.; Posch, A.; Weiss, W. *Electrophoresis*, 1995, **16**, 1079. (b) Humphery-Smith, I.; Cordwell, S. J.; Blackstock, W. P. *Electrophoresis*, 1997, **18**, 1217. (c) Lamond, A. I.; Mann, M.; *Trends Cell Biol.* 1997, **7**, 139. (d) Patterson, S. D.; Aebersold, R. *Electrophoresis*, 1995, **16**, 1791. (e) Pennington, S. R.; Wilkins, M. R.; Hochstrasser, D. F.; Dunn, M. J. *Trends Cell Biol.*, 1997, **7**, 168. (f) Wilkins, M. R.; Williams, K. L.; Appel, R. D.; Hochstrasser D. F., eds., *Proteome Research: New Frontiers in Functional Genomics*, Springer, Berlin, 1997. (g) Yates, J. R.; *J. Mass Spectrom.* 1998, **33**, 1.
10. Henry, C. M. *C&E News*, 2001, **April 2**, 47–49.
11. Yao, X.; Freas, A.; Ramirez, J.; Demirev, P. A.; Fenselau, C. *Anal Chem.*, 2001, **73**, 2836–2842.
12. Saenger, W. *Principles of Nucleic Acid Structure*, Springer-Verlag, New York, 1984.
13. Cowan, J. A. *Inorganic Biochemistry, An Introduction*, 2nd ed., Wiley-VCH, New York, 1997.
14. Nar, H.; Messerschmidt, A.; Huber, R.; van de Kamp, M.; Canters, G. W. *J. Mol. Biol.*, 1991, **221**, 427–447.
15. Borman, S. *C&E News*, 2001, **Feb. 12**, 9.
16. International Human Genome Sequencing Consortium, *Nature*, 2001, **409**, 860–921.
17. Venter, J. C., et al. *Science*, 2001, **291**, 1304–1351.
18. Hanas, J. S.; Hazuda, D. J.; Bogenhagen, D. R.; Wu, F. Y.-H.; Wu, C.-W. *J. Biol. Chem.*, 1983, **258**, 14120.
19. Klug, A.; Rhodes, D. *Trends Biochem. Sci.*, 1987, **12**, 464.
20. (a) Vallee, B. L.; Auld, D. S. *Acc. Chem. Res.*, 1993, **26**, 543–551. (b) Frankel, A. D.; Bredt, D. S.; Pabo, C. O. *Science*, 1998, **240**, 70.
21. Berg, J. M. *Proc. Natl. Acad. Sci. USA*, 1988, **85**, 99–102.
22. Wolfe, S. A.; Nekludova, L.; Pabo, C. O. *Annu. Rev. Biophys. Biomolec. Struct.* 2000, **29**, 183–212.
23. Berg, J. M. *Curr. Opin. Struct. Biol.*, 1993, **3**, 11.
24. Bertini, I.; Gray, H. B.; Lippard, S. J.; Valentine, J. S. *Bioinorganic Chemistry*, University Science Books, Mill Valley, CA, 1994, pp. 491–493.

25. (a) Marmorstein, R.; Harrison, S. C. *Genes Dev.*, 1994, 2504–2512. (b) Marmorstein, R.; Carey, M.; Ptashne, M.; Harrison, S. C. *Nature*, 1992, **356**, 408.
26. Parraga, G.; Horvath, S. J.; Eisen, A.; Taylor, W. E.; Hood, L.; Young, E. T.; Klevit, R. E. *Science*, 1988, **241**, 1489.
27. Pavletich, N. P.; Pabo, C. O. *Science*, 1991, **252**, 809.
28. Pavletich, N. P.; Pabo, C. O. *Science*, 1993, **261**, 1701–1707.
29. Dutnall R. N.; Neuhaus D.; Rhodes D. *Structure*, 1996, 599–611. (PDB: 1NCS)
30. Desjarlais, J. R.; Berg, J. M. *Proc. Natl. Acad. Sci. USA*, 1993, **90**, 2256–2260.
31. Greisman, H. A.; Pabo, C. O. *Science*, 1997, **275**, 657–660.
32. Fairall, L.; Schwabe, J. W. R.; Chapman, L.; Finch, J. T.; Rhodes, D. *Nature*, 1993, **366**, 483–487. (PDB: 2DRP)
33. Shi, Y.; Berg, J. M. *Science* 1995, **268**, 282–284.

3

INSTRUMENTAL AND COMPUTER-BASED METHODS

3.1 INTRODUCTION

3.1.1 Analytical Instrument-Based Methods

Lawrence Que, Jr., has stated the following in the preface to *Physical Methods in Bioinorganic Chemistry*: "By piecing together the various clues derived from the physical methods, bioinorganic chemists have been able to form a coherent picture of the metal binding site and to deduce the role of the metal ion in a number of biological processes."[1] Physical methods are used for analysis of systems under study by all chemists, whether or not they are trained analytical chemists. Therefore all researchers must become familiar with the capabilities and limitations of the analytical methods they use. In addition to instrumental and physical methods, chemists have additional investigational capabilities based on computer methodology. This chapter contains an introduction to some physical (instrumental) and computer-based methods. The discussion is not intended as an exhaustive description or even listing of available methods but rather concentrates on those mentioned frequently in the following chapters of this text. The methods include X-ray absorption spectroscopy (XAS and EXAFS), X-ray crystallography, nuclear magnetic resonance, electron paramagnetic resonance (including ENDOR), and Mössbauer spectroscopies. The last method mentioned is especially relevant for characterization of bioinorganic iron-containing species. Other important methods used to analyze bioinorganic species such as ultraviolet-visible, infrared, resonance Raman, circular dichroism, magnetic circular dichroism spectroscopies, and many others are not discussed here. Students are referred to analytical and instrumental texts[2] for more information on all methods, discussed here or not. Reference 1 is especially helpful because of its emphasis on instrumental methods applied to the analysis of bioinorganic systems.

3.1.2 Spectroscopy

Students will be familiar with the absorption or emission of electromagnetic radiation as the basis for spectroscopic methods. Electromagnetic radiation itself is perceived as mutually perpendicular oscillating electric and magnetic fields. The total energy of the radiation, which has a number of components, is determined by the relationship shown in equation 3.1:

$$E_{\text{total}} = h\nu = \frac{hc}{\lambda} = E_{\text{translation}} + E_{\text{rotation}} + E_{\text{vibration}} + E_{\text{electron spin}} + E_{\text{nuclear spin}} + E_{\text{nuclear levels}} + \cdots \tag{3.1}$$

where

h = Planck's constant = 6.626×10^{-34} Js
c = the speed of light = 3×10^{8} m s^{-1}
ν = the frequency of light in s^{-1} (Hz)
λ = the wavelength of light in meters (m)

Atomic and molecular energy levels represent specific quantum states (ground and excited) illustrated in Figure 3.1. Transitions between these states, which may be caused either by energy absorption or by energy emission, are responsible for physical method observations. Examples include:

1. Moving an electron from a ground state to an excited state ($E_{\text{electronic}}$), leading to observations in the ultraviolet-visible spectroscopic region
2. Reorienting an electron in a magnetic field ($E_{\text{electron spin}}$) as seen in electron paramagnetic resonance (EPR) spectroscopy

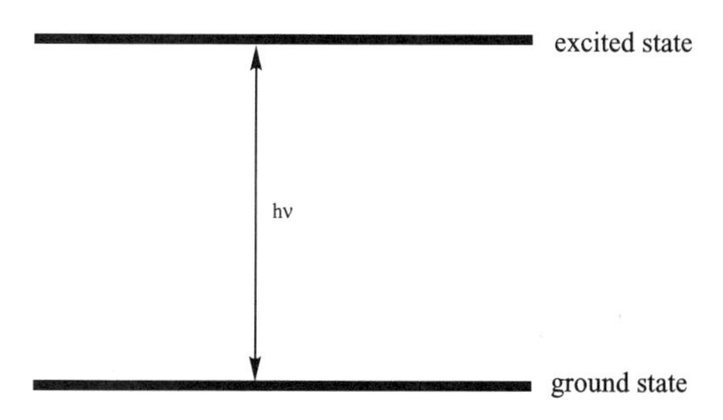

Figure 3.1 Energy transition (absorption or emission) between ground and excited states.

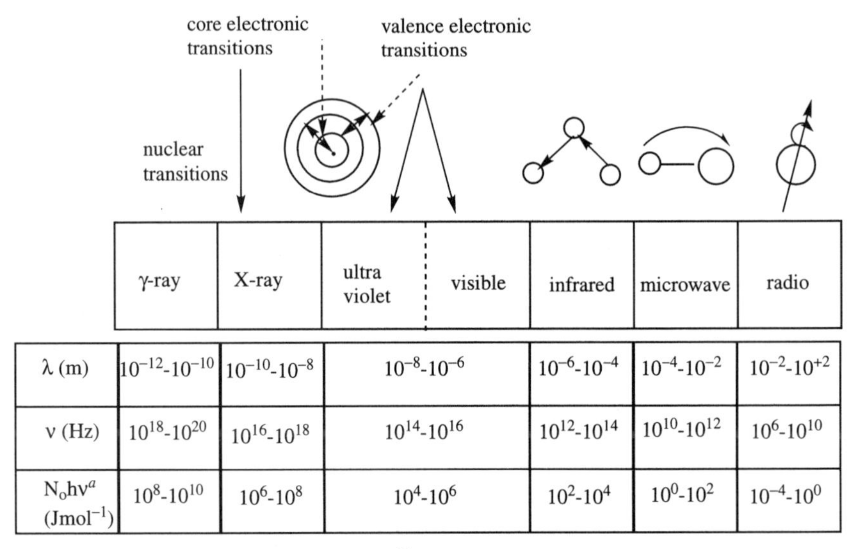

	γ-ray	X-ray	ultra violet / visible	infrared	microwave	radio
λ (m)	10^{-12}-10^{-10}	10^{-10}-10^{-8}	10^{-8}-10^{-6}	10^{-6}-10^{-4}	10^{-4}-10^{-2}	10^{-2}-10^{+2}
ν (Hz)	10^{18}-10^{20}	10^{16}-10^{18}	10^{14}-10^{16}	10^{12}-10^{14}	10^{10}-10^{12}	10^{6}-10^{10}
$N_o h\nu^a$ ($Jmol^{-1}$)	10^{8}-10^{10}	10^{6}-10^{8}	10^{4}-10^{6}	10^{2}-10^{4}	10^{0}-10^{2}	10^{-4}-10^{0}

[a] N_o = Avogradro's number = 6.023×10^{23}

Figure 3.2 Illustrated energy transitions for several useful regions of the electromagnetic spectrum. (Adapted with permission from Figure 2.2 of Cowan, J. A. *Inorganic Biochemistry, An Introduction*, 2nd ed., Wiley-VCH, New York, 1997. Copyright 1997, Wiley-VCH.)

3. Reorienting a nuclear moment in a magnetic field ($E_{\text{nuclear spin}}$) as observed in nuclear magnetic resonance (NMR) spectrometry
4. Detecting energy absorbed or emitted in deforming bonds between atoms in molecules ($E_{\text{vibration}}$) as observed in infrared or resonance Raman spectroscopy

The entire spectrum of electromagnetic resonance, as shown in Figure 3.2 adapted from Cowan,[3] is available for various physical methods. High-energy, short-wavelength transitions occur at the left end of Figure 3.2, whereas low-energy, long-wavelength transitions occur at the right.

3.2 X-RAY ABSORPTION SPECTROSCOPY (XAS) AND EXTENDED X-RAY ABSORPTION FINE STRUCTURE (EXAFS)

3.2.1 Theoretical Aspects and Hardware

The use of X-ray diffraction from crystalline samples can result in a complete three-dimensional crystal structure of a molecule, but requires a single crystal suitable for proper diffraction (see Section 3.3). X-ray absorption spectroscopy (XAS) can yield limited molecular structural information on noncrystalline (amorphous) solid samples, frozen solutions, liquid solutions, and gases. A tunable X-ray source is required because of the energy-dependent absorption coefficient of the material under study. Synchrotron radiation has provided a high-intensity tunable source of X-rays, and its use has been increasingly responsible for the popularity of this technique over the last three decades.[4] The following discussion summarizes the material found in reference 4.

Figure 1 of reference 4 shows a typical X-ray absorption spectrum. A sharp rise, called the X-ray absorption edge, occurs at a well-defined X-ray photon energy, which is unique to the absorbing element. The absorption edge is due to electron dissociation from a core level of the absorbing atom to valence energy levels. Spectral features in the edge region, sometimes called the near edge, are related to the electronic structure of the absorbing atom and can often be used to identify the geometric arrangement of its ligands. For instance, one might be able to distinguish between octahedral and tetrahedral geometric arrangements about the absorbing atom through analysis of XAS edge spectral features. For the same absorbing atom, differences in absorption energy at the edge are related to its valence or oxidation state. Usually, higher oxidation states will result in absorption at higher energies. This analysis is often given the acronym XANES (X-ray absorption near-edge structure).

In the region above (to the right of) the edge, variation in the X-ray absorption coefficient known as extended X-ray absorption fine structure (EXAFS) may be analyzed to yield structural information about atoms in the ligand sphere (within 4- to 5-Å radius) of the absorbing atom. The EXAFS phenomenon arises through interaction of photoelectrons from the absorbing atom, symbolized by an "a," with electron density of surrounding atoms. Scattering atoms or scatterers are symbolized by an "s." Scattering atoms contribute damped sine waves of measurable frequency (related to distance between a and s), amplitude (related to coordination number about a and s atom types), and phase (related to s atom type). Analysis of EXAFS data can answer the question, How many of what type of atom are at what distance from the absorbing atom?

The XAS spectrometer is similar to a UV-visible system in that it consists of a source, a monochromator, and a detector. The most favorable XAS source, synchrotron radiation, is tunable to different wavelengths of desirable high intensity. A laboratory instrument for analysis of solids and concentrated solutions may use a rotating anode source (further described in Section 3.3). The monochromator for X-ray radiation usually consists of silicon single crystals. The crystals can be rotated so that the wavelength (λ) of the X rays produced depends on the angle of incidence (θ) with a Bragg lattice plane of the crystal and the d spacing of the crystal according to Bragg's law as shown in equation 3.2:

$$n\lambda = 2d \sin\theta \tag{3.2}$$

Detectors for quantitative measurement of X-ray absorption spectra must measure the flux (photons s^{-1}) of the X-ray beam. Ionization chambers consisting of X-ray transparent windows on each end of a chamber holding an inert gas work well as transmission detectors for concentrated samples. For transmission detectors, $\ln(I_0/I)$ is proportional to absorption coefficient (I_0 = incident X-ray photonin tensity, I = transmitted intensity). Fluorescence excitation techniques provide a more sensitive detection system in which fluorescent X-ray photons (a fraction of the ionized absorbing atoms relax by emission of a fluorescent X-ray photon) are counted as the photon energy is scanned. The signal generated is proportional to the absorption coefficient.

The data collected are subjected to Fourier transformation yielding a peak at the frequency of each sine wave component in the EXAFS. The sine wave frequencies are proportional to the absorber–scatterer (a–s) distance R_{as}. Each peak in the display represents a particular shell of atoms. To answer the question of how many of what kind of atom, one must do curve fitting. This requires a reliance on chemical intuition, experience, and adherence to reasonable chemical bond distances expected for the molecule under study. In practice, two methods are used to determine what the back-scattered EXAFS data for a given system should look like. The first, an empirical method, compares the unknown system to known models; the second, a theoretical method, calculates the expected behavior of the a–s pair. The empirical method depends on having information on a suitable model, whereas the theoretical method is dependent on having good wave function descriptions of both absorber and scatterer.

While extended X-ray absorption fine structure (EXAFS) in general does not give direct information about the geometry of ligand atoms about the metal center, it is possible to relate specific X-ray absorption edge behaviors to geometry for nickel(II) compounds for instance. In the examples discussed by Scott in Figure 13 of reference 4, Ni(II) complexes with square-planar (D_{4h}) symmetry give rise to a characteristic K edge spectra with a large characteristic pre-edge peak about 5 eV below the main edge. The peak is assigned to a $1s \rightarrow 4p_z$ transition with simultaneous ligand → metal change transfer (LMCT) to a Ni(II) $3d$ orbital. Because neither octahedral nor tetrahedral Ni(II) complexes show this peak (but have their own characteristic spectral behavior), one can, in combination with information about ligands and bond lengths from EXAFS, assign metal complex geometry. Figure 13 of reference 4 also shows another typical behavior: Soft ligands result in lower edge energy. Information about the oxidation state of a metal ion can also be gathered from edge and pre-edge behavior. Cu(I) two- and three-coordinate complexes (tetrahedral Cu(I) complexes do not show the behavior) exhibit a pre-edge peak that disappears upon oxidation to Cu(II), allowing for quantitative determination of Cu(I) content in these systems.[5]

3.2.2 Descriptive Examples

Quantitative information about the first coordination sphere structure depends on analysis of EXAFS data. From analytical data or knowledge of common ligands in metalloenzymes (N, O, S, Se), one can decide which ligands are likely to be present in the coordination sphere. An example discussed by Scott[4] tests the hypothesis of a Cu(II)–S bond being present in the compound shown in Figure 3.3.

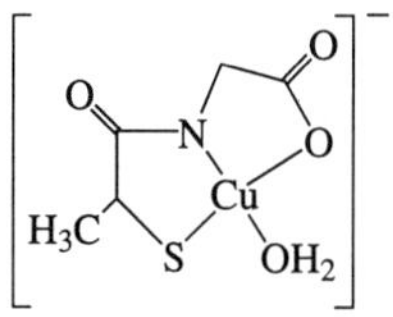

$[Cu(MPG)(H_2O)]^-$

Figure 3.3 Cu(II) complex with potential chelating ligand mercaptopropionylglycine (MPG). (Adapted from reference 4.)

Fourier transformation of Cu EXAFS data gathered on the Cu(MPG) complex reveals two separate peaks representing shells at distances of 1.9 and 2.3 Å. When tested for N_s (coordination number), metal–ligand distance (R_{as}), and Debye–Waller parameter difference ($\Delta\sigma^2_{as}$) followed by comparison to known model compounds, results show that the presence of both a Cu–(N, O) and Cu–S shell is necessary to obtain an adequate fit to the EXAFS data. Therefore it was concluded that a Cu–S bond is present in the compound.

The Fe and Mo EXAFS of the enzyme nitrogenase's MoFe–protein in partially oxidized and singly reduced forms has been studied.[6] Chapter 6 includes detailed discussion of this important and complex enzyme that reduces N_2 to NH_3 in sequential reduction and protonation steps. Figure 6.1 indicates nitrogenase's major components: dimeric complexes called Fe–protein and tetrameric complexes called MoFe–protein. Fe–protein dimers bridge to MoFe–protein tetramers by way of an $[Fe_4S_4]$ cluster through which electrons are shuttled. Metal sulfur clusters are discussed in more detail in Section 6.3. Electrons arriving at MoFe–protein pass through an unusual $[Fe_8S_7]$ iron sulfur cluster (called the P–cluster) and eventually arrive at a third metal sulfur cluster of formula $[Fe_7MoS_8$(homocitrate)] (called the FeMo cofactor or M center) where it is believed that N_2 is complexed, reduced, and protonated to produce ammonia. Each metal sulfur cluster exhibits several metal oxidation states that change during the catalytic cycle as electrons are shuttled through the enzyme toward N_2. Beginning in 1992, X-ray crystallography of nitrogenase components established metal–metal and metal–sulfur ligand bond distances for the resting enzyme (the E_0 state). These data are discussed in Chapter 6, and Figure 6.5 outlines a catalytic cycle for nitrogenase that illustrates the "E" terminology.

As an example of the kind of information obtained from EXAFS studies, in reference 6 the authors study EXAFS spectra of MoFe protein to establish bond lengths for the one-electron reduced form E_1 (Figure 6.5) and also for an oxidized form of the enzyme. The results are compared to information on the resting form E_0 and to X-ray crystallographic data. The one-electron reduced MoFe protein EXAFS spectrum was obtained by examining a steady-state mixture of resting (symbolized as P/M) and one-electron reduced MoFe protein (symbolized as P/M^-). Specifically, they measured the Fe K-edge EXAFS of the $E_0 + E_1$ steady-state mixtures and subtracted the 50% E_0 contribution thus isolating an E_1 spectrum. EPR spectroscopy verified the $E_0 + E_1$ steady-state mixture instrumentally as the mixture exhibited a 50% reduction in the $S = 3/2$ FeMo-cofactor EPR signal compared to the control E_0 state sample. MoFe–protein with oxidized P–clusters were prepared by oxidation with indigodisulfonate (IDS). EPR spectroscopy again indicated the FeMo cofactor was unchanged while signals attributed to P–cluster oxidation appeared. The IDS-oxidized clusters are symbolized P^{2+}/M. To obtain appropriate values for EXAFS threshold energy shifts, the researchers then fitted spectra of model compounds containing iron–sulfur and iron–molybdenum–sulfur clusters with theoretical shift and amplitude functions while constraining bond distances to crystallographic values. Parameters taken from the model-compound-fit results were then fixed for the fits on MoFe protein EXAFS data.

The two clusters—P–cluster and FeMo cofactor or M center—found in MoFe protein present a challenging problem in separating and defining metal–metal and metal–sulfur distances. To solve this problem, a "split shell model" was used, in which the average M center short Fe–Fe distance for a thionine-oxidized (P^{2+}/M^{+}) MoFe protein was explored. The Fe–Fe short distance was found to be 2.61 Å compared to a P–cluster short Fe–Fe distance of 2.74 Å. Resting enzyme EXAFS bond distances of 2.58 Å for the M center and 2.67 Å for the P–cluster bonds were compared to X-ray crystallographic average short Fe–Fe distances of 2.60 Å for the M center and 2.71 Å for the P–cluster.[7] I find an average short Fe–Fe distance for P_N from reference 7 to be 2.66 Å rather than the 2.71 Å found by the reference 6 authors. This bond length is found by averaging the data of Table 6.2 (data taken from Table 3 of reference 7). These data and others are collected in Table 3.1. Long Fe–Fe bond distances or cross-cluster bond distances (such as from Fe_2–Fe_5 or Fe_3–Fe_6 in Figure 6.9) are also listed along with Fe–S and Fe–Mo distances.

Table 3.1 Metal–Metal and Iron–Sulfur Bond Distances in Oxidized, Resting, and Reduced Forms of Nitrogenase's MoFe Protein

Enzyme State	Experimental Technique	Fe–S (Å)	Fe–Fe (Å) Short M Center	Fe–Fe (Å) Short P–Cluster	Fe–Fe′ (Å) Long Cross-Cluster	Fe–Mo (Å)
Thionine oxidized (P^{2+}/M^{+})	EXAFS	2.29	2.61	2.74	3.74	2.72
IDS oxidized (P^{2+}/M)	EXAFS	2.31	2.59	2.70	3.74	2.71
Resting (P/M)	EXAFS	2.31	2.58	2.67	3.74	2.71
Resting (P/M)	X-ray diffraction		2.56	2.66	3.60	2.65
Reduced (P/M^{-})	EXAFS	2.33	2.54	2.66	3.72	2.66

Sources: Reference 6 (EXAFS) and 7 (X-ray diffraction).

It was concluded that a contraction of metal–metal distances for short Fe–Fe couples takes place as the nitrogenase enzyme is reduced. These data contrast with those for the great majority of documented $[Fe_4S_4]^{n+}$ metal clusters (as well as for most known metal–ligand clusters) where metal–metal distances expand upon electron addition. In EXAFS studies reported here, a slight expansion in Fe–S bond distances takes place during reduction while long or cross cluster Fe–Fe distances do not appear to change appreciably. More recent X-ray crystallographic studies of oxidized and reduced M centers and P–clusters of MoFe-protein (reference 7 of this chapter and those discussed in Section 6.5) confirm the decrease in Fe–Fe bond lengths for the P–cluster, while changes found for the M center are much less. These authors speculate that the structural changes accompanying oxidation of the P–cluster may be involved in coupling the necessary electron and proton transfer to FeMo cofactor as required for nitrogenase activity.

3.3 X-RAY CRYSTALLOGRAPHY

3.3.1 Introduction

X-ray crystallographic molecular structures of proteins have been available since the 1960s and 1970s when pioneering work by Kendrew[8] and Perutz[9] produced X-ray diffraction structures of myoglobin and hemoglobin. These oxygen carrying metalloproteins are discussed in Chapter 4. Since that time the introduction of sophisticated computer hardware and software has made the solution of protein structure in the solid state using X-ray crystallography more accurate and less time-consuming. The field continues to evolve as hardware and instrument design improvements are implemented and as crystallographers discover more powerful software algorithms for solving structures after the necessary data has been collected. At the time of this writing, 175+ X-ray crystallographic data sets were deposited in the Research Collaboratory for Structural Bioinformatics' Protein Data Bank (RCSB-PDB at http://www.rcsb.org/pdb/) for hemoglobin and hemoglobin mutants as well as 191+ data sets for myoglobin and myoglobin mutant species. Nuclear magnetic resonance protein structure determination in solution provides a complementary structural technique that does not require the production of single crystals necessary for X-ray diffraction studies. However, at this time, NMR solution structures are limited to smaller proteins of molecular weights less than 30,000. In contrast, X-ray crystallography can produce structures of proteins of up to 1×10^6 molecular weight. Recombinant DNA technology has aided the X-ray crystallographic study of proteins by allowing large amounts of a protein of interest to be produced through expression of its cloned gene in a microorganism. Site-directed mutagenesis of a selected protein's gene has allowed researchers to study three-dimensional structural changes brought about by amino acid replacement in the protein's primary amino acid sequence. These techniques are discussed in Sections 2.3.4 and 2.3.5. Much of the discussion in this section on X-ray crystallography has been taken from a recent text written by author and crystallographer

Jan Drenth.[10] Readers are referred to the Department of Crystallography site at Würzburg University (http://www.uni-wuerzburg.de/mineralogie/crystal/teaching/teaching.html) for tutorials on X-ray diffraction methodology. The site includes interactive tutorials describing basic examples, reciprocal space, the crystallographic phase problem, and diffuse scattering and defect structures. Tutorials on convolution theorem, modification of a structure, solving a simple structure, anomalous scattering, and powder diffraction are also found on this site.

3.3.2 Crystallization and Crystal Habits

The first requirement for protein structure determination using X-ray crystallography is to grow suitable crystals. Protein crystallization is mainly a trial-and-error procedure complicated by impurities contaminating the selected protein, by the poorly understood process of crystal nucleation, and by other unknown factors. Protein purity means not only that other compounds must be absent but also that all molecules of the protein have the same surface properties (charge distribution) as the latter affects crystal packing of the molecules. After purification the protein is dissolved in a suitable solvent (usually water–buffer mixtures for pH control) and the solution is brought to supersaturation so that nuclei for crystal growth appear. To avoid continued nucleation that will result in crystals too small for diffraction and to foster crystal growth, supersaturation must be decreased at this point. This is usually achieved by raising the temperature. The next step involves addition of a precipitant such as polyethyleneglycol (PEG), salt, or an organic solvent and perhaps adjustment of pH.

Crystallization techniques include liquid–liquid diffusion, several types of vapor diffusion, and dialysis through a membrane. Liquid–liquid diffusion of protein and precipitant solutions may conveniently take place in a melting point capillary tube using approximately 5 μl of each solution. The solution of higher density, which may be either the protein solution or the precipitant solution, is added to the capillary tube first using a syringe. The other solution is layered onto the top of the first, forming a sharp boundary. Slow diffusion of the layers may produce suitable crystals, although it may be necessary to test many variations of solvent, solute concentration, and diffusing solvent before the desired crystal size and purity is attained. The most common vapor diffusion method is called the *hanging drop method*. Drops containing 3–10 μl of protein and precipitant solutions are placed on a siliconized (to prevent spreading of the drop) microscope glass cover slip. The slip is placed upside down over a depression in a tray. The depression is partially filled with precipitant solution. Diffusion of vapors from or to the precipitant solution may result in crystal formation given the same caveats as mentioned above for solvent diffusion. Many different dialysis techniques are available. The advantage of this technique is that the protein solution is confined within a membrane allowing different precipitant solutions outside the membrane to be tried sequentially. Gilliland and Ladner have published a review of crystallization techniques (reference 11a), and Gilliland has assembled a database of macromolecule crystal and crystallization data (reference 11b).[11] To be suitable for X-ray crystallography, a crystal must reach dimensions of ≥ 0.2 mm, although smaller crystals may be used with newer diffraction and data collection techniques.

The flat faces of crystals reflect the regular packing of atoms, molecules, or ions within the crystalline structure, resulting in anisotropic (directional) behavior. Observation of this property is easily carried out by examining the crystals between crossed polarizers in a polarizing microscope. Single crystals will display birefringence—that is, beautiful colors in one orientation extinguished to darkness by rotation of the crystal 90°. The anisotropic birefringent behavior is caused by the crystal having different refractive indices in different crystal orientations. Although one cannot achieve atomic resolution with an electron microscope (resolution to 10–20 Å is possible, whereas the C–C distance in ethane, for instance, is 1.54 Å), it is possible to show the regular packing of molecules in crystals. Because of the regular arrangement of the crystal's atoms, molecules, or ions in planes with fixed angles and distances between the planes, single crystals will diffract X-rays. As an example, placing a crystal having a four-fold symmetry axis in an X-ray beam path and observing the X-ray pattern as the crystal oscillates over a small angle will produce reflection spots arranged in concentric circles. The concentric circle pattern of diffraction spots can be thought of as the intersection of a series of parallel planes with a sphere as shown in Figure 3.4. The planes represent the three-dimensional lattice that is not actually the crystal lattice but its reciprocal lattice. The unit distances in the lattice are reciprocally related to unit distances in the crystal, hence the name reciprocal lattice. The direction in which the X-ray beams are diffracted depends on two factors: the unit cell distances in the crystal and the X-ray wavelength. To be detected, the diffraction spots must be on, or pass through, the surface of the sphere represented by the circle in Figure 3.4. The radius of the sphere, called the sphere of reflection or the "Ewald sphere," is reciprocal to the

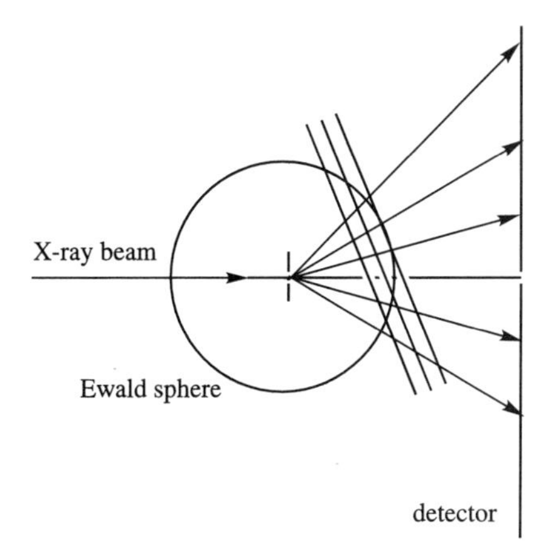

Figure 3.4 X-ray beam passing through the "Ewald sphere" and diffracted by planes in a single crystal produces reflection spots. (Adapted with permission from Figure 1.13 of Drenth, J. *Principles of Protein X-ray Crystallography*, 2nd ed., Springer-Verlag, New York, 1999. Copyright 1999 Springer-Verlag, New York.)

X-ray radiation wavelength—that is, $1/\lambda$. When the crystal is rotated, the reciprocal lattice rotates with it and different points within the lattice are brought to diffraction. The diffracted beams are called "reflections" because each of them can be regarded as a reflection of the primary X-ray beam against planes in the crystal.

It is beyond the scope of this text to discuss crystal habits in great detail, and the student is referred to discussion in reference 10. A short discussion follows here. Molecules of organic, organometallic, or inorganic materials, when precipitating from solution, attempt to reach their lowest free energy state. Frequently, this is accomplished by packing the molecules in a regular way in a crystalline habit. Regular packing of molecules in a crystal will define a unit cell through generation of three repeating vectors **a** (the x axis), **b**, (the y axis), and **c** (the z axis), with angles α, β, and γ between them. Planes can be constructed through lattice points, and these repeat regularly throughout the crystal. Parallel planes are equidistant with distance d between them. Lattice planes cut the x, y, and z axes into equal parts having whole numbers called indices. A set of lattice planes is determined by three indices h, k, and l if the planes cut the x axis as a/h, the y axis as b/k, and the z axis as c/l. In Figure 3.5A, lattice planes are shown in a two-dimensional lattice. In this figure, taken from Figure 3.5 of reference 10, $h = 2$ and $k = 1$. The lattice plane distance d is the projection of a/h, b/k, and c/l on the line perpendicular to the corresponding lattice plane $(h\,k\,l)$. If a set of planes is parallel to an axis, that particular index is 0 (the plane intercepts the axis at infinity). Therefore the unit cell is bounded by planes (100), (010), and (001), with the parentheses indicating $(h\,k\,l)$. Line segments are given in brackets; that is, [100] is the line segment from the

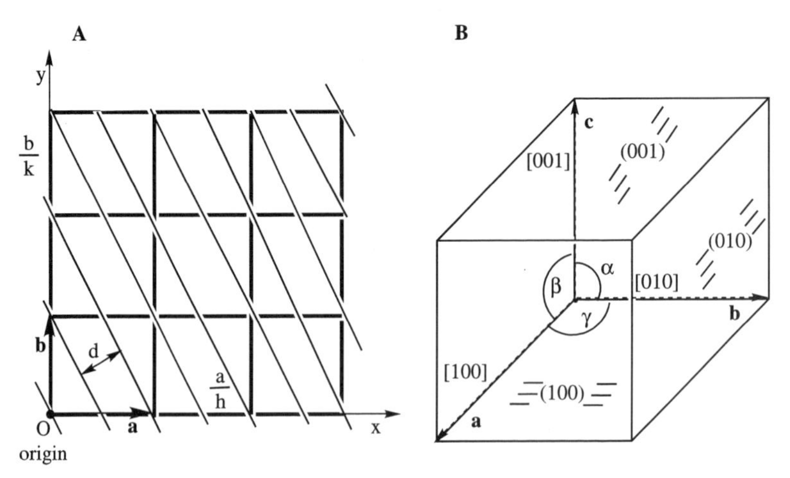

Figure 3.5 (A) Lattice planes in a two-dimensional lattice as taken from Figure 3.5 of reference 10 ($h = 2$, $k = 1$). (B) Unit cell bounded by planes (100), (010), (001). Directions along **a**, **b**, and **c** are indicated by [100], [010], and [001]. (Adapted with permission from Figure 3.6 of Drenth, J. *Principles of Protein X-Ray Crystallography*, 2nd ed., Springer-Verlag, New York, 1999. Copyright 1999, Springer-Verlag, New York.)

origin of the unit cell to the end of the *a* axis and [111] is the body diagonal from the origin to the opposite corner. These properties are illustrated in Figure 3.5B, adapted from Figure 3.6 of reference 10.

The unit cell considered here is a primitive (P) unit cell; that is, each unit cell has one lattice point. Nonprimitive cells contain two or more lattice points per unit cell. If the unit cell is centered in the (010) planes, this cell becomes a B unit cell; for the (100) planes, an A cell; for the (001) planes a C cell. Body-centered unit cells are designated I, and face-centered cells are called F. Regular packing of molecules into a crystal lattice often leads to symmetry relationships between the molecules. Common symmetry operations are two- or three-fold screw (rotation) axes, mirror planes, inversion centers (centers of symmetry), and rotation followed by inversion. There are 230 different ways to combine allowed symmetry operations in a crystal leading to 230 space groups.[12] Not all of these are allowed for protein crystals because of amino acid asymmetry (only L-amino acids are found in proteins). Only those space groups without symmetry (triclinic) or with rotation or screw axes are allowed. However, mirror lines and inversion centers may occur in protein structures along an axis.

Seven crystal systems as described in Table 3.2 occur in the 32 point groups that can be assigned to protein crystals. For crystals with symmetry higher than triclinic, particles within the cell are repeated as a consequence of symmetry operations. The number of asymmetric units within the unit cell is related but not necessarily equal to the number of molecules in a unit cell, depending on how the molecules are related by symmetry operations. From the symmetry in the X-ray diffraction pattern and the systematic absence of specific reflections in the pattern, it is possible to deduce the space group to which the crystal belongs.

Table 3.2 The Seven Crystal Systems

Crystal System	Conditions Imposed on Cell Geometry	Minimum Point Group Symmetry
Triclinic	None	1
Monoclinic	$\alpha = \gamma = 90°$ (*b* is the unique axis; for proteins this is a twofold axis or screw axis) or $\alpha = \beta = 90°$ (*c* is the unique axis; for proteins this is a twofold axis or screw axis)	2
Orthorhombic	$\alpha = \beta = \gamma = 90°$	222
Tetragonal	$a = b$; $\alpha = \beta = \gamma = 90°$	4
Trigonal	$a = b$; $\alpha = \beta = 90°$; $a = b$; $\gamma = 120°$ (hexagonal axes) or $a = b = c$; $\alpha = \beta = \gamma$ (rhombohedral axes)	3
Hexagonal	$a = b$; $\alpha = \beta = 90°$; $\gamma = 120°$	6
Cubic	$a = b = c$; $\alpha = \beta = \gamma = 90°$	23

Source: Adapted from reference 12.

In summary, it is important to know crystal quality, unit cell dimensions of the crystal (a larger crystal absorbs X rays more strongly, 0.3–0.5 mm is considered the optimal size), the crystal's space group, and how many protein molecules are in the unit cell and in one asymmetric unit. The great majority of crystals useable for X-ray crystallography are not ideal but contain lattice defects. This is true for protein crystals, which are also weak scatterers because the great majority of the component atoms are light atoms, namely, C, N, and O. One deduces the space group from the symmetry in the crystal's diffraction pattern and the systematic absence of specific reflections in that pattern. The crystal's cell dimensions are derived from the diffraction pattern for the crystal collected on X-ray film or measured with a diffractometer. An estimation of Z (the number of molecules per unit cell) can be carried out using a method called V_M, proposed by Matthews.[13] For most protein crystals the ratio of the unit cell volume and the molecular weight is a value around 2.15 $Å^3$/Da. Calculation of Z by this method must yield a number of molecules per unit cell that is in agreement with the decided-upon space group.

3.3.3 Theory and Hardware

The mathematics necessary to understand the diffraction of X rays by a crystal will not be discussed in any detail here. Chapter 4 of reference 10 contains an excellent discussion. The arrangement of unit cells in a crystal in a periodic manner leads to the Laue diffraction conditions shown in equations 3.3 where vectors **a**, **b**, and **c** as well as lattice indices h, k, and l have been defined in Figure 3.5 and **S** is a vector quantity equal to the difference between the resultant vector **s** after diffraction and the incident X-ray beam wave vector $\mathbf{s_0}$ so that $\mathbf{S} = \mathbf{s} - \mathbf{s_0}$.

$$\begin{aligned} \mathbf{a} \cdot \mathbf{S} &= h \\ \mathbf{b} \cdot \mathbf{S} &= k \\ \mathbf{c} \cdot \mathbf{S} &= l \end{aligned} \tag{3.3}$$

The same crystalline arrangement leads to the expression of Bragg's law applied to X-ray diffraction with incident X-ray beam of wavelength λ as shown in equation 3.4 and where the terms are defined as in Figures 3.5 and 3.6.

$$\lambda = 2d \sin \theta \tag{3.4}$$

As mentioned above, the formalism of the reciprocal lattice is convenient for constructing the directions of diffraction by a crystal. In Figure 3.4 the Ewald sphere was introduced. The radius of the Ewald sphere, also called the sphere of reflection, is reciprocal to the wavelength of X-ray radiation—that is, $1/\lambda$. The reciprocal lattice rotates exactly as the crystal. The direction of the beam diffracted from the crystal is parallel to MP in Figure 3.7 and corresponds to the orientation of the reciprocal lattice. The reciprocal space vector $\mathbf{S}(h\,k\,l) = \mathbf{OP}(h\,k\,l)$ is perpendicular to the reflecting plane $h\,k\,l$, as defined for the vector **S**. This leads to the fulfillment of Bragg's law as $|S(h\,k\,l)| = 2(\sin\theta)/\lambda = 1/d$.

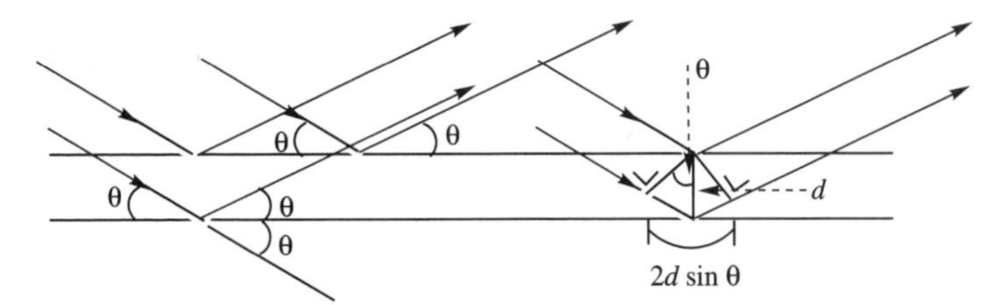

Figure 3.6 Two lattice planes separated by distance *d*. Incident and reflected X-ray beams make the angle θ with the lattice planes. (Adapted with permission from Figure 4.17 of Drenth, J. *Principles of Protein X-Ray Crystallography*, 2nd ed., Springer-Verlag, New York, 1999. Copyright 1999, Springer-Verlag, New York.)

Molecules and atoms within a crystal are not in static positions but vibrate around an equilibrium position. Atoms around the periphery of a molecule will vibrate to a greater extent, whereas central atoms will have relatively fixed positions. The resultant weakening of X-ray beam intensity, especially at high scattering angles, is expressed as the temperature factor. In the simplest case, components of the atom's vibration are all in the same direction, called the isotropic case. The component perpendicular to the reflecting plane and thus along **S** is equal for each $(h\,k\,l)$, and the temperature correction factor for isotropic atomic scattering is given by equation 3.5.

$$T_{(\mathrm{iso})} = \exp\left[-B\frac{\sin^2\theta}{\lambda^2}\right] = \exp\left[-\frac{B}{4}\left(\frac{2\sin\theta}{\lambda}\right)^2\right] = \exp\left[-\frac{B}{4}\left(\frac{1}{d}\right)^2\right] \tag{3.5}$$

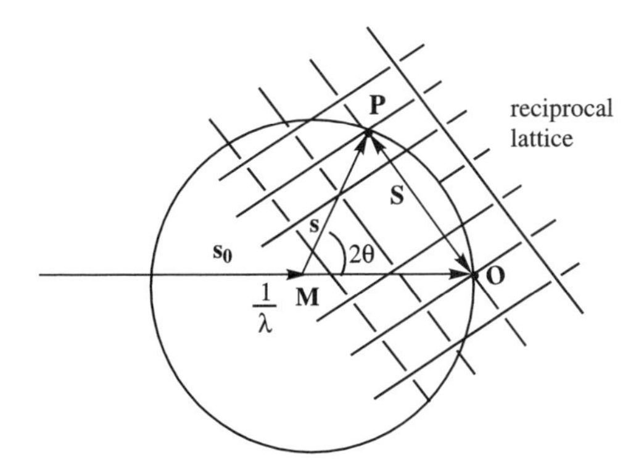

Figure 3.7 The Ewald sphere used to construct the direction of the scattered beam. The sphere has radius $1/\lambda$. The origin of the reciprocal lattice is **O**. The incident X-ray beam is labeled $\mathbf{s_0}$ and the scattered beam is labeled **s**. (Adapted with permission from Figure 4.19 of Drenth, J. *Principles of Protein X-Ray Crystallography*, 2nd ed., Springer-Verlag, New York, 1999. Copyright 1999, Springer-Verlag, New York.)

Assuming isotropic and harmonic vibration, the thermal parameter B becomes the quantity shown in equation 3.6, where $\bar{u}^2$ is the mean square displacement of the atomic vibration:

$$B = 8\pi^2 \times \bar{u}^2 \tag{3.6}$$

For anisotropic vibration the temperature factor is more complex because $\bar{u}^2$ now depends on the direction of **S**. The anisotropic temperature factor is often displayed in the form of an ellipsoid of vibration, commonly so that the vibrating atom has a 50% chance of being within the ellipsoid. This is illustrated in Figure 3.8 for a Kitajima group copper compound, which is a Cu(I) product of the reaction of $[Cu(II)(HB(3,5\text{-}i\text{-}Pr_2pz)_3]_2(O_2)$ with carbon monoxide. $[Cu(II)(HB(3,5\text{-}i\text{-}Pr_2pz)_3]_2(O_2)$ was designed as model for dioxygen binding in hemocyanin ($[HB(3,5\text{-}i Pr_2pz)_3]$ = hydrotris(3,5-diisopropylpyrazolyl)borate anion).[14] This so-called ORTEP view of $[Cu(I)(CO)(HB(3,5\text{-}i\text{-}Pr_2pz)_3]$ is drawn using 30% probability ellipsoids. Note that the outer atoms show greater uncertainty in position as seen in their elliptical rather than spherical shape. Section 5.3 contains a lengthy discussion of copper enzyme model compounds.

For proteins the X-ray structures usually are not determined at high enough resolution to use anisotropic temperature factors. Average values for B in protein structures range from as low as a few $Å^2$ for well-ordered structures to 30 $Å^2$ for structures involving flexible surface loops. Using equation 3.6, one can calculate the root mean square displacement $\sqrt{u^2}$ for a well-ordered protein structure at approximately 0.25 Å (for $B = 5\ Å^2$) and for a not-so-well-ordered structure at

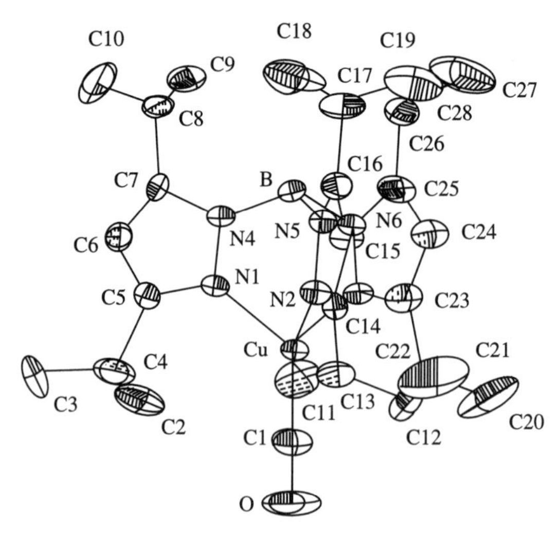

Figure 3.8 Ellipsoids of vibration for a small molecule. (Reprinted with permission from Kitajima, N.; Fujisawa, K; Fujimoto, C.; Moro-oka, Y.; Hashimoto, S.; Kitagawa, T.; Toriumi, K.; Tatsumi, K.; Nakamura, A. *J. Am. Chem. Soc.*, 1992, **114**, 1277–1291. Copyright 1992, American Chemical Society.)

0.62 Å (for $B = 30\,\text{Å}^2$). These seemingly small errors in atomic positions of C, N, and O atoms derive from the fact that the bond distances and angles for individual amino acids in small compounds are well known, and it is assumed that these do not change when the amino acids are incorporated into large protein molecules. In fact the limited resolution of a protein X-ray diffraction pattern does not permit calculation of an electron density map at atomic resolution, although amino acid residues can be distinguished from differences in their side chains. Usually these are displayed in stereo diagrams such as seen in Figure 4.23 of reference 10. Figure 6.4 presents a stereodiagram of the Cα trace for Av2 Fe–protein discussed in Section 6.4.

In addition to the dynamic disorder caused by temperature-dependent vibration of atoms, protein crystals have static disorder due to the fact that molecules, or parts of molecules, do not occupy exactly the same position or do not have exactly the same orientation in the crystal unit cell. However, unless data are collected at different temperatures, one cannot distinguish between dynamic and static disorder. Because of protein crystal disorder, the diffraction pattern fades away at some diffraction angle θ_{max}. The corresponding lattice distance d_{min} is determined by Bragg's law as shown in equation 3.7:

$$d_{min} = \frac{\lambda}{2 \sin \theta_{max}} \tag{3.7}$$

The end result of an X-ray structural determination reports the electron density in the crystal. The fundamental equation for its calculation follows in equation 3.8:

$$\rho(x\,y\,z) = \frac{1}{V}\sum_h \sum_k \sum_l |F(h\,k\,l)| \exp[-2\pi i(hx + ky + lz) + i\alpha(h\,k\,l)] \tag{3.8}$$

The electron density at every position x, y, z in the unit cell, $\rho(x\,y\,z)$, is the Fourier transform of the structure factor $\mathbf{F}(h\,k\,l)$, which is in turn a function of the electron density distribution in the unit cell and the integrated intensity of the reflected beam, called $I(h\,k\,l)$. Values of $I(h\,k\,l)$, the integrated reflected beam intensity, are obtained from the diffraction pattern after the application of correction factors. The term $|F(h\,k\,l)|$ is the structure factor amplitude of reflection $(h\,k\,l)$ including the temperature factor, $\alpha(h\,k\,l)$ is the phase angle, and x, y, and z are coordinates in the unit cell. The factor i is the mathematical imaginary term. The summation occurs over all the discrete directions in which diffraction by the crystal occurs. Sophisticated computer programs have been designed to calculate the electron density in the above equation except for the phase angles $\alpha(h\,k\,l)$, which cannot be derived in a straightforward manner from the diffraction pattern. Several methods have been developed to solve the phase angle problem as listed below.

1. The isomorphous replacement method requires attachment of heavy atoms to protein molecules in the crystal. In this method, atoms of high atomic number are attached to the protein, and the coordinates of these heavy atoms in the unit cell are

determined. The X-ray diffraction pattern of both the native protein and its heavy atom derivative are determined. Application of the so-called *Patterson* function determines the heavy atom coordinates. Following the refinement of heavy atom parameters, the calculation of protein phase angles proceeds. In the final step the electron density of the protein is calculated.

2. The multiple-wavelength anomalous diffraction method relies upon sufficiently strong anomalously scattering atoms in the protein structure itself. In this method, diffraction data must be collected at a number of different wavelengths, usually requiring data collection with synchrotron radiation. Anomalous scattering by an atom arises from the fact that its electrons cannot be considered as completely free electrons. The effect is negligible for light atoms C, N, and O but becomes useful for heavier atoms (S onwards). Anomalous scattering can be exploited for protein phase angle determination by the multiple-wavelength anomalous dispersion (MAD) method. Determination of the absolute configuration of a protein structure is also possible.

3. The molecular replacement method assumes similarity of the unknown structure to a known one. This is the most rapid method but requires the availability of a homologous protein's structure. The method relies on the observation that proteins which are similar in their amino acid sequence (homologous) will have very similar folding of their polypeptide chains. This method also relies on the use of Patterson functions. As the number of protein structure determinations increases rapidly, the molecular replacement method becomes extremely useful for determining protein phase angles.

4. The so-called direct methods rely on the principles that phase information is included in the intensities, that electron density is always positive, and that the crystal contains atoms that are or may be considered equal. Phase relations based on probability theory have been formulated and applied to clusters of reflections. Direct methods are still under development for proteins, although they are standard techniques for determining phase angles in smaller molecules.

A preliminary structural model of a protein is arrived at using one of the methods described above. Calculated structure factors based on the model generally are in poor agreement with the observed structure factors. The agreement is represented by an R-factor defined as found in equation 3.9 where k is a scale factor:

$$R = \frac{\sum_{hkl} ||F_{\text{obs}}| - k|F_{\text{calc}}||}{\sum_{hkl} |F_{\text{obs}}|} \times 100 \tag{3.9}$$

Refinement takes place by adjusting the model to find closer agreement between the calculated and observed structure factors. For proteins the refinements can yield R-factors in the range of 10–20%. An example taken from reference 10 is instructive. In a refinement of a papain crystal at 1.65-Å resolution, 25,000 independent X-ray reflections were measured. Parameters to be refined were the positional parameters (x, y, and z) and one isotropic temperature factor parameter

(*B*) for each of the 2000 nonhydrogen atoms in the molecule. Four times 2000 yields 8000 parameters. With 25,000 measurements the ratio of observations to parameters is slightly more than 3, a poor (low) overdetermination. The number of "observations" is increased by incorporating bond length and angle data from small molecules. Another technique, called "solvent flattening," which adjusts for disordered solvent molecules in the channels between protein molecules, is imposed. In any case, because protein structures are large and complex, their refinement is a large project computationally requiring fast computers and application of fast Fourier transform methods. Application of refinement to the papain structure would probably result in an *R*-factor of about 16% (usually reported as a decimal, i.e., $R = 0.16$). Application of refinement methods cannot completely rule out wrong interpretations of electron density maps and consequent entirely or partially incorrect protein structures. A number of checks for gross errors are available, including stereochemistry checks for detailed analysis of protein geometry. One of these, PROCHECK,[15] conducts a detailed analysis of all geometric aspects of proteins using data extracted from the Cambridge Crystallographic Data Centre (http://www.ccdc.cam.ac.uk) for bond lengths, angles, and planarity in peptide structures. For torsion angles, PROCHECK uses comparison data from high-resolution protein structures in the Protein Data Bank (PDB; http://www.rcsb.org/pdb/).

The hardware necessary for collection of X-ray diffraction data include an X-ray source and X-ray detector. Most commonly, the radiation is emitted from a copper source and has a wavelength of 1.5418 Å. X-ray sources include: (1) sealed X-ray tubes; (2) a more powerful system having a tube with a rotating anode but requiring maintenance of high vacuum; and (3) the most powerful source, a particle accelerator such as a synchrotron or storage ring, an extremely large and expensive facility. Protein crystallography benefits from use of the rotating anode tube because the X rays emitted are of higher intensity. Extremely high intensity X-ray radiation from a synchrotron is of value when collecting data from weakly diffracting crystals. Synchrotron radiation is also tunable to wavelengths at or below 1 Å, an advantage for certain detector types. A crystal diffracts shorter wavelength radiation more weakly, but the crystal suffers less radiation damage because it absorbs less radiation at the shorter wavelength. X-ray detectors also come in several varieties: (1) single-photon counters that give accurate results but require up to several weeks to acquire the 10,000–100,000 (10^4–10^5) reflections necessary to compile a complete data set for a protein crystal; (2) image plates that operate much like photographic film but are 10 times more sensitive; and (3) area detectors, electronic devices that detect X-ray photons on a two-dimensional surface. Both fluorescent-type detectors, image plates and area detectors, are more sensitive at the shorter wavelengths of X-ray radiation from synchrotron sources.

3.3.4 Descriptive Examples

X-ray crystallographic structures of the enzyme nitrogenase first became available in 1992 with refinements of the structures continuing to the present time. As of this

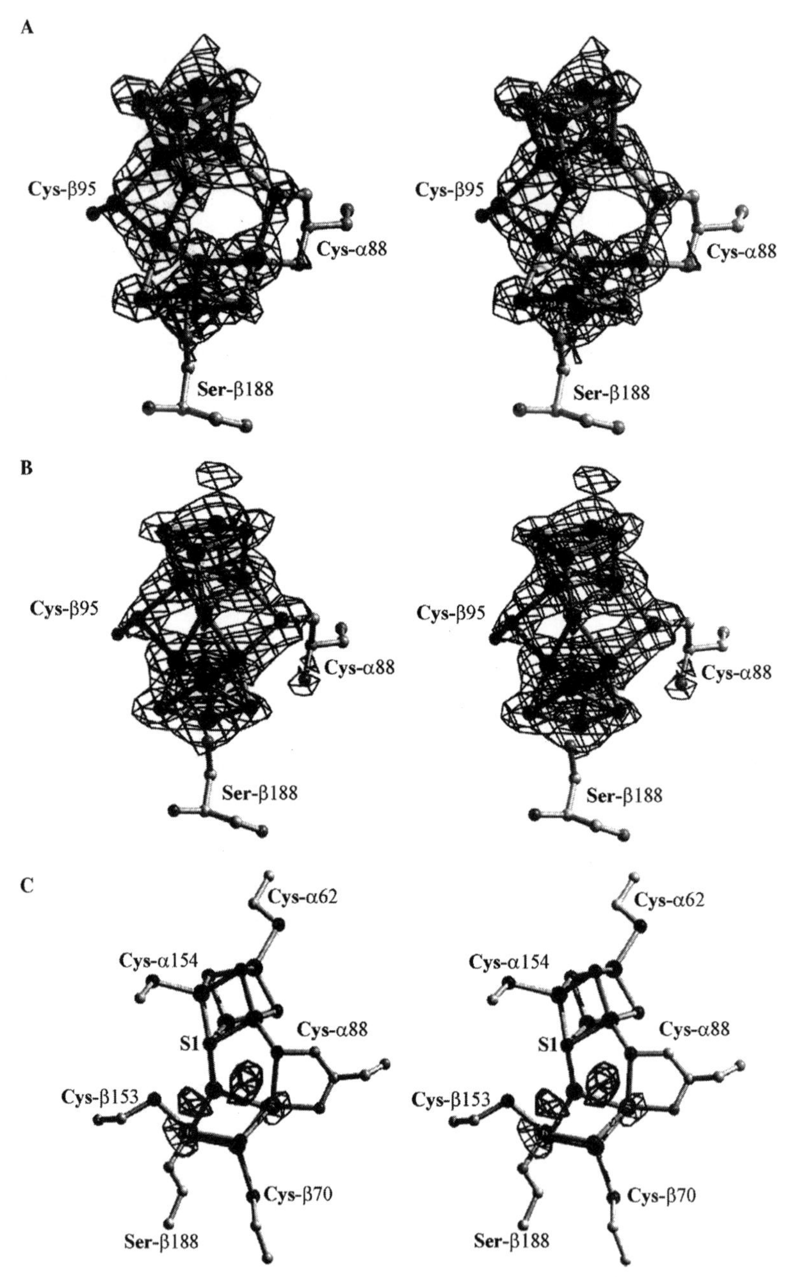

Figure 3.9 Stereoview of 2 $F_{obs} - F_{calc}$ electron density maps at 2-Å resolution for (A) P^{OX} and (B) P^{N}. (C) Stereoview of the $F_{obs} - F_{obs}$ (reduced minus oxidized) electron density maps showing differences in the vicinity of the P cluster. (Reprinted with permission from Figure 3 of Peters, J. W.; Stowell, M. H. B.; Soltis, S. M.; Finnegan, M. G.; Johnson, M. K.; Rees, D. C. *Biochemistry*, 1997, **36**, 1181–1187. Copyright 1997, American Chemical Society.)

writing, data on 16 nitrogenase crystallographic structures have been deposited in the PDB. Typical purification, crystallization, and data collection parameters are summarized here as taken from reference 7. Wild-type *azobacter vinelandii* was cultured and large-scale growth accomplished using 20 L carboys. Purification by gel filtration chromatography was accomplished under anaerobic conditions. MoFe–protein was crystallized by microcapillary batch diffusion using a combination of reagents and buffers as precipitating solution. Dark brown MoFe–protein crystals grew in 1–1.5 months to an average size of $0.1 \times 0.3 \times 0.8\,\text{mm}^3$ in space group $P2_1$. These were considered to be the MoFe–protein oxidized state having both the P–cluster and the FeMo cofactor in an oxidized state (P^{OX}/M^{OX}). The P^{OX}/M^{OX} crystals showed no EPR resonances, indicating that both the P–cluster and FeMo cofactor spin states corresponded to their oxidized state. The MoFe–protein crystals were assayed for nitrogenase acetylene reduction activity. Some crystals treated with sodium dithionite as a reducing agent were considered to be in a reduced state equivalent to the native (resting) enzyme (P^N/M^N). In the P^N/M^N state, EPR studies indicated an $S = 3/2$ state typical of the FeMo cofactor in its reduced state. At approximately $-180°\text{C}$, crystals were flash-cooled in liquid nitrogen and data were collected under a continuous nitrogen stream. Data were collected using synchrotron radiation processed using DENZO[15d] and SCALEPACK,[15d] refined in parallel with X-PLOR,[15e] and structurally analyzed using PROCHECK.[15a,b] Stereoviews of the P^N and P^{OX} molecules are shown in Figure 3.9 as reprinted from Figure 3 of reference 7. The stereoviews are obtained from 2 $F_{obs} - F_{calc}$ electron density maps at 2-Å resolution. The electron density map is contoured at 3.5 σ for Figures 3.9A and 3.9B ($\sigma =$ the standard deviation of the measurements). Section 3.13 contains some useful hints for productive viewing of stereodiagrams such as those of Figure 3.9.

3.4 ELECTRON PARAMAGNETIC RESONANCE

3.4.1 Theory and Determination of *g*-Values

In electron paramagnetic resonance (EPR) spectroscopy or electron spin resonance (ESR), radiation of microwave frequency is absorbed by molecules, ions, or atoms (organic or inorganic) containing a paramagnetic center—that is, a system with one or more unpaired electrons. In EPR, the unpaired electron spin moment ($m_s = \pm 1/2$ for a free electron) interacts with an applied magnetic field producing the so-called Zeeman effect. The effect is illustrated in Figure 3.10 for the electron spin functions α and β corresponding to $m_s = +1/2$ and $m_s = -1/2$.

Three basic equations (3.10–3.12) are needed to describe the technique. In the equations, μ is the magnetic moment of the electron, sometimes also written as μ_e, g is called the g factor or spectroscopic splitting factor, **S** is defined as the total spin associated with the electron (in bold type because it is considered as a vector), **B** is the imposed external magnetic field (also defined as a vector quantity), and

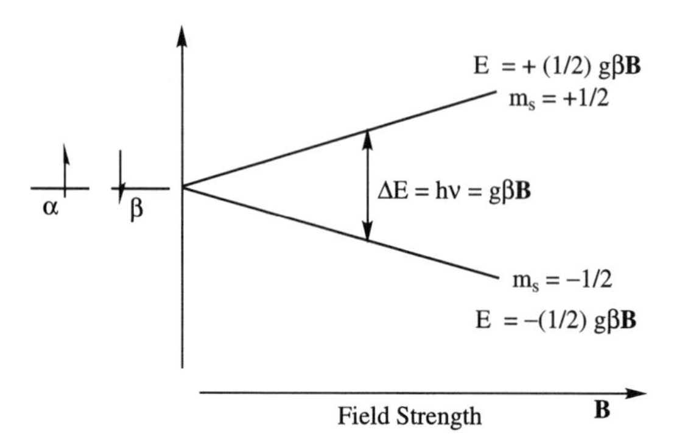

Figure 3.10 Removal of degeneracy of the α and β electron spin states by a magnetic field. (Adapted with permission from Figure 2.16 of Cowan, J. A. *Inorganic Biochemistry, An Introduction*, 2nd ed., Wiley-VCH, New York, 1997. Copyright 1997, Wiley-VCH.)

$\beta = (e/2m) \times (h/2\pi)$ and is called the Bohr magneton.

$$\mu = -g\beta\mathbf{S} \tag{3.10}$$

$$E = \mu \cdot \mathbf{B} \tag{3.11}$$

$$\Delta E = h\nu \tag{3.12}$$

Replacing μ by its equivalent operator from equation 3.10 and replacing *E* by *H*, the Hamiltonian, equation 3.11, becomes

$$H = g\beta\mathbf{S} \cdot \mathbf{B} \tag{3.13}$$

or by replacing the dot product by m_s, the projection of **S** onto **B**, and multiplying by the magnitude of **B**, equation 3.14 is written as

$$E = g\beta m_s B \tag{3.14}$$

Giving m_s its two values, $\pm 1/2$, one obtains equation 3.15:

$$\Delta E = E_{1/2} - E_{-1/2} g\beta \mathrm{B} \tag{3.15}$$

If one then defines the resonance condition, B_R (alternately written as B_0 or as B_L when referring to the laboratory field associated with a particular EPR instrument system), as the magnetic field at which the energy of the transition comes into resonance with the field, one finds equation 3.16 or, more usefully, equation 3.17:

$$\Delta E = h\nu = g\beta B_R \tag{3.16}$$

$$g = \frac{h\nu}{\beta B_R} \tag{3.17}$$

From the knowledge of the spectrometer's operating frequency (held constant) and the magnetic field intensity at which maximum EPR absorption occurs as one varies the magnetic field, one easily calculates g from equation 3.17.

One can calculate the ratio of populations of spin-up to spin-down electron orientations at room temperature ($T = 300\,\mathrm{K}$) from the Boltzmann formula finding that N^+/N^- is approximately equal to one (0.999), indicating that there is about a 0.1% net excess of spins in the more stable, spin-down orientation at room temperature. Using the same mathematical expression, this difference in populations can be shown to increase as the temperature is lowered. Actually, the EPR signal will be linearly dependent on $1/T$, and this linear dependence is called the Curie law. Because of the excited state population's temperature dependence, most EPR spectra are recorded at temperatures between 4 and 77 K.

The magnetic field strength at which ΔE (proportional to the microwave frequency as seen from equation 3.16) comes into resonance with the transition illustrated in Figure 3.10 will produce an intensity spike along the magnetic field abscissa. Because of line-broadening phenomena the typical absorption curve will look like Figure 3.11A. The absorption curve is routinely displayed as its first derivative (Figure 3.11B) because details of the spectrum are more easily detected.

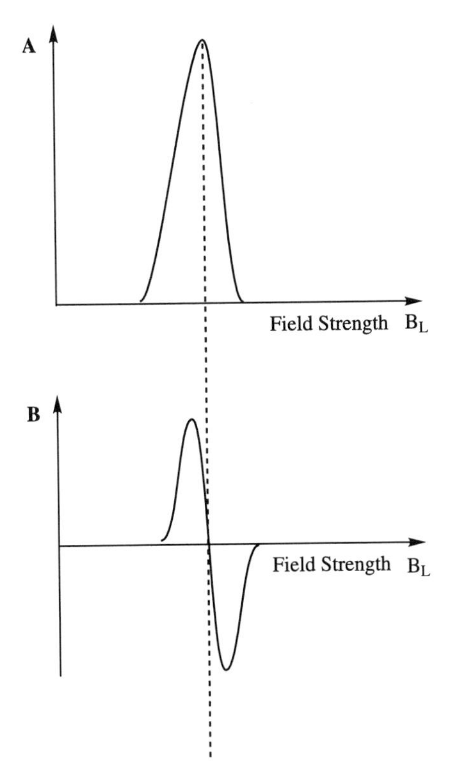

Figure 3.11 (A) Typical EPR absorption curve. (B) First derivative EPR absorption curve.

This chapter's discussion does not treat inorganic and organic free radicals and triplet states (such as dioxygen, O_2), which produce EPR spectra. Rather, the focus here will be on EPR behavior of transition metal centers that occur in biological species. An excellent presentation of the subject, written by Graham Palmer, is found in Chapter 3 of reference 1.[16] The discussion here is summarized mostly from that source.

When an electron is exposed to a magnetic field, B_L, the electron can be either stabilized ($m_s = -1/2$) or destabilized ($m_s = +1/2$), with the magnitude of the effect varying linearly with the intensity of B_L. This interaction of magnetic moments with B_L is called the Zeeman effect or Zeeman interaction. Figure 1 on page 126 of reference 16, along with Figures 3.10 and 3.11, relates the energy level diagram for the electron to the absorption and first-derivative modes of EPR spectral presentation. Standard EPR instrumentation utilizes a fixed frequency (usually the 9-GHz "X band") and a variable magnetic field. Other frequencies may also be used to enhance spectra, thereby increasing resolution of unresolved hyperfine structure (3 GHz) or resolution associated with g anisotropy (35 GHz).

If one calculates g from equation 3.17, the measurable experimental quantity would appear to be a single number of approximately 2. Observed g factors for paramagnetic metal ions range from <1 to 18 (measured for some lanthanide ions). Two phenomena, known as spin–orbit interactions (spin–orbit coupling) and zero-field splitting, are responsible for g factor deviations from the free electron value. Spin–orbit coupling arises because the magnetic dipole associated with the orbital momentum of the electrons (L) tends to align itself with the magnetic dipole due to the electrons' intrinsic spin (S). Spin–orbit coupling tends to be quenched if the metal ion exists in a ligand field that lifts the degeneracy of the d orbitals, and in practice the g factor value will lie somewhere between the free ion value (favored by spin–orbit coupling) and the free electron value (quenched spin–orbit coupling). The greater the lifting of the degeneracy of the d orbitals, the more effective is the quenching of spin–orbit coupling, and the closer g will be to free electron value. A simple spin–orbit interaction is illustrated in Figure 3.12 as adapted from Figure 2.18 of reference 3.

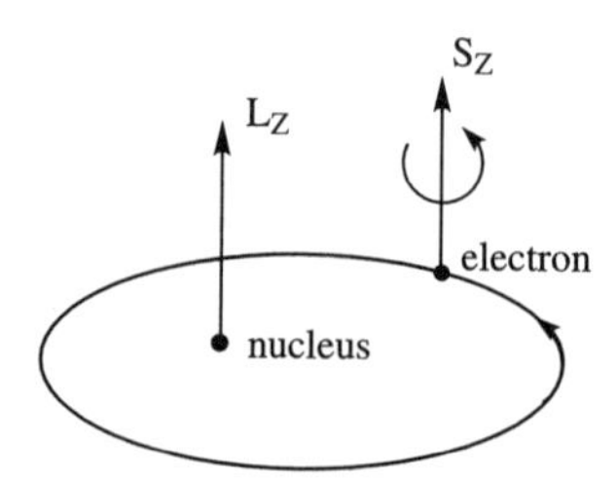

Figure 3.12 The motion of an electron in orbit about a nucleus generates an orbital momentum (L) adding a component to the magnetic field experienced by the electron spin (S). (Adapted with permission from Figure 2.18 of Cowan, J. A. *Inorganic Biochemistry, An Introduction*, 2nd ed., Wiley-VCH, New York, 1997. Copyright 1997, Wiley-VCH.)

For the simple case of one unpaired electron ($S = 1/2$), the associated magnetic moment is not a simple number but is directionally oriented—that is, is anisotropic. Taking spin–orbit interactions into account for EPR spectra leads to four limiting cases:

1. $g_x = g_y = g_z$. The magnetic moment is independent of orientation. This case is called isotropic, and a single symmetric EPR absorption is obtained. The paramagnet can be represented by a sphere. The EPR spectrum will resemble Figures 3.11A, 3.11B, and 3.13A.

2. $g_x = g_y < g_z$. The paramagnet is represented by a football shape exhibiting a minor feature at low field (from g_z, often called $g_{\parallel}$ or g parallel) and a major feature at high field (from g_x and g_y, often called $g_{\perp}$ or g perpendicular). The two common g values (x, y) are usually referred to as $g_{\perp}$ (g perpendicular), while the unique g value is usually called $g_{\parallel}$ (g parallel). The spectrum is said to be axial. If Cu(II) d^9 ions in octahedral ligand fields exhibit z-axis elongated d-orbital splitting because of the Jahn–Teller effect (see Section 1.6 in Chapter 1 for a discussion of these topics), the EPR envelope will appear as shown in Figure 3.13B. All g values will be >2.

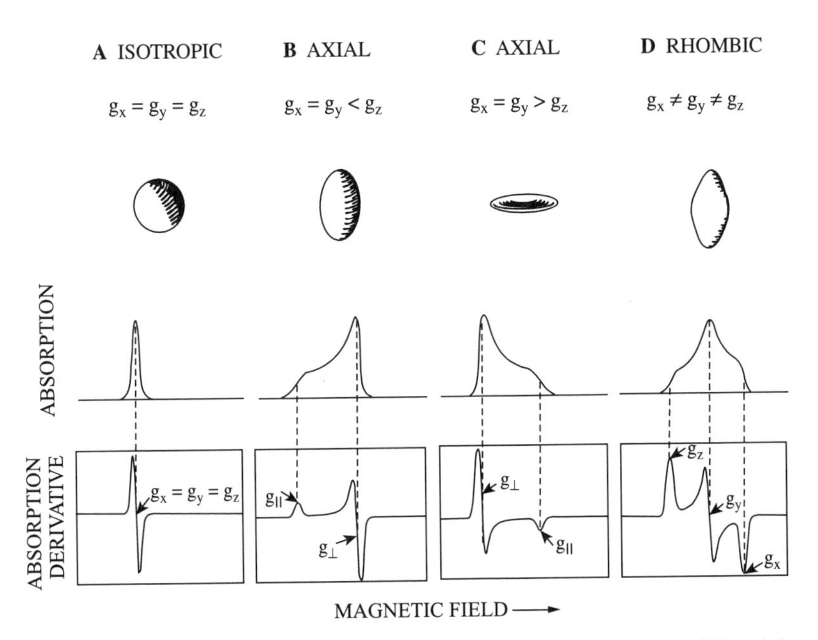

Figure 3.13 EPR absorption curves. (A) Isotropic spectrum, $g_x = g_y = g_z$ (B) Axial spectrum, $g_x = g_y < g_z$. (C) Axial spectrum, $g_x = g_y > g_z$. (D) Rhombic spectrum, $g_x \neq g_y \neq g_z$. (Reprinted with permission from Figure 4 of Palmer, G., in Que, L., ed. *Physical Methods in Bioinorganic Chemistry: Spectroscopy and Magnetism.* University Science Books, Sausalito, CA, 2000, pp. 121–185. Copyright 2000, University Science Books.)

3. $g_x = g_y > g_z$. The paramagnet is represented by a discus shape exhibiting a minor feature at high field (from g_z, $g_\parallel$ or g parallel) and a major feature at low field (from g_x and g_y, often called $g_\perp$ or g perpendicular). The spectrum is also said to be axial. It will appear as shown in Figure 3.13C.

4. $g_x \neq g_y \neq g_z$. The spectrum is said to be rhombic. Three different EPR values are recorded. Figure 3.13D shows this behavior.

When there is more than one unpaired electron in the paramagnetic center, zero-field splitting (zfs) will occur. Zero-field splitting is the separation in energy of the various m_s states in the absence of an applied magnetic field. It is the result of interelectronic interactions and ligand fields of low symmetry. The hamiltonian for zfs is written as equation 3.18, where D is the axial zfs parameter and E/D indicates the degree of rhombic distortion in the electronic environment. The zfs is applied as a correction to the energies of the individual spin states arising from spin–orbit coupling.

$$H_{\text{zfs}} = D\left[S_z^2 - \frac{1}{3}S^2 + \frac{E}{D}(S_x^2 - S_y^2)\right] \tag{3.18}$$

The case of high-spin Fe(III) ($S = 5/2$) is important because it is found in high-spin heme systems of hemoglobin and myoglobin (see Chapter 4). The parameter D is a directed quantity; that is, it characterizes the magnitude and direction of axial distortion. In the high-spin Fe(III) heme case the direction is normal to the heme plane. Actually, three cases are of interest for the $S = 5/2$ system. When $D = 0$ there is no zfs, the separation between all levels is the same, and each m_s level converges to a common origin at zero field. Therefore each level comes into resonance at the same value of B_L, and a single EPR line is seen. This simple behavior has never been observed in a biological system. When $0 < D < h\nu$, the levels are split in zero field because the paramagnet experiences a small asymmetry in its environment which separates the m_s values. The levels converge to different origins. When the zero-field separation is small, the levels can come into resonance with B_L and five separate resonances are observed. In the third case, $D \gg h\nu$, the zero-field splitting is very large. The separation between m_s $\pm 3/2$ and $\pm 5/2$ is too large for the transitions to be observed. Only transitions within the $m_s = \pm 1/2$ levels are observed, and these are very sensitive to the orientation of the paramagnet with respect to the applied field. The EPR spectrum would look like that of Figure 3.13C with greater separation of $g_\parallel$ from $g_\perp$ and a lower relative intensity for $g_\parallel$. It can be calculated that the $m_s = \pm 1/2$ transitions correspond to a $g_\parallel$ value of 2.0 and $g_\perp$ values of 6.0 ($2S + 1$ with $S = 5/2$). This is the circumstance found for hemes where the square-planar array of four porphyrin nitrogen ligands gives rise to large D values. A more complete discussion is found on pages 145–152 of reference 16; Figures 11 and 14 are particularly instructive, and the interested student is referred there.

3.4.2 Hyperfine and Superhyperfine Interactions

Hyperfine interactions in EPR spectra arise when the paramagnet finds itself in the vicinity of a nucleus having a nuclear spin and consequently a nuclear magnetic moment. The interactions may be of two types: (1) the hyperfine interaction—the unpaired electron interacting with the nucleus having a nuclear spin belongs to the same atom; and (2) the superhyperfine interaction—the unpaired electron interacting with the nucleus having a nuclear spin belongs to a different atom in the molecule.

The resonance expression relating the magnetic field and energy (frequency) as seen in equation 3.16, $\Delta E = h\nu = g\beta B_R$, has its analog for hyperfine interactions written as equation 3.19, where A is the hyperfine coupling constant in Hz:

$$hA = g\beta a \tag{3.19}$$

For g equal to 2, A will equal $2.8a$ (MHz). Usually A is expressed in reciprocal centimeters using the relation that $30\,\text{GHz} = 1\,\text{cm}^{-1}$. For $g = 2$, a hyperfine splitting of $10\,\text{mT}$ can be expressed as $280\,\text{MHz}$ or $0.0093\,\text{cm}^{-1}$. (Use $\beta = 9.2741 \times 10^{-21}\,\text{erg}\,\text{G}^{-1}$, $h = 6.626 \times 10^{-27}\,\text{erg s}$, $1\,\text{G} = 10^{-4}\,\text{T}$ to calculate that $10\,\text{mT} = 0.0093\,\text{cm}^{-1}$). Provided that z is parallel to the magnetic field, one can write

$$E = -\mu_z B_L \tag{3.20}$$

Assuming that B_{HF} (where HF = hyperfine) is much smaller than B_L, one can write equation 3.21 from equations 3.10, 3.14 and 3.20. The constant h is often omitted in this expression.

$$E = -\mu_z(B_L - \text{am}_\text{I}) = g\beta m_s B_L + A(h)m_s m_I \tag{3.21}$$

The hyperfine interaction is shown in Figure 21 of reference 16. The $M_s = \pm 1/2$ states of an $S = 1/2$ paramagnet interact with an $I = 1/2$ nuclear moment to create the hyperfine interaction. Interactions from $m_s = -1/2$ to $M_I = -1/2$ and $m_s = -1/2$ to $M_I = +1/2$, for instance, create the magnetic field specified as the hyperfine interaction A. Figure 21 of reference 16 describes the behavior for an $I = 1/2$ nuclear moment. The number of hyperfine lines will be equal to $2I + 1$ for nuclear moments greater than 1/2. Each hyperfine line will be of equal intensity when the electron is interacting with its own nucleus. For instance the $Cu^{2+}, I = 3/2$ nucleus will produce four hyperfine lines as described in the next section.

3.4.3 Descriptive Examples

As described in the previous section, four hyperfine lines will be found in the EPR spectrum of Cu,Zn-superoxide dismutase (CuZnSOD) because of the $I = 3/2$

nuclear spin of the copper nucleus. This behavior has been studied by many researchers, and one typical spectrum is shown in Figure 3.14A as reported by Valentine.[17] A second EPR spectrum, that for spinach plastocyanin, was reported in a *Chemical Reviews* article by Solomon and coworkers and is reprinted in Figure 3.14B.[18] Additional information on superoxide dismutases will be found in Section 5.2.3, and more data on blue copper proteins such as plastocyanin will be found in Section 5.2.2.

The superhyperfine interaction is observed for metal complexes in cases where the metal ligands have a nuclear moment. For instance, the nitrosyl (NO) complexes of iron(II) heme proteins have two inequivalent axial nitrogen ligands. The $^{14}N(I = 1)$ NO couples strongly to the unpaired electron, yielding a widely split triplet with each component of equal intensity and separated by 2.1 mT. The second

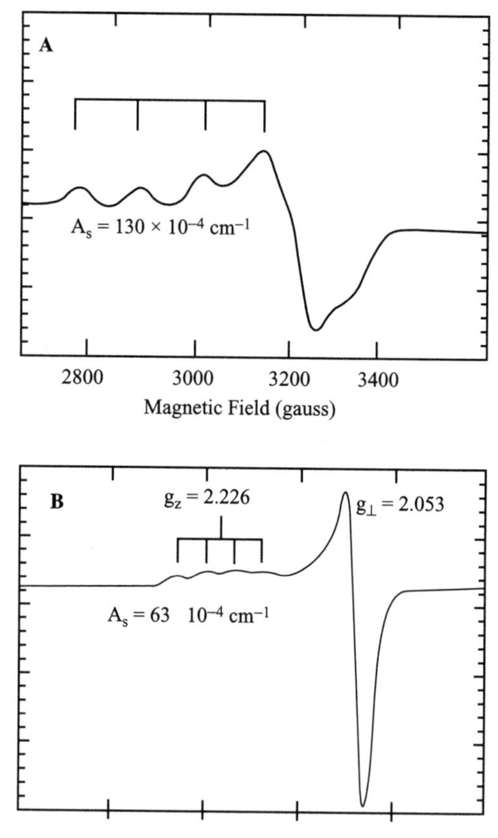

Figure 3.14 (A) EPR spectrum for superoxide dismutase. (Reprinted with permission from Figure 5 of Valentine, J. S.; DeFreitas, D. M. *J. Chem. Ed.*, 1985, **62**(11), 990–997. Copyright 1985, Division of Chemical Education, Inc.) (B) EPR spectrum for spinach plastocyanin. (Reprinted with permission from Figure 3b of Solomon, Edward I.; Baldwin, Michael J.; Lowery, Michael D. *Chem. Rev.*, 1992, **92**, 521–542. Copyright 1992, American Chemical Society.)

nitrogen of the histidine proximal ligand (see Figure 4.7 of Chapter 4) shows weaker coupling with the splitting about 0.7 mT. Difficulties arise in trying to distinguish between hyperfine and superhyperfine coupling in a bioinorganic system where both may be present. Palmer discusses the solutions to this problem on pages 163–165 of reference 16.

A rhombic EPR spectrum is found for the iron–molybdenum cofactor contained in the nitrogenase enzyme. Nitrogenase has been briefly described in Section 3.2.2, while Section 6.5.3 contains detailed information on FeMoco or the M center of nitrogenase.) The description of nitrogenase's EPR spectrum will be found in Section 3.6.4, along with information gained from ENDOR and Mössbauer studies of FeMo cofactor. The ENDOR technique just mentioned is one of a variety of pulsed EPR techniques that are available. Electron nuclear double resonance (ENDOR) and electron spin echo envelope modulation (ESEEM) are discussed in Chapter 4 of reference 1. The techniques can define electron–nuclear hyperfine interactions too small to be resolved within the natural width of the EPR line. For instance, because a paramagnetic transition metal center in a metalloprotein interacts with magnetic nuclei such as ^{1}H, ^{2}H, ^{13}C, ^{14}N, ^{15}N, ^{17}O, ^{31}P, or ^{33}S, these interactions may be detected by ENDOR or ESEEM analysis. Using these techniques, a particular ligand nuclei complexed to the metal may be identified. In favorable circumstances, metal–ligand bond distances and bond angles may be determined as well. ENDOR and ESEEM techniques will not be further discussed here. In the section on Mössbauer spectroscopy later in this chapter, ENDOR will be described as a complementary technique used to identify the seven iron ions in the enzyme nitrogenase's iron–molybdenum cofactor.[30]

3.5 NUCLEAR MAGNETIC RESONANCE

3.5.1 Theoretical Aspects

While chemists are usually concerned with the behavior of electrons orbiting the nucleus, the instrumental method called nuclear magnetic resonance (NMR) is a technique based on the properties of the atom's nucleus. Interaction of the atom's electrons with the nucleus yields important structural and chemical information that can be gathered using NMR techniques. Nuclear magnetic resonance spectroscopy can provide information on structure, dynamics, kinetics, binding processes, electronic structure, and magnetic properties of bioinorganic molecules in solution. Basic information about the NMR technique as summarized here is taken from reference 19.

Nuclei of natural isotopes (atoms of chemical elements differing in the number of neutrons in their nuclei) may possess angular momentum or spin and therefore magnetic moments. One defines spin by the following equation:

$$\text{spin} = \hbar[I(I+1)]^{1/2} \tag{3.22}$$

where $\hbar = h/2\pi$ and I = nuclear angular momentum $= 0, 1/2, 1, 3/2, \ldots$.

The nuclear angular momentum, I, is quantized with magnitude $\hbar m$, where $m = I, I-1, I-2, \ldots -I$, yielding $2I+1$ equally spaced spin states with angular quantum numbers I. Additionally, a nucleus with spin has a magnetic moment, μ. Components of μ having different spin states $\mu m/I$ yield $2I+1$ components for μ. Application of an external magnetic field splits spin states into different potential energy states, accounting for the origin of the nuclear magnetic resonance phenomenon.

The angular momentum and magnetic moment of nuclei act as parallel or antiparallel vectors, and the ratio between these is known as the magnetogyric ratio, γ. The magnetogyric ratio is given in terms of equation 3.23:

$$\gamma = \frac{2\pi}{h}\frac{\mu}{I} = \frac{\mu}{I\hbar} \tag{3.23}$$

For $I > 1/2$, nuclei also possess an electric quadrupole moment, Q. Quadrupolar nuclei exchange energy with electric fields in the rest of the molecule in which they are located causing profound effects on NMR spectra. Table 3.3 lists properties for some nuclei found in bioinorganic systems.

Receptivity, or natural signal strength, of a nucleus as listed in Table 3.3 depends on the intrinsic sensitivity of the nucleus weighted by its natural abundance. Usually the receptivity of a nucleus is high if the magnetic moment, μ, is high. In Table 3.3 the ^{13}C nucleus, which has a fairly weak signal and natural abundance of 1.108%, is arbitrarily given a receptivity = 1.00 with the receptivity of all other nuclei calculated relative to it. The resonant frequency in a particular magnetic field (2.348 T in Table 3.3) is proportional to the magnetogyric ratio and will vary slightly according to the chemical and electronic environment in which the nucleus finds itself. This variation results in the "chemical shift effect" (to be explained further below), one of the most useful results of nuclear magnetic resonance spectroscopy.

Nuclei possessing even numbers of both protons and neutrons have angular momentums equal to $0 (I = 0)$ and are magnetically inactive. Two major isotopes, ^{12}C and ^{16}O, belong to this group, a fact that simplifies the NMR spectra of organic molecules containing them. Odd numbers of either (or both) protons or neutrons lead to nonzero spins, with the actual number of spin states being dependent on the number of possible arrangements. Because s bonding electrons have charge density within the nucleus and will perturb nuclear spin states, ultimately the NMR technique leads to information about molecular arrangements about the nucleus under study. This is the information being sought in the NMR experiment.

Placing the nucleus with $I = 1$ and a positive magnetogyric ratio in a particular magnetic field $\mathbf{B}_0$ (aligned along a z axis) leads to $2I+1$ arrangements at differing potential energies split by $(m\mu/I)\mathbf{B}_0$, as shown in Figure 3.15A (adapted from Figure 1.1a of reference 19).

Alignment of vectors relative to $\mathbf{B}_0$ for each value of m results in vectors of length $\hbar[I(I+1)]^{1/2}$ and a z component $\hbar[I(I+1)]^{1/2}$ where $\cos\theta = m/[I(I+1)]^{1/2}$, as shown in Figure 3.15B. Differing energy states lead to interaction with

Table 3.3 Magnetic Properties of Selected Nuclei

Element	Atomic Weight	Spin, I	Natural Abundance (%)	Receptivity (^{13}C =1.00)	Quadrupole Moment (10^{-30} m^2)	Resonant Frequency (MHz) at 2.348 T (23.48 kG)
Hydrogen	1	1/2	99.985	5670	None	100.00
Carbon	13	1/2	1.108	1.00	None	25.15
Nitrogen	14	1	99.63	5.70	1.67	7.23
Nitrogen	15	1/2	0.37	0.022	None	10.14
Oxygen	17	5/2	0.037	0.061	−2.6	13.56
Phosphorus	31	1/2	100	377	None	40.48
Manganese	55	5/2	100	1014	40	24.84
Iron	57	1/2	2.19	0.00425	None	3.24
Cobalt	59	7/2	100	1560	42	23.73
Nickel	61	3/2	1.19	0.24	16	8.93
Copper	63	3/2	69.09	368	−22	26.51
Zinc	67	5/2	4.11	0.67	15	6.25
Molybdenum	95	5/2	15.72	2.92	−1.5	6.55
Iodine	127	5/2	100	541	−79	20.15
Platinum	195	1/2	33.8	19.9	None	21.50
Praseodymium	141	5/2	100	1620	−4.1	29.03
Europium	151	5/2	47.82	464	114	24.28
Dysprosium	163	5/2	24.97	1.79	251	4.77
Ytterbium	171	1/2	14.31	4.5	None	17.70
Gold	197	3/2	100	0.153	55	1.75
Bismuth	209	9/2	100	819	−37	16.36

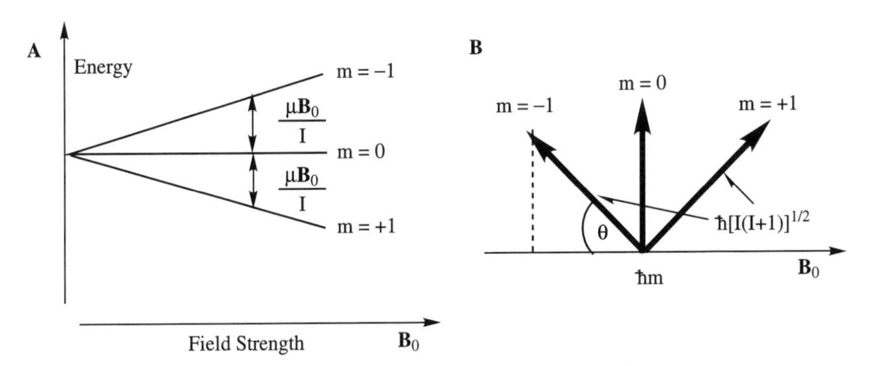

Figure 3.15 (A) Nuclear spin energy for a nucleus having $I = 1$ (e. g., ^{14}N) plotted as a function of $\mathbf{B}_0$. (B) Nuclear vectors relative to $\mathbf{B}_0$ with vector length $\hbar[I(I+1)]^{1/2}$ and z component $\hbar m$ so that $\cos\theta = m/[I(I+1)]^{1/2}$. (Adapted with permission of Nelson Thornes Ltd. from Figures 1a and 1b of Akitt, J. W. *NMR and Chemistry*, 3rd ed., 1992.)

electromagnetic radiation of the correct frequency according to $\Delta E = h\nu$. For nuclear magnetic resonance observations, equation 3.24 must hold because the selection rule for energy transitions between adjacent states ($\Delta m = \pm 1$) operates. In terms of the magnetogyric ratio one obtains equation 3.25:

$$h\nu = \mu \frac{\mathbf{B}_0}{I} \tag{3.24}$$

$$\gamma = \frac{2\pi}{h} \frac{\mu}{I} \tag{3.25}$$

Substitution and rearrangement lead to equation 3.26:

$$\nu = \frac{\gamma}{2\pi} \mathbf{B}_0 \tag{3.26}$$

Values of the magnetogyric ratio, γ, are such that frequencies lie within the radio range (frequencies, $\nu = < 1000$ MHz, see Figure 3.2).

The low-energy, low-frequency range for NMR transitions corresponds to a small change in energy, ΔE. This has implications for the population of excited states, the Boltzmann distribution. For a spin-1/2 nuclei with $\Delta E = \mu\mathbf{B}_0/I$ and $I = 1/2$, equation 3.27 applies. Because $N^+ \approx N^-$, one can write equation 3.28:

$$\frac{N^+}{N^-} = \exp\left(-\frac{\Delta E}{kT}\right) = \exp\left(-\frac{2\mu\mathbf{B}_0}{kT}\right) \tag{3.27}$$

$$\frac{N^+}{N^-} = 1 - \left(\frac{2\mu\mathbf{B}_0}{kT}\right) \tag{3.28}$$

For the proton ($I = 1/2, \Delta E = 2\mu\mathbf{B}_0$) in a magnetic field of 9.39 T (tesla)—resonance frequency of 400 MHz—and at a temperature of 300 K, the quantity $2\mu\mathbf{B}_0/kT$ has a value of about 6×10^{-5}. In words, the excess population in the lower energy state is extremely small, approximately one nucleus in 300,000. As will be seen in the following, the NMR experiment monitors the *relaxation* of molecules from an excited state to a ground state. This is in contrast to the EPR technique and many other instrumental techniques that detect *absorption* of energy in a species moving from a ground state to an excited state.

If one defines the length of the vector as in Figure 3.15B as $[I(I+1)]^{1/2}$, the angle θ is given by equation 3.29:

$$\cos\theta = \frac{m}{[I(I+1)]^{1/2}} \tag{3.29}$$

It is possible to translate these, in three dimensions, as the motion of a magnet of moment μ in an applied magnetic field. The magnet axis becomes inclined to the field, precesses around it, and is defined as the Lamour precession. The half-apex angle of the cone equals θ and the angular velocity is $\gamma\mathbf{B}_0$, so that one finds equation 3.26 now rewritten as

$$\nu = \left(\frac{\gamma}{2\pi}\right)\mathbf{B}_0 \tag{3.30}$$

Equation 3.30 defines the nuclear resonant frequency. The precession cone of excess low-energy nuclei is shown in Figure 3.16 (as adapted from Figure 1.2b on p. 10 of reference 19), with the magnetic axis indicated as the z axis leading to a net magnetization $\mathbf{M}_z$ along the z axis. The x- and y-axis components average to zero, that is, $\mathbf{M}_x = \mathbf{M}_y = 0$.

To produce an observable magnetic effect, the NMR technique resorts to an additional perturbation for detection of a resonance. The additional perturbation results from the application of a sinusoidally oscillating magnetic field with frequency $= \gamma\mathbf{B}_0/2\pi$ along the y axis of Figure 3.16. The perturbing field is generated by passing a radio-frequency (RF) alternating current through a coil wrapped around the sample space. The field generates a vector (called $\mathbf{B}_1$) around which the nuclei precess. The precession yields a resultant magnetization vector $\mathbf{M}_{xy}$ and consequently yields a net absorption of energy by the nuclei. When $\mathbf{B}_1$ is cut off, the precession stops; however, the signal output due to the loss of the energy previously absorbed by the nuclei is observable for between 10 ms and 10s. The point at which maximum current has been induced in the coil during $\mathbf{B}_1$'s application is known as a 90° pulse. If one observes only the resultant nuclear magnetization, this laboratory frame of reference is known as the rotating frame. A simplified version of this concept is shown in Figure 3.17, as adapted from Figure 1.6 of reference 19. In this scheme the resultant nuclear magnetization

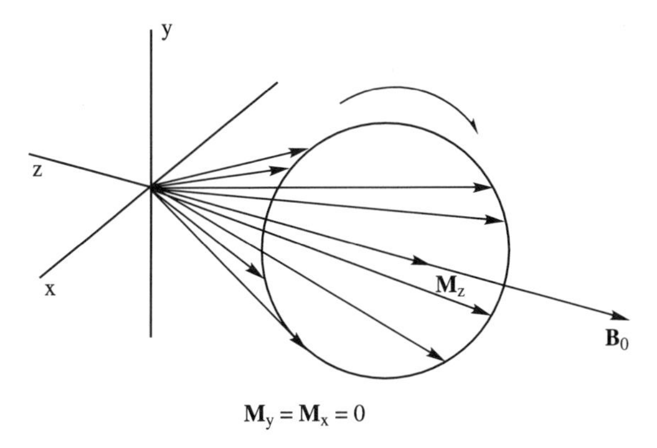

Figure 3.16 Freely precessing nuclei in a magnetic field. The figure represents low-energy nuclei in a sample arising from different atoms (drawn with the same origin). (Adapted with permission of Nelson Thornes Ltd. from Figure 1.2b of Akitt, J. W. *NMR and Chemistry*, 3rd ed., 1992.)

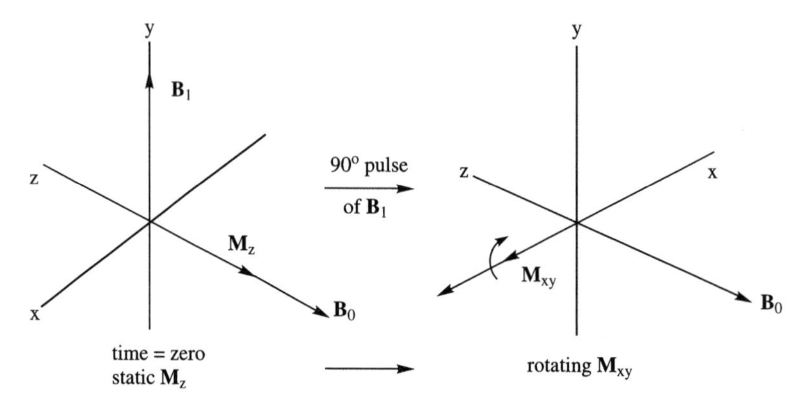

Figure 3.17 Rotating Frame when $\mathbf{B}_1$ is applied in the form of a 90° pulse. (Adapted with permission of Nelson Thornes Ltd. from Figure 1.6 of Akitt, J. W. *NMR and Chemistry*, 3rd ed., 1992.)

$\mathbf{M}_{xy}$ is illustrated with $\mathbf{B}_1$'s position frozen. The fact that all low-energy nuclei act in concert and at a single frequency leads to the name "resonance spectroscopy."

A nuclear magnetic resonance spectrometer measures the frequency of "nuclear resonance" with sufficient accuracy to provide useful information. The spectrometer features a strong stable magnet containing a gap within which the sample is placed. The sample is surrounded by a transmitter–receiver coil. The usual modern system contains a superconducting cryomagnet for generation of magnetic field strengths between 200 and 750 to 1000 MHz.

3.5.2 Nuclear Screening and the Chemical Shift

If all the nuclei being detected in an NMR experiment (all protons in an organic ligand molecule, for instance) resonated at the same frequency, chemists would not be very interested because little information about structure, and so on, would be gained. However, when a magnetic field is applied during an NMR experiment, electrons surrounding nuclei in the molecules under study set up a secondary magnetic field. The secondary field opposes the main field, reducing the nuclear frequency. The magnitude of the frequency change is proportional to $\mathbf{B}_0$. This is important in that there will be larger separations between resonant frequencies at higher magnetic field strengths, allowing one to detect finer differences between the different protons in any liquid sample. The effect of electrons surrounding the nucleus on the nucleus in the applied magnetic field is termed screening (or shielding). Taking equation 3.30 and introducing the screening constant, σ, one finds equation 3.31:

$$\nu = \left(\frac{\gamma}{2\pi}\right)\mathbf{B}_0(1 - \sigma) \tag{3.31}$$

The screening constant, σ, is dimensionless and usually recorded in parts per million (ppm). Contributors to σ, opposite in sign, are σ_d (the diamagnetic term) and σ_p (the paramagnetic term). The diamagnetic term depends upon the density of circulating electrons. The paramagnetic effect in this context does *not* imply the presence of unpaired electrons (to be discussed below) but is substantial, and dominates, for heavier atoms with many electrons in outer orbitals involved in chemical bonding. Several factors affect σ_p:

1. The inverse of the energy separation, ΔE, between ground and excited electronic states of the molecule. This means that there will be a correlation between NMR spectra and absorption in the visible and ultraviolet spectral regions.
2. The relative electron density in p orbitals involved in bonding.
3. The value of $\langle 1/r^3 \rangle$, the average inverse cube distance from the nucleus to the electronic orbitals involved.

The paramagnetic screening constant becomes disproportionately larger for heavier elements; thus while ^{1}H, the proton, exhibits screening for its compounds within a range of 20 ppm, thallium (^{205}Tl) compound screening constants range over 5500 ppm. Changes in screening of each nucleus do not increase continuously with atomic number but are periodic, following the value of $\langle 1/r^3 \rangle$, increasing along each period and then falling markedly at the beginning of the next. Screening constants change in complex manners dependent upon a number of factors including charge density near the nucleus (^{14}N nucleus is 25 ppm more shielded in NH_3 than in NH_4^+), the influence of neighboring π systems, and oxidation states or coordination number of the nucleus being observed (^{31}P screening increases in the series $PCl_3 < PCl_4^+ < PCl_5 < PCl_6^-$). Usually, screening increases for substituted main group elements as the electronegativity of the substituent increases. The "normal" halogen effect, increased screening for the series $AlCl_4^- < AlBr_4^- < AlI_4^-$, is found to be a decreased screening effect for certain transition metals. The nephelauxetic effect (expansion of the electron cloud and increasing electron delocalization in ligand–metal bonding) changes the screening effect down the halogen group; thus while the difference between $AlCl_4^-$ and $AlBr_4^-$ is 22 ppm, that between $AlBr_4^-$ and AlI_4^- is 47 ppm.

Anisotropic magnets may be formed in chemical bonds within a molecule so that nuclei in the vicinity may be screened or descreened. Anisotropic behavior would be found in the vicinity of a carbonyl bond, for instance. The benzene ring exhibits ring current anisotropy, leading to large descreening (downfield shifts) of benzene protons. Molecules containing electric dipoles perturb molecular orbitals and therefore perturb the screening of a nuclei. The closer the nucleus is to the bond generating the electric field, the more they are descreened. In 1-chloropropane the descreening shifts, compared to CH_4, are α-CH_2 3.24 ppm, β-CH_2 1.58 ppm, and CH_3 0.83 ppm.

The electron ($s = 1/2$) has a very large magnetic moment that affects the NMR spectrum of any molecule possessing a paramagnetic center. If paramagnetic transition metal ions are present in the molecule, large effects are observed. The screening constants cover a much larger range than is normal for the nucleus under study because the unpaired electrons apparently can delocalize throughout the molecule and appear at or "contact" nuclei. The large resonance frequency shifts that result are called contact shifts. For proton spectra these shifts may have magnitudes of several hundred parts per million. The behavior has great utility in simplifying complex NMR spectra, as illustrated here for some lanthanide elements. Octahedral complexes of lanthanide ions such as europium (Eu), dysprosium (Dy), praseodymium (Pr), and ytterbium (Yb), complexed with organic ligands to render the metals soluble in organic solvents normally used for NMR samples, are added to the system being studied. Because the lanthanides may assume higher coordination numbers, the so-called shift reagent may react with suitable donor sites (such as O or N) within the target molecule. The interaction produces a pseudocontact shift caused by the anisotropic magnetic moment of the shift reagent (similar to the neighbor anisotropy effect described above). The shift not only may move the resonance of protons to different locations in the spectrum but also affects each proton differently, causing larger separation of the resonances. Both effects are useful in revealing more information about the structure of the molecule. The effect of paramagnetic transition metals on NMR spectra have been used to great advantage in the analysis of bioinorganic systems. One example of this utility will be given later in this chapter after a discussion of two-dimensional NMR spectroscopy.

Normally, one speaks of "chemical shifts" rather than "screening" when discussing NMR spectra. For two nuclei in different environments with screening constants σ_1 and σ_2, from equation 3.31, the corresponding nuclear frequencies will result in equation 3.32. Applying equation 3.32 to σ_2, subtracting, eliminating $\mathbf{B}_0$, and setting $\sigma_1 \ll 1$, one obtains equation 3.33:

$$\nu_1 = \left(\frac{\gamma}{2\pi}\right)\mathbf{B}_0(1 - \sigma_1) \tag{3.32}$$

$$\left(\frac{\nu_1 - \nu_2}{\nu_1}\right) = (\sigma_2 - \sigma_1) \tag{3.33}$$

The quantity σ_2–σ_1 is given the symbol δ and is called the chemical shift. Its value is expressed in parts per million (ppm). It is normal to establish a chemical shift scale for a given nucleus choosing a standard and defining its chemical shift arbitrarily at zero ppm. ^{29}Si tetramethylsilane, $(CH_3)_4Si$, or TMS is used as the standard for the common nuclei, ^{1}H and ^{13}C. In recording spectra using TMS, it is normal to depict the descreened (deshielded) nuclei to the left of the standard. One calls the deshielded region "low field" or "high frequency." A useful diagram detailing the terminology adapted from Figure 2.12 of reference 19 is shown in Figure 3.18.

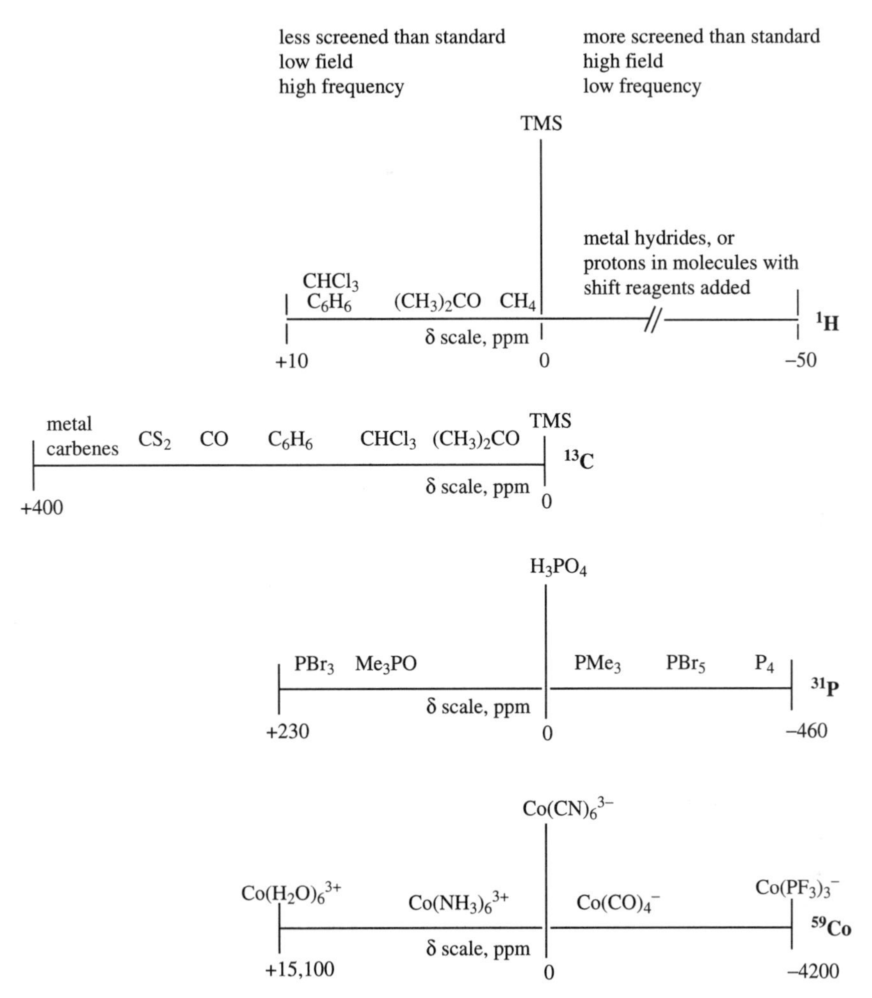

Figure 3.18 Chemical shift scales and chemical shifts of some compounds. (Adapted with permission of Nelson Thornes Ltd. from Figure 2.12 of Akitt, J. W. *NMR and Chemistry*, 3rd ed., 1992.)

3.5.3 Spin–Spin Coupling

A nucleus under study by nuclear magnetic resonance techniques is affected by other nuclei in the same molecule. This phenomenon is known as *spin–spin coupling*. The effect arises (in adjacent nuclei) from the two electrons joining the nuclei in a covalent bond. Suppose the energy of states in which the electrons in the bond have opposing spins is lower than the state in which the electron spins are parallel. Then the ΔE between the two states (in this case a negative number) is called the coupling constant, J, expressed in frequency units, Hz. Internuclear spin–spin coupling constants may be either positive or negative and depend on a number of factors:

1. The number and bond order of bonds intervening between the nuclei as well as the bond angles. Usually the interaction is observed only through one to four bonds, and the effect is attenuated (the J value becomes smaller) as the number of intervening bonds increases.
2. The magnetic moments of the two interacting nuclei. These are directly proportional to the product of the magnetogyric ratios $(\gamma_A\gamma_B)$ of the interacting nuclei.
3. The valence s electron density at the nucleus. This is affected by the s character of the bonding orbitals between the interacting nuclei.

Nuclei coupling to each other through spin–spin interactions may have very similar or very different chemical shifts. The difference or similarity will affect the appearance of the resonances associated with the coupled nuclei. Nuclei separated by small chemical shifts are denoted by the letters A, B, C while sets of nuclei separated by large chemical shifts are designated A, M, and X. The number of nuclei in each letter category is indicated by a subscript. Using the proton as an example, CH_3CH_2Cl (chloroethane) is an example of an A_3X_2 system while CH_2CHCl (vinyl chloride) is an example of an ABX system. When chemical shifts differences are large, coupling between protons on adjacent atoms will follow the simple $n+1$ multiplicity rule for the number of peaks in a multiplet (the general rule is $2nI+1$, where I is the nuclear spin). This is named a first-order pattern. The ABX system is almost first-order, but $A_aM_bX_x$ or $A_aB_bC_c$ systems exhibit complex spin-coupling multiplet patterns.

An example of spin–spin coupling between the ^{195}Pt nucleus ($I = 1/2$, abundance = 33.8%) and the proton (^{1}H, $I = 1/2$, abundance = 99.985%) is shown schematically in Figure 3.19 for the complex *trans*-$MeBrPt(PMe_2R)_2$ (where R is a 2,4-dimethoxyphenyl group).[19] The two major methyl proton resonances are indicated and are connected to the responsible peaks. The height of the central downfield methyl resonance indicates that it corresponds to the protons of four methyl groups attached to phosphorus, and thence to the magnetically inactive

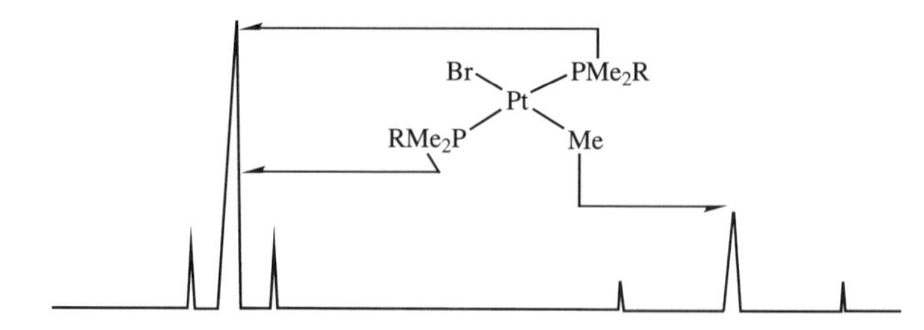

Figure 3.19 ^{1}H spectrum of the complex *trans*-$MeBrPt(PMe_2R)_2$. (Adapted with permission of Nelson Thornes Ltd. from Figure 3.13 of Akitt, J. W. *NMR and Chemistry*, 3rd ed., 1992.)

platinum nucleus. The two smaller satellite peaks at one-quarter intensity on either side of the major downfield peak originate from the same methyl protons coupled to the magnetically active ^{195}Pt nucleus. The longer coupling path from ^{195}Pt through ^{31}P to ^{1}H results in a weaker, smaller coupling constant (a so-called ^{3}J coupling) when compared to the upfield pattern for methyl protons of the methyl group directly attached to platinum. The 1 : 4 : 1 pattern for the upfield peak again indicates that the coupling corresponds to the 33.8% abundant platinum nucleus. The upfield resonance corresponds to the protons of the methyl group directly attached to the platinum atom, and thus the satellite peaks exhibit an appreciably stronger coupling and consequently a larger J value.

A more complete discussion of spin–spin coupling may be found in Chapter 3 of reference 19 and many instrumental chemistry texts.[2]

3.5.4 Techniques of Spectral Integration and Spin–Spin Decoupling

Two other important NMR concepts and related NMR techniques are introduced in Figure 3.19. The first of these is that the size of the peak is proportional to the number of protons resonating at a given frequency. One can integrate the area under the different peaks in a spectrum containing a number of proton resonances and determine how many protons resonate at each given frequency. Frequently, this can lead to confirmation of a proposed molecular structure. The second concept might have occurred to the reader when considering the phosphorus nucleus in the compound. ^{31}P nuclei are 100% abundant and should couple to both the magnetically active platinum and hydrogen nuclei. This fact should complicate the spin–spin couplings in Figure 3.19. This has not taken place because the effect of the ^{31}P spin-1/2 nucleus has been removed by a technique called double irradiation or spin-decoupling. In this technique an NMR sample is irradiated at a resonant frequency for one particular nucleus (in this case the ^{31}P nucleus). This irradiation causes the orientation of the nucleus to become indeterminate, and the resonance of adjacent nuclei will not show splitting due to spin–spin coupling with the irradiated nucleus. Thus the spectrum is simplified. This technique is applied in many different ways in NMR spectroscopy. For instance, heteronuclear spin-decoupling is essential for ^{13}C NMR spectroscopy. The large number of ^{31}C–^{1}H couplings in an organic molecule or ligand would make the ^{13}C spectrum extremely complex if the spins were not decoupled. However, because many protons in different electronic environments in organic molecules or ligands result in many resonances, a range of frequencies must be used for spin-decoupling. This has the favorable result of increasing the intensity of ^{13}C nuclei (1.108% abundant) and has the unfavorable result of not allowing integration of peaks to count numbers of C atoms of specific types. Spin–spin coupling is discussed further in the section on the nuclear Overhauser effect (NOE), Section 3.5.6.

3.5.5 Nuclear Magnetic Relaxation

If one perturbs a physical system from equilibrium and then removes the perturbing influence, the system will return to its original equilibrium condition. This does not happen instantaneously but occurs over some time, according to the equation

$$(n - n_e)_t = (n - n_e)_0 \exp\left(\frac{-t}{T}\right) \tag{3.34}$$

where $(n - n_e)_t$ is the displacement from equilibrium, n_e, at time t, $(n - n_e)_0$ is the displacement from equilibrium, n_e, at time zero, and T is the relaxation time.

Two types of relaxation processes are known with possibly different relaxation times. These are known as T_1 and T_2. Looking at Figure 3.17, one can imagine a 180° pulse that inverts the magnetization. Following the end of the pulse, relaxation processes begin to return magnetization to its initial state. This process is called T_1 or longitudinal relaxation because it takes place in the direction of $\mathbf{B}_0$. If one uses a 90° $\mathbf{B}_1$ pulse, the magnetization is moved to the xy plane ($\mathbf{M}_{xy}$) as in the rotating frame figure, Figure 3.17. This transverse magnetization rotates at the nuclear Larmor frequency; and because some nuclear spins are faster and some are slower, the xy magnetization starts to fan out and lose coherence, eventually resulting in $\mathbf{M}_{xy} = 0$. The characteristic time for this process is called transverse relaxation or T_2. The equations governing the behavior of transverse ($\mathbf{M}_{xy}$) or longitudinal ($\mathbf{M}_z$) magnetization and their return to equilibrium following a $\mathbf{B}_1$ pulse are given by equation 3.35 (in which $\mathbf{M}_z$ increases from zero to its equilibrium value) and equation 3.36 (in which the transverse magnetization falls from its maximum (equal to $\mathbf{M}_z$) to zero when sufficient time has elapsed).

$$(\mathbf{M}_z)_t = (\mathbf{M}_z)_\infty \left[1 - \exp\left(\frac{-t}{T_1}\right)\right] \tag{3.35}$$

$$(\mathbf{M}_{xy})_t = (\mathbf{M}_{xy})_0 \exp\left(\frac{-t}{T_2}\right) \tag{3.36}$$

Methods for measuring T_1 and T_2 are discussed in Chapter 5 of reference 19. Suffice it to say here that understanding the method for measuring T_2 (the Carr–Purcell pulse sequence or spin-echo method) becomes important for discussing two-dimensional NMR spectra. When spin–spin coupling is present, a modulation of spin echoes is produced, and it is this fact that is important in 2-D NMR spectroscopy. Nuclear relaxation rates and mechanisms become important when discussing the effect of paramagnetic metal centers on NMR spectroscopy. These will be further discussed for a bioinorganic system after the discussion of two-dimensional NMR.

3.5.6 The Nuclear Overhauser Effect (NOE)

The double resonance experiment can be used to simplify a spectrum as discussed above, or to probe correlations between different nuclei. Two types of double resonance experiments are described. In the homonuclear double resonance experiment the nuclei irradiated are the same isotope as those observed: Shorthand notation for this is, for example, ^{1}H{^{1}H}. In heteronuclear double resonance, the nucleus irradiated may differ from that observed: Observing ^{13}C while irradiating ^{1}H has the notation ^{13}C{^{1}H}.

As stated previously, for a spin = 1/2 nuclei so that $I = 1/2$ and with $\Delta E = \mu \mathbf{B}_0 / I$ one obtains equation 3.37. Rewriting this in terms of the magnetogyric ratio, where $\gamma = 2\pi/h(\mu/I)$, yields equation 3.38.

$$\frac{N_+^I}{N_-^I} = \exp\left(\frac{-\Delta E}{kT}\right) = \exp\left(\frac{-2\mu \mathbf{B}_0}{kT}\right) \tag{3.37}$$

$$\frac{N_+^I}{N_-^I} = \exp\left(\frac{-\gamma_I \hbar \mathbf{B}_0}{kT}\right) \tag{3.38}$$

The number of nuclei in the upper energy state, N_+^I, is less than that in the lower energy state, N_-^I, and the probabilities of upwards and downwards transitions are different. The spin transitions are caused by the spins S of a nucleus, and the influence of these occurs directly through space. The transitions and their relaxations may be between the same or different type of nucleus, but in either case they are chemically shifted from the spin I. At equilibrium, some relative spin I populations exist and the fractional difference in populations between the two energy states can be written (remembering that e^x is approximated by $1 + x$) as

$$\frac{(N_-^I - N_+^I)}{N_-^I} = \frac{\gamma_I \hbar \mathbf{B}_0}{kT} \tag{3.39}$$

A similar expression for the difference of the two spin S populations can be written as

$$\frac{(N_-^S - N_+^S)}{N_-^S} = \frac{\gamma_I \hbar \mathbf{B}_0}{kT} \tag{3.40}$$

Now if S is strongly irradiated, then it is saturated and S is no longer at its Boltzmann equilibrium. Therefore it cannot maintain the Boltzmann equilibrium of spins I, and the intensity of the I signal is changed. Equalizing S populations produces a proportional change in I populations such that equation 3.41 can be written in which η_{IS} is called the nuclear Overhauser enhancement (NOE) factor.

$$\frac{\text{Intensity } I \text{ with } S \text{ irradiated}}{\text{Normal intensity } I} = \frac{\dfrac{\gamma_I \hbar \mathbf{B}_0}{kT} + \dfrac{\phi \gamma_S \hbar \mathbf{B}_0}{kT}}{\dfrac{\gamma_I \hbar \mathbf{B}_0}{kT}} = 1 + \frac{\phi \gamma_S}{\gamma_I} = 1 + \eta_{\mathrm{IS}} \tag{3.41}$$

Two factors contribute to η_{IS}. One is the ratio of the magnetogryric ratios of the two different spins, and the other depends on relaxation mechanisms. Provided that the relaxation mechanism is purely dipole–dipole, ϕ has the value 1/2. If other relaxation mechanisms affect spin I, then ϕ may approach zero. Assuming that the dipolar mechanism is operational (no quadrupolar nuclei with $I > 1/2$ are present), η has the value $\gamma_S/2\gamma_I$ and is regarded as $\eta_{\max}$. In the homonuclear case we have $\eta_{\max} = 1/2$. Usually one chooses nuclei where $\gamma_S > \gamma_I$ to ensure that the NOE is significant. For observation of ^{13}C for instance, if the protons in the molecule are double irradiated, the ratio is 1.99 and $1 + \eta_{\max}$ equals approximately 3. To repeat a statement made above, proton broad-band irradiation enhances the intensity of the ^{13}C nucleus, which otherwise has very low receptivity.

3.5.7 Obtaining the NMR Spectrum

Three parameters affect an NMR spectrum: the chemical shift, coupling, and nuclear relaxation. These must be accounted for when obtaining the NMR spectrum from the spectrometer's output. Obtaining the NMR spectral plot from the output (the free induction decay, FID) of a modern NMR spectrometer involves the analysis of the mathematical relationship between the time (t) and frequency (ω) domains known as the Fourier relationship:

$$F(\omega) = \int_{-\infty}^{\infty} = f(t)\exp(-i\omega t)\,dt \tag{3.42}$$

which is also written as

$$F(\omega) = \int_{-\infty}^{\infty} = f(t)[\cos(\omega t) - i\sin(\omega t)]\,dt \tag{3.43}$$

The Fourier transform (FT) relates the function of time to one of frequency—that is, the time and frequency domains. The output of the NMR spectrometer is a sinusoidal wave that decays with time, varies as a function of time and is therefore in the time domain. Its initial intensity is proportional to $\mathbf{M}_z$ and therefore to the number of nuclei giving the signal. Its frequency is a measure of the chemical shift and its rate of decay is related to T_2. Fourier transformation of the FID gives a function whose intensity varies as a function of frequency and is therefore in the frequency domain.

When one applies the perturbing field $\mathbf{B}_1$, the nuclei in the sample precess as discussed above. The point at which maximum current has been induced during $\mathbf{B}_1$'s application is known as a 90° pulse. Short $\mathbf{B}_1$ pulses—less than 100 μs—ensure that all nuclei in a sample, whatever their chemical shift, are swung around $\mathbf{B}_1$ by an appropriate angle. Long $\mathbf{B}_1$ pulses choose nuclei of a particular chemical shift to precess around $\mathbf{B}_1$ without affecting other nuclei in the sample. The ideal Fourier transform experiment, allowing spins to relax to equilibrium before successive pulses are applied, is illustrated in Figure 3.20 as adapted from Figure 5.12 of reference 19.

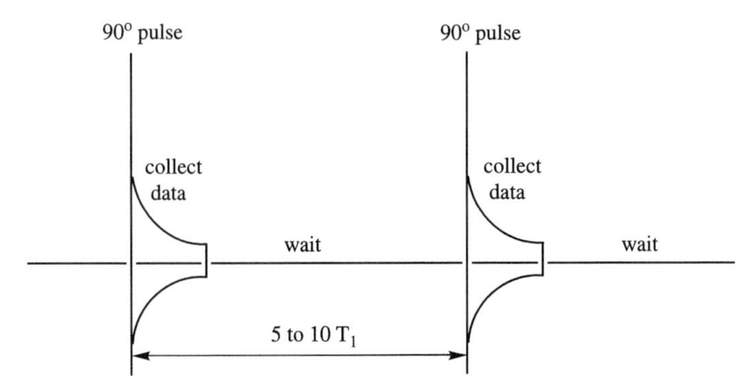

Figure 3.20 The ideal Fourier transform experiment. (Adapted with permission of Nelson Thornes Ltd. from Figure 5.12 of Akitt, J. W. *NMR and Chemistry*, 3rd ed., 1992.)

Assuming the spectrometer is stable, a series of isolated 90° pulses will each give an identical nuclear response and these can be added together in computer memory to yield a strong total response. Actually most pulsing sequences omit the waiting time and have pulse sequences in the 40° to 30° range, depending on the nuclear isotope observed, the chemical shift range, the relaxation time T_1, and the computer memory size. Different nuclei in different parts of the molecule may have different relaxation times, so that pulse lengths for each are slightly different. Some distortion of signal intensity results from this factor; thus if precise quantitative data are required, the pulse sequence of Figure 3.20 is necessary. The intensity of an FID is proportional to the number of nuclei contributing to the signal. When transformed to the so-called absorption spectrum, integration of the area under the peaks relates to the number of nuclei resonating at a given frequency.

Each FID signal is accompanied by noise; however, the noise is incoherent—sometimes positive, sometimes negative—so that it increases more slowly than the desired nuclear signal. A series of N FID's has a signal-to-noise ratio $\sqrt{N}$ times better than a single FID, allowing spectroscopists to obtain useful chemical information from otherwise unreceptive nuclei or from dilute solution samples having few of the nuclei of interest.

The output of the NMR spectrometer must be transformed from an analog electrical signal into digital information that can be stored in the computer's dedicated computer. The minicomputers used in NMR spectroscopy have memory used for data accumulation, programs for manipulating the data, and storage devices to store large collections of data for future or additional manipulation into useful spectral results.

3.5.8 Two-Dimensional (2D) NMR Spectroscopy

The two dimensions of two-dimensional NMR spectroscopy are those of time. In one time domain, FIDs containing frequency and intensity information about the

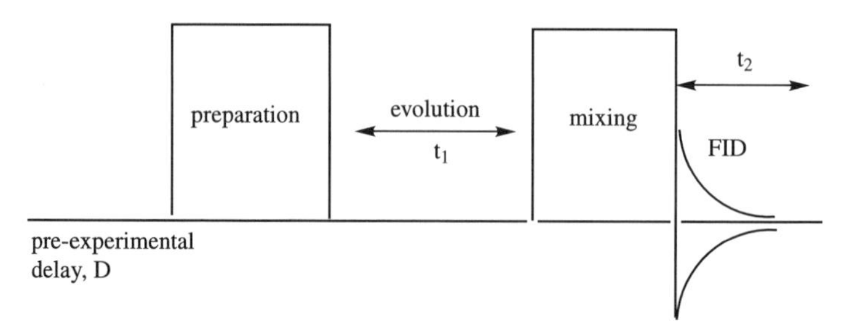

Figure 3.21 Conceptual diagram of a two-dimensional (2D) NMR experiment. (Adapted with permission of Nelson Thornes Ltd. from Figure 8.1 of Akitt, J. W. *NMR and Chemistry*, 3rd ed., 1992.)

observed nuclei are collected. The second time dimension refers to the time that elapses between some perturbation of the system and the onset of data collection in the time domain. The second time period is varied, and a series of FID responses are collected for each of the variations. Following this, each FID is Fourier-transformed to produce perturbed spectra. These are stacked, and the sets of data at each frequency point is transformed once again. The results give data that are displayed on two frequency axes, with the peaks represented by contours on a relief map. The two-dimensional plot contains connectivities between the different nuclei. The connectivities or correlations may involve spin–spin coupling of nuclei or show spatial proximity through NOE effects. A simple conceptual diagram of the 2D experiment is shown in Figure 3.21 as adapted from Figure 8.1 of reference 19.

3.5.9 Two-Dimensional Correlation Spectroscopy (COSY)

Two-dimensional techniques continue to evolve. Several examples will be discussed here to give the reader some idea of the information that can be gained. COSY or correlation spectroscopy is a technique with many variations, all depending upon the existence of spin–spin coupling between nuclei. The coupling provides responses relating to the chemical shift positions of coupled nuclei. Its equivalent in one-dimensional spectroscopy would be a series of double resonance experiments at each multiplet in the spectrum. One can perform homonuclear (probing couplings between the same nuclei, i.e., ^{1}H–^{1}H) or heteronuclear (probing connectivities between different nuclei, i.e., ^{1}H–^{13}C) COSY experiments.

In the pulse sequence for a homonuclear COSY experiment, the first 90° pulse flips the z magnetization into the x direction and into the xy plane. Considering an AX spin system (one in which the nuclei have very different chemical shifts) having two doublets due to spin–spin coupling, the magnetization will include four components precessing at different frequencies. During a time t_1 the magnetization separates into two pairs of vectors due to the chemical shift difference, with the separation in each pair being due to the coupling interaction. The second 90° pulse swings the magnetization into the yz plane, where it starts to precess around the z axis. If the vectors were in the right half of the xy plane, they are swung into the $-z$ direction; if they were in the left half of the xy plane, they are swung into the $+z$ direction. Inversion of the components causes transfer of polarization, and the signal of either doublet will be a function of t_1. For each value of t_1 the FID will contain the four frequencies of the two AX doublets distorted in both intensity and phase. The FIDs are collected at a number of t_1 values, and then they are transformed to yield a stack of spectra with different degrees of distortion. These are then transformed at each frequency to obtain a two-dimensional (2D) map. Where there is no correlation between spins, the frequency is the same in both dimensions. These signals appear on a diagonal plot, which is the normal 1D spectrum. Where there is a correlation, the chemical shift and coupling constant mix in each resonance and a signal (peak) appears off the diagonal. The off-diagonal peaks (cross peaks) appear at the intersection of the A and X chemical shift resonances. Instantaneously, one can identify two peaks in the 1D spectrum that correlate to each other, that exhibit spin–spin coupling, and that are therefore located next to each other.

The heteronuclear COSY experiment investigates the connectivities imparted by coupling paths between two different nuclei, most commonly those between ^{1}H and ^{13}C. The experiment is facilitated by the fact that one-bond CH coupling constants are much larger than the two- or three-bond (CCH or CCCH) coupling constants. Decoupling of protons from ^{13}C is accomplished in the usual way by broad-band decoupling during the t_2 period while accumulating the ^{13}C FID. Other, more complex pulse sequences are capable of decoupling the proton resonances as well. See Figure 3.22 for a simplified COSY pulse sequence as adapted from Figure 8.9 of reference 19. The first 90° pulse at the ^{1}H frequency produces transverse y magnetization. This evolves with t_1 and depends on the frequency of the proton's

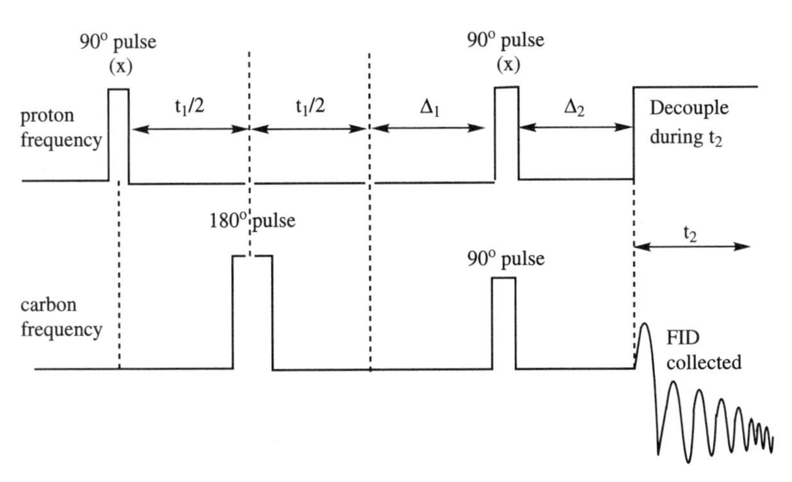

Figure 3.22 Pulse sequence used to produce ^{1}H–^{13}C heteronuclear COSY spectrum. (Adapted with permission of Nelson Thornes Ltd. from Figure 8.9 of Akitt, J. W. *NMR and Chemistry*, 3rd ed., 1992.)

chemical shift. Protons attached to carbon nuclei will have two magnetization components precessing at different frequencies. These components are refocused halfway through t_1 by inverting the carbon magnetization with a 180° pulse. The proton magnetization at the end of t_1 is a function of t_1. This magnetization is transferred to the carbon magnetization by a second proton 90° pulse placing the y magnetization into the z direction and permitting polarization transfer. The period Δ_1 allows the two magnetization components to come into phase before the second proton 90° pulse is applied. This delay is timed to coincide with $1/2J(\mathrm{CH})$, one-half $^1J(\mathrm{CH})$, the coupling constant between ^{13}C and ^1H nuclei. The 90° pulse applied at the carbon frequency at the same time as the second proton pulse creates transverse carbon magnetization and produces the output FID. The refocusing delay Δ_2 is usually the same as Δ_1. The experiment is repeated for many values of t_1, and the stacks of FIDs obtained (which contain both carbon and proton chemical shift information) are transformed and plotted. Usually the carbon spectrum appears on the x axis of the 2D plot with the proton spectrum plotted vertically. Contours appear at the intersection of carbon and proton resonances corresponding to the carbon and proton atoms bonded to each other.

3.5.10 Nuclear Overhauser Effect Spectroscopy (NOESY)

NOESY NMR spectroscopy is a homonuclear two-dimensional experiment that identifies proton nuclei that are close to each other in space. If one has already identified proton resonances in one-dimensional NMR spectroscopy or by other methods, it is then possible to determine three dimensional structure through NOESY. For instance, it is possible to determine how large molecules such as proteins fold themselves in three-dimensional space using the NOESY technique. The solution structures thus determined can be compared with solid-state information on the same protein obtained from X-ray crystallographic studies. The pulse sequence for a simple NOESY experiment is shown in Figure 3.23 as adapted from Figure 8.12 of reference 19.

The first 90° pulse places magnetization in the xy plane, and the two components then precess at their individual frequencies for t_1. At the end of t_1, each component will have its own x and y component of magnetization. The second 90° pulse places

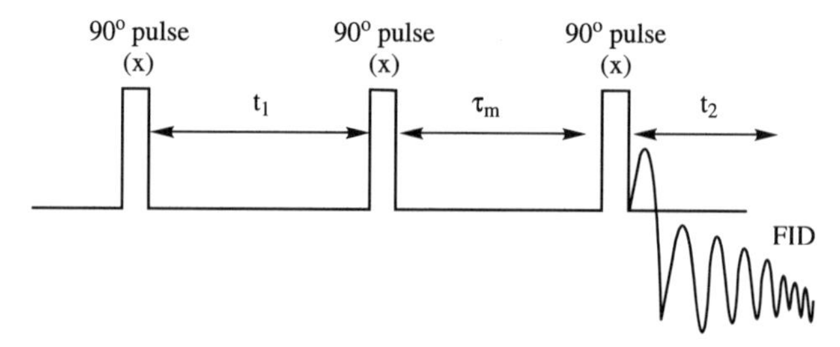

Figure 3.23 The pulse sequence used for NOESY. (Adapted with permission of Nelson Thornes Ltd. from Figure 8.12 of Akitt, J. W. *NMR and Chemistry*, 3rd ed., 1992.)

the magnetization into the xz plane, where the y components continue to precess about the z axis. A short mixing time, τ_m (typically 0.05 s), follows, allowing an exchange of magnetization to take place. At the end of τ_m the two component frequencies have been modulated through the process of exchange going on between the nuclei (this in turn is a function of t_1). An FID is produced by a third 90° pulse, which moves the magnetization back into the xy plane. Transformation of the FID stacks yields the two-dimensional trace with cross peaks between exchanging nuclei. The pulse sequence is set up to detect two uncoupled spins having different chemical shifts but that undergo slow exchange. In nonexchanging nuclei, through-space relaxation can also produce an exchange of magnetization through the NOE effect. In either case, nuclei that are close together in space may produce cross peaks in the NOESY experiment that, along with additional structural information, may lead to three-dimensional structure determination of large molecules in solution.

3.5.11 Descriptive Examples

An example utilizing two-dimensional NMR spectroscopy for a paramagnetic transition metal system is taken from Chapter 8 of reference 1. The chapter, entitled "Nuclear Magnetic Resonance of Paramagnetic Metal Centers in Proteins and Synthetic Complexes," is written by Li-June Ming.[20] The system to be analyzed is CuNiSOD (SOD = superoxide dismutase) containing an antiferromagnetically coupled Cu^{2+}–Ni^{2+} pair. The enzyme superoxide dismutase is discussed in Section 5.2.3. As discussed previously in this chapter, paramagnetic metal centers can extend the range of chemical shifts for nearby nuclei greatly, simplifying spectra by shifting important resonances away from congested regions of the spectrum. They also cause enhanced nuclear relaxation that may lead to significant increase in line width (line broadening). As with chemical shifts, paramagnetic centers cause great changes in nuclear relaxation rates. Longitudinal (T_1), spin-lattice, and transverse (T_2), spin–spin, nuclear relaxations were discussed in an earlier section of the NMR discussion and described in equations 3.35 and 3.36. The most quantifiable paramagnetic nuclear relaxation mechanism is a through-space dipolar interaction between the nuclear and electron magnetic moments. In its simplest approximation the rate of dipolar relaxation is directly proportional to τ_e, the electronic relaxation time of the paramagnetic center, and inversely proportion to $r^6_{\text{M-H}}$, the sixth power of the distance between the paramagnetic center and the resonating nucleus as shown in equation 3.44:

$$T_{\text{M}}^{-1}(\text{dipolar}) \propto r^6_{\text{M-H}}\tau_e$$

$$T_{1\text{M}}^{-1}(\text{dipolar}) = T_{2\text{M}}^{-1}(\text{dipolar}) = \frac{4}{3}\left(\frac{(\gamma_N^2 g_e^2 \beta_e^2 S(S+1))}{r^6_{\text{M-H}}}\right)\tau_e \qquad (3.44)$$

Dipolar relaxation is usually assumed to be the predominant mechanism in a paramagnetic molecule, although this is a simplification. For instance, if the contribution of contact relaxation time (a function of correlation time, τ_c) is ignored, and is sizable as it would be when there is significant metal–ligand covalency, then the dipolar relaxation rates would be too large, providing estimates of metal–ligand distances that are too short. Continuing with the CuNiSOD example, the Cu^{2+} ($d^9, S = 1/2$) ion is a slowly relaxing paramagnetic center and its isotropically shifted ^1H NMR signals are often broadened beyond detection in a mononuclear complex. However, the Ni^{2+} ($d^8, S = 1$) ion exhibits a short $\tau_e(<10^{-11}s)$ for its unpaired electrons, leading to a fast nuclear relaxation rate and sharper NMR signals. Antiferromagnetic coupling of the ions may result in two spin levels for the pair: $S' = 1/2$ and $3/2$. The ^1H NMR spectrum of the magnetically coupled Cu–Ni pair is shown in Figure 29 of reference 20. In this spectrum, one is able to distinguish and assign histidine protons of three his ligands attached to Cu from those attached to the nickel ions three histidine ligands.

Assignment of the isotropically shifted signals observed for the CuNiSOD example discussed in the previous paragraph has been achieved by means of anion titrations (not discussed here) and nuclear Overhauser enhancement spectroscopy (NOESY), to be discussed next. In Figure 3.24B the CuNiSOD active site is depicted with histidine nitrogens and protons identified for the discussion of the NOESY results. The copper(II) ion is coordinated to the N^{ε} ligand atoms of his46

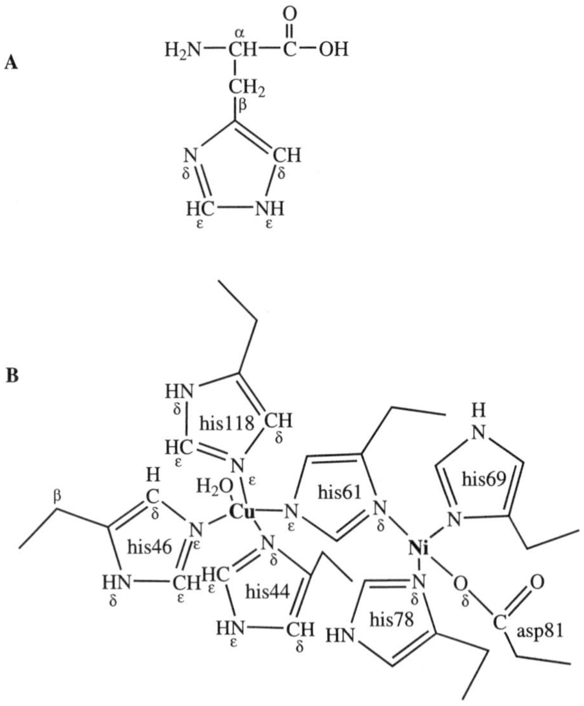

Figure 3.24 (A) Labeling of side-chain atoms in histidine residues for identification purposes. (B) CuNiSOD coordination sphere for discussion of NOESY spectrum.

and his118, the N^{δ} ligand atom of his44, and bridges to the Ni(II) ion using the N^{ε} atom of his61. His44 and his46 actually are in trans position to each other in the distorted square-planar CuN_4 coordination sphere although this is not apparent in Figures 3.24 and 3.25. The nickel ion's geometry is distorted tetrahedral with ligand atoms N^{δ} of his78 and his69, the N^{δ} of the bridged his61 side chain, and the carboxyl oxygen (O^{δ}) of asp81.

The pulse sequence for the NOESY experiment follows that of Figure 3.23, with a mixing time τ_m chosen so as to be not much greater than about $2\,T_{1av}$ (where T_{1av} is the average T_1 for the paramagnetically coupled pair). In the case of CuNiSOD, T_1 is approximately 4–30 ms for protons on the coordinated ligands.[21] The coordination modes of two histidine ligands can be unambiguously assigned in

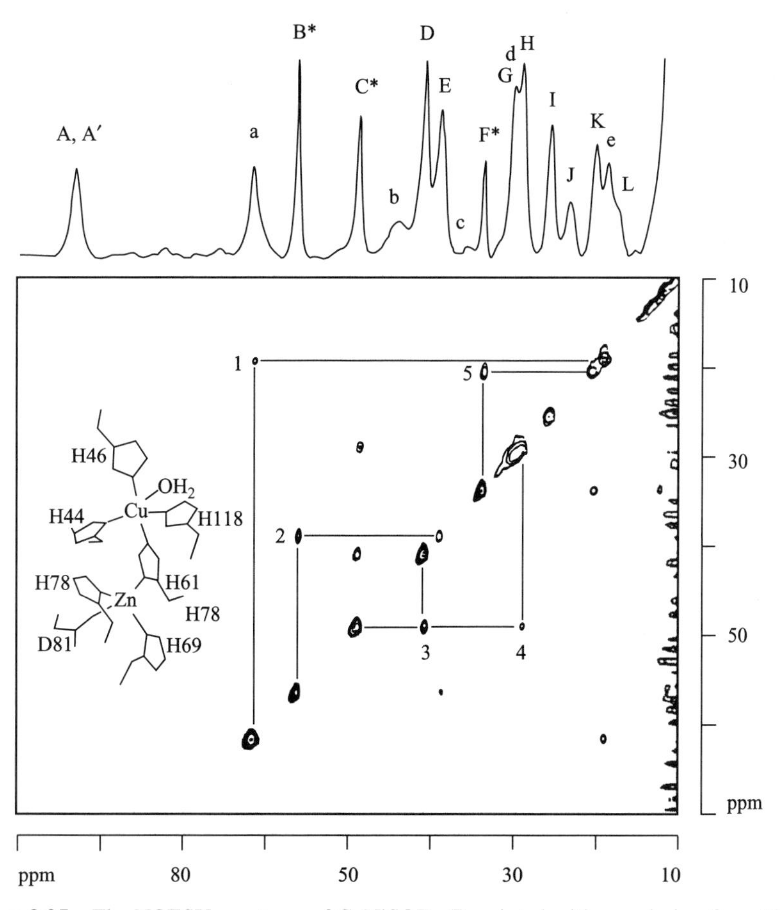

Figure 3.25 The NOESY spectrum of CuNiSOD. (Reprinted with permission from Figure 45 of Ming, L.-J., in Que, L., ed. *Physical Methods in Bioinorganic Chemistry: Spectroscopy and Magnetism*, University Science Books, Sausalito, CA, 2000, pp. 375–464. Copyright 2000, University Science Books.)

the following manner as illustrated in Figure 3.25, as reprinted from Figure 45 of reference 20. The signals A–L are attributable to his ring protons at the Cu^{2+} site, and the signals a–e are attributable to his ring protons on the Ni^{2+} site. The asterisked signals are solvent exchangeable his ring NH protons. The most downfield-shifted overlapped signals are assigned to the bridging his61 ring protons. The copper ion's coordination to N_ε or N_δ histidine atoms is indicated by the following reasoning process. The $N_\varepsilon H$ proton (C* in Figure 3.25) of the N_δ–Cu coordinated his44 shows cross-relaxation with both adjacent $C_\delta H$ (D in Figure 3.25) and $C_\varepsilon H$ (H in Figure 29a of reference 20) protons (cross peaks 3 and 4 in Figure 3.25), whereas the $N_\delta H$ protons on the N_ε–Cu coordinated his118 and his46 residues show cross peaks only with the adjacent $C_\varepsilon H$ protons (cross peaks 2 and 5) but not with the more distant $C_\delta H$ protons. The NOESY spectrum was acquired on a 360-MHz instrument at 298 K with the CuNiSOD dissolved in 50 mM phosphate/H_2O buffer at pH 6.5. NOESY techniques have found applications in structure determination for other paramagnetic metalloproteins, including Fe(III)–heme proteins as well as Co^{2+}- and Ni^{2+}-substituted proteins.

Over the last 10–15 years, multidimensional (3D and 4D) NMR spectroscopic and computational techniques have been developed to solve structures of large proteins in solution. This technology has the advantage of resolving the severe overlap in 2D NMR spectra for proteins having >100 residues. Currently, the upper limit of applicability of the new methods is probably around 60–70 kDa (approximately 250 residues). Using these methods, one is able to resolve protein structure at the same level as X-ray crystallographic data (resolved to approximately 2.5 Å). The reference 22 authors state that a carefully refined X-ray structure of a given protein (a solid-state picture) may not be identical to the "true" solution structure but in most cases can be quite close. Evidence to suggest the match comes from calculations of three-bond coupling ($^3J_{\mathrm{Hn}\alpha}$) constants from good crystallographic data that have been found to be in excellent agreement with experimentally determined values in solution. Methods that improve structure refinement in NMR solution studies include the use of a conformational database potential and direct refinement against three-bond coupling constants (^3J values), secondary ^{13}C shifts, ^1H shifts, T_1/T_2 ratios, and residual dipolar couplings. The article by Clore and Gronenborn gives an excellent overview of the new refinement strategies that increase the accuracy of the resulting solution structures.[22]

3.6 MÖSSBAUER SPECTROSCOPY

3.6.1 Theoretical Aspects

The Mössbauer effect as a spectroscopic method probes transitions within an atom's nucleus and therefore requires a nucleus with low-lying excited states. The effect has been observed for 43 elements. For applications in bioinorganic chemistry, the ^{57}Fe nucleus has the greatest relevance and the focus will be exclusively on this nucleus here. Mössbauer spectroscopy requires (a) the emission of γ rays from the source element in an excited nuclear state and (b) absorption of these by the same element in the sample under study. The Mössbauer phenomenon requires that the emission and absorption of the γ radiation take place in a recoil-free manner; this is accomplished by placing the nucleus in a solid or frozen solution matrix. Three main types of interaction of the nuclei of interest with the chemical environment surrounding it cause detectable changes in the energy required for absorption: (1) resonance line shifts from changes in electron environment—see discussion of the isomer shift, δ, below; (2) quadrupole interactions—see discussion of ΔE_Q below; and (3) magnetic interactions. The last type, magnetic interactions, is especially important in studying bioinorganic systems such as iron–sulfur clusters.

The source for the 14.41-keV γ radiation used in Mössbauer experiments is indicated by the boldface arrow in Figure 3.26.[3] Origin of the isomer shift and quadrupole splitting phenomena are indicated at the right-hand side of the diagram.

The electronic environment about the sample's nucleus influences the energy of the γ ray necessary to cause the nuclear transition from the ground to the excited state. The energies of the γ rays from the source can be varied by moving the source relative to the sample. In order to obtain the Mössbauer spectrum, the source is moved relative to the fixed sample, and the source velocity at which maximum absorption of γ rays occurs is determined. For a ^{57}Fe source emitting a 14.4-keV γ ray, the energy is changed by 4.8×10^{-8} eV for every mm s^{-1} of velocity imposed on the source.[23] Detectors will be similar to those used for X-ray crystallography.

Figure 3.27 shows the Mössbauer spectrum that results from splitting of the ^{57}Fe excited state, a quadrupole doublet, for a sample containing randomly oriented molecules such as found in polycrystalline solids or frozen solutions. The two doublets are separated in energy by the quadrupole splitting, ΔE_Q, defined by the

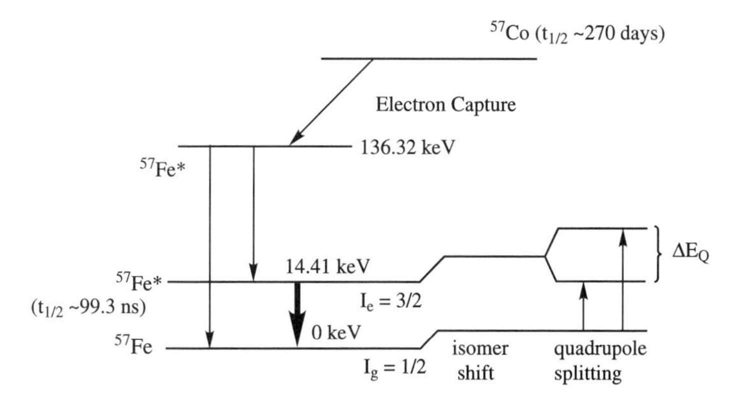

Figure 3.26 Cobalt-57 source of 14.41-keV γ radiation used in Mössbauer experiments. Isomer shift and quadrupole splitting characteristics are shown at right. (Adapted from Figure 2.26 of reference 3 and Figure 1 of reference 24.)

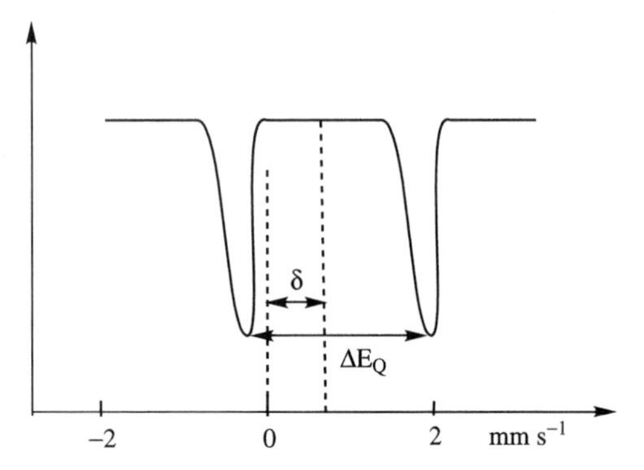

Figure 3.27 Typical Mössbauer spectrum for a sample containing randomly oriented molecules.

following equation:

$$\Delta E_Q = \left(\frac{eQV_{zz}}{2}\right)\sqrt{\left[\frac{1-\eta^2)}{3}\right]} \tag{3.45}$$

where V_{zz} is the electric field gradient (EFG) tensor (defined along the z axis by convention) and η is the asymmetry parameter (defined by $(V_{xx} - V_{yy})/V_{zz}$.[24]

Asymmetry in the ligand environment, either geometric or in charge distribution (or both), affect the asymmetry parameter, η. An $\eta = 0$ value corresponds to complete axial symmetry, whereas $\eta = 1$ corresponds to pure rhombic symmetry. Electric monopole interactions between the nuclear charge distributions and the electrons at the nucleus cause a shift of the nuclear ground and excited states. These interactions are known as the isomer shift, δ. Both the Mössbauer source and the absorber (the sample of interest) experience an isomer shift, and it is customary to quote δ relative to a standard, usually Fe metal or $Na_2[Fe(CN)_5NO] \cdot 2H_2O$ at 298 K.

3.6.2 Quadrupole Splitting and the Isomer Shift

The isomer shift, δ, is a measure of the s-electron density at the iron nucleus. Influences on this parameter include changes in the s population of a valence shell or shielding effects caused by increasing or decreasing p or d electron density. Because the radial distributions of s and d electrons overlap, δ can be a good measure of iron's oxidation state in the sample. Values for the isomer shift may also yield information on the spin state (several high- or low-spin states exist for Fe(IV) d^4, Fe(III) d^5, or Fe(II) d^6), the coordination sphere of the metal ion, and the degree of covalency in the metal–ligand bond being studied. High-spin Fe(II) has a unique range of δ values higher than those for low-spin Fe(II) and higher than those for high- or low-spin Fe(III) or Fe(IV). The trend can be understood through the

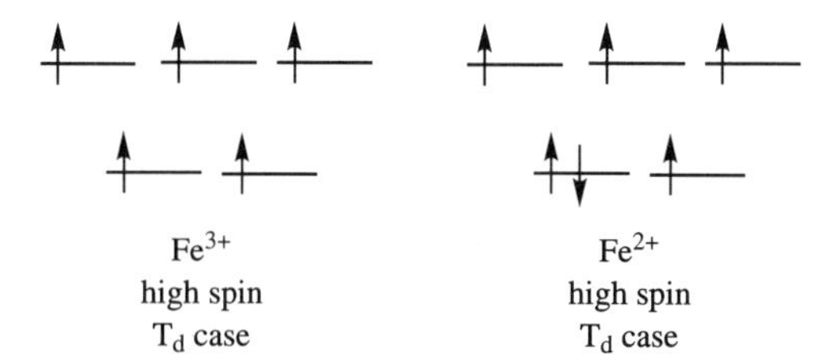

Figure 3.28 Spherical symmetry and asymmetry of electric field gradients affecting ΔE_Q.

following reasoning. A decrease in the number of *d* electrons with increasing iron oxidation state decreases the screening of *s* electron density at the nucleus (increases *s* electron density at the nucleus). This, in turn, tends to decrease the value of the isomer shift. ΔE_Q values depend on the spherical symmetry or asymmetry of the electric field gradients at the nucleus as explained in more detail below. In general, the higher the symmetry of electric field about the nucleus, the smaller the value of ΔE_Q. For instance, $\Delta E_Q(Fe^{2+}) > \Delta E_Q(Fe^{3+})$ for the high-spin case because the d^5 high-spin $Fe^{3+}(S = 5/2)$ ion has spherical symmetry and the d^6 high-spin $Fe^{2+}(S = 2)$ does not, as indicated in Figure 3.28. Comparison of ΔE_Q values for heme ligands in Table 3.4 illustrates the trend: For Fe^{3+}–heme $(S = 5/2)$ we have $\Delta E_Q = 0.5$–1.5, while for Fe^{2+}–heme (S = 2) we have $\Delta E_Q = 1.5$–3.0.

Table 3.4 lists values for ΔE_Q and δ for some important oxidation and spin states found in bioinorganic molecules. Data are taken from reference 24 and from Table 1 of reference 25 for hemoglobin, myoglobin, and the picket-fence porphyrin model compound, FeTpivPP(1-MeIm).[25] The myoglobin and hemoglobin model compounds are discussed in Section 4.8.2. Reference 26 provides the Table 3.4 data on iron sulfur clusters found in many bioinorganic species.[26] The unusual iron–sulfur and iron–molybdenum–sulfur clusters found in the enzyme nitrogenase are discussed more fully below and in Chapter 6.

3.6.3 Magnetic Hyperfine Interactions

A nucleus with spin quantum number *I* has $(2I + 1)$ magnetic energy levels. The energy gaps between these levels cause splitting in Mössbauer spectra. This effect is known as the *magnetic hyperfine interaction.* At higher temperatures of observation (195 K) the splitting of energy levels may not be observed for paramagnetic species because the electrons may change spins so rapidly that the time-averaged effect is zero. Below a certain temperature, the Curie temperature, thermal agitation is insufficient to prevent the alignment of unpaired electrons so that magnetic hyperfine splitting will be exhibited for the paramagnetic centers. For diamagnetic species $(S = 0)$, magnetic hyperfine splitting may be induced by the application of an external magnetic field which is often done experimentally. Figure 3.29 illustrates the magnetic hyperfine splittings of the nuclear ground state to the 14.4-keV level (see Figure 3.26), assuming $\Delta E_Q = 0$. In the presence of a nuclear

Table 3.4 Mössbauer Parameters for Some Biological Species[a]

Oxidation State	Spin State	Ligands	ΔE_Q (mm s^{-1})	δ (mm s^{-1})
Fe(IV)	$S = 2$	Fe–(O, N)[b]	0.5–1.0	0.0–0.1
Fe(IV)	$S = 1$	Hemes	1.0–2.0	0.0–0.1
		Fe–(O, N)	0.5–4.3	−0.20–0.10
Fe(III)	$S = 5/2$	Hemes	0.5–1.5	0.35–0.45
Fe(III)	$S = 5/2$	Fe–S[c]	<1.0	0.20–0.35
Fe(III)	$S = 5/2$	$[FeS_4]$ cluster (rubredoxin)		0.25
Fe(III)	$S = 5/2$	$[Fe_2S_2]^{2+}$ cluster		0.27
Fe(III)	$S = 5/2$	Fe(III) in $[Fe_2S_2]^+$ cluster		0.30
Fe(III)	$S = 5/2$	$[Fe_3S_4]^+$ cluster		0.27
Fe(III)	$S = 5/2$	$[Fe_3S_4]^0$ cluster		0.32
Fe(III)	$S = 5/2$	Fe–(O, N)	0.5–1.5	0.40–0.60
Fe(III)	$S = 3/2$	Hemes	3.0–3.6	0.30–0.40
Fe(III)	$S = 1/2$	Hemes	1.5–2.5	0.15–0.25
		$Hb(O_2)$	2.19	0.26
		FeTpivPP(1-MeIm)(O_2)[d]	2.04	0.27
Fe(III)	$S = 1/2$	Fe–(O, N)	2.0–3.0	0.10–0.25
Fe(II)	$S = 2$	Hemes	1.5–3.0	0.85–1.0
		Hb	2.22	0.92
		Mb	2.17	0.91
		FeTpivPP(1-MeIm)[d]	2.32	0.88
Fe(II)	$S = 2$	Fe–S	0.0–3.0	0.60–0.70
Fe(II)	$S = 2$	$[FeS_4]$ (rubredoxin) cluster		0.70
Fe(II)	$S = 2$	Fe(II) in $[Fe_2S_2]^+$ cluster		0.72
Fe(II)	$S = 0$	Fe–(O, N)	1.0–3.2	1.1–1.3
Fe(II)	$S = 0$	Hemes	<1.5	0.30–0.45
Delocalized $Fe^{2.5+}$–$Fe^{2.5+}$	$S = 1/2$ or $S = 9/2$ coupled to $S = 5/2$ Fe^{3+} site	$[Fe_3S_4]^0$		0.46

[a] Typical values at 4.2 K. Isomer shifts are taken relative to the standard value for Fe metal at 298 K.
[b] Fe–(O, N) hexacoordinate or pentacoordinate sites.
[c] Fe–S tetrahedral sulfur ligation.
[d] TpivPP = 5, 10, 15, 20-tetrakis-[*o*-(pivalamido)phenyl]porphyrinate($^{2-}$); 1-MeIm = 1-methylimidazole.
Source: References 24–26.

quadrupole the hyperfine splittings will be unequally spaced, as can be seen in Figure 2b of reference 24. Other splitting patterns are observed for $S = 1/2$ systems that have isotropic magnetic properties; these are illustrated in Figure 3 of reference 24.

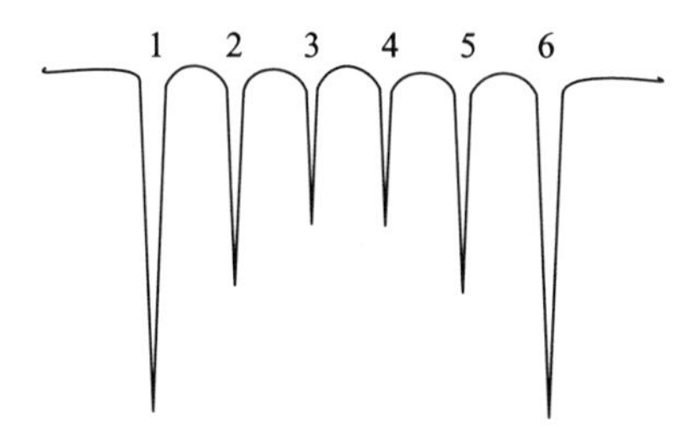

Figure 3.29 Magnetic hyperfine splittings of the nuclear ground state assuming $\Delta E_Q = 0$. (Adapted from Figure 2 of reference 24.)

For Figure 3.29, note that lines 1, 3, 4, and 6 obey the selection rule $|\Delta m_I| = 1$ for the allowed γ transitions between the nuclear sublevels, whereas lines 2 and 5 obey the $\Delta m_I = 0$ selection rule. For an isotropic $(g_x = g_y = g_z = g)$ sample in which the effective magnetic field is parallel to the observed γ rays, the intensity of the $\Delta m_I = 0$ lines vanishes so that only four lines are seen in the spectrum (Figure 3a of reference 24). The same lines that are missing in the isotropic case will be maximized when the effective magnetic field is perpendicular to the γ rays (Figure 3b of reference 24). For a uniaxial case ($g_x = g_y = 0$ and $g_z \neq 0$, Figure 3c of reference 24) or the extreme anisotropic case $(g_z \gg g_x, g_y)$, the intensities of the absorption lines are independent of the orientation of the applied field. In other words, the parallel versus perpendicular dependence of the Mössbauer spectrum becomes less pronounced as g becomes more anisotropic $(g_x \neq g_y \neq g_z \neq 0)$ and the spectrum looks more like Figure 3.29.

The six-line magnetic hyperfine splitting spectrum seen for the anisotropic cases is averaged over all molecular orientations and will have a 3 : 2 : 1 : 1 : 2 : 3 intensity pattern as shown in Figure 3.29. Because g values are determined by the electron paramagnetic resonance (EPR) technique as discussed in Section 3.4, one can make reasonable predictions about the shape of the Mössbauer spectrum if the results of an EPR study are known. For instance, a magnetic-field-independent Mössbauer spectrum suggests an EPR silent state; conversely, a magnetic field-dependent Mössbauer spectrum implies that an EPR spectrum should be observed.

3.6.4 Descriptive Examples

Mössbauer spectroscopy along with evidence gathered from EPR and ENDOR have identified the behavior of iron atoms in the M center or FeMo cofactor of the enzyme nitrogenase from *Azobacter vinelandii*. Figure 6.9 shows a schematic diagram of nitrogenase's M center. At the time of the EPR study, it was known that two M center clusters were contained in each nitrogenase MoFe–protein subunit, that each was of composition $MoFe_{6-8}S_{9\pm1}$, and that each cluster contained three unpaired electrons in an $S = 3/2$ system.[27] The rhombic EPR signal with g factors $g_1 (= g_x) = 4.32$, $g_2 (= g_y) = 3.68$ and $g_3 (= g_z) = 2.01$ would have the appearance of the rhombic spectrum of Figure 3.13D in Section 3.4.1. (Also see Figure 16 of reference 29.)

The electron–nuclear double resonance (ENDOR) technique mentioned in the previous paragraph is used to study electron–nuclear hyperfine interactions (see EPR discussion in Section 3.4.2) that are too small to be resolved within the natural width of the EPR line. Usually the hyperfine splitting values, A, detected by ENDOR range from 2 to 40 MHz. In transition metal complexes and metalloproteins, magnetic nuclei such as ^{1}H, ^{2}H, ^{13}C, ^{14}N, ^{15}N, ^{17}O, ^{31}P, and ^{33}S can be detected by ENDOR as being in the vicinity of paramagnetic metal ions such as high-spin Fe(II) or high- or low-spin Fe(III). In the current example the ^{57}Fe ENDOR spectra were obtained with protein from *Azobacter vinelandii* grown on ^{57}Fe-enriched media.[28] The high-field g_3 and low-field g_1 edges of the EPR spectrum (see previous paragraph) were used to generate the ENDOR spectra assuming that, in different ENDOR experiments, molecules were oriented with their g_3 and g_1 axes approximately parallel to the applied magnetic field.[29] The spectra revealed that there were five distinguishable (inequivalent) iron sites designated A^1, A^2, A^3, B^1, and B^2 for the *Azobacter vinelandii* M center. It is not possible to count the total number of irons in the spectrum using the ENDOR technique; however, subsequent X-ray crystallography (see Section 6.5.3) indicated seven iron atoms in the M center cluster (five of which are inequivalent).

In reference 30, the authors take the ENDOR data described above and extend its conclusions using Mössbauer spectroscopy.[30] Specifically, ^{57}Fe enrichment of nitrogenase's M center is used to identify all seven irons present. Between the time of the EPR and ENDOR studies described in the previous paragraphs and the Mössbauer study presented in reference 30, the X-ray crystallographic structure of nitrogenase was published (see Section 6.5.3). The X-ray studies revealed that the M center consists of two cuboidal fragments, [Mo–3Fe–3S] and [4Fe–3S], these being linked by three sulfide bridges and attached to the nitrogenase protein via a cysteine and a histidine. The M center ($S = 3/2$) is further classified as being in the M^N state to distinguish it from the diamagnetic M^{OX} ($S = 0$) moiety not found in vivo and the M^R EPR-silent ($S > 1$) reduced state found in vivo in the presence of nitrogenase's Fe–protein subunit and MgATP (see Section 6.4). The reference 30 authors identified the seventh Fe site (A^4) as having an unusually small and anisotropic magnetic hyperfine coupling constant, $A \cong -4$ MHz. They also identified the previously identified (by ENDOR) B^1 site as representing two equivalent Fe sites having the same hyperfine interactions. The following values for the isomer shifts, δ, in mm/s are reported in reference 30 for M^N: A^1, 0.39; A^2, 0.48; A^3, 0.39; A^4, 0.41; B^1, 0.33; and B^2, 0.50; this yielded a δ_{avg} of 0.41 mm/s. The δ_{avg} of 0.41 mm/s for M^N differs by only 0.02 mm/s from that found for M^R, while M^N and M^{OX} differ by 0.06 mm/s. Because M^N and M^R have very similar high-field Mössbauer spectra, the authors believe that the iron oxidation and spin states remain the same in both forms while the Mo ion becomes reduced in the M^R state. The conclusion is supported by the fact that the magnetic hyperfine interactions for M^R and M^N are quite similar. The δ_{avg} difference for the M^N versus the M^{OX} pair led the researchers to conclude that the redox events (leading to nitrogenase's production of ammonia from dinitrogen) center on the iron ions of FeMo cofactor. Putting together their Mössbauer analysis with previous ENDOR data, EXAFS analyses, and X-ray crystallographic evidence, along with comparison to a model Fe(II) complex with trigonal sulfur coordination, allowed the reference 30 authors to describe the M^N center as (Mo^{4+}–$3Fe^{3+}$–$4Fe^{2+}$).

As the preceding discussion of nitrogenase metal–sulfur clusters indicate, analysis of complex bioinorganic systems requires the use of multiple analytical techniques and the cooperative exchange of data and ideas of many researchers. The descriptions in this chapter have attempted to give students some idea of the scope and complexity of instrumental techniques available to the bioinorganic chemist. It has not been intended to be either comprehensive or theoretical in presentation. Students are encouraged to acquaint themselves further with the theory and practice of instrumental techniques, especially those that are important to their particular research interests.

3.7 OTHER INSTRUMENTAL METHODS

3.7.1 Atomic Force Microscopy

Atomic force microscopy (AFM) is part of a range of emerging microscopic methods for chemists and biologists that offer the magnification range of both the light and electron microscope, but allow imaging under the "natural" conditions usually associated with the light microscope. AFM offers the prospect of high-resolution images of biological material, images of molecules and their interactions even under physiological conditions, and the study of molecular processes in living systems. Applications of AFM in the biosciences include analysis of (1) DNA and RNA, (2) protein–nucleic acid complexes, (3) chromosomes, (4) cellular membranes, (5) proteins and peptides, (6) molecular crystals, (7) biopolymers and biomaterials, and (8) ligand–receptor binding.

The atomic force microscope is one of about two dozen types of scanned-proximity probe microscopes. All of these microscopes work by measuring a local property—height, optical absorption, or magnetism—with a probe or "tip," typically made from Si_3N_4 or Si, placed very close to the sample. The small probe–sample separation (on the order of the instrument's resolution) makes it possible to take measurements over a small area. To acquire an image, the microscope raster-scans the probe over the sample while measuring the local property in question. The resulting image resembles an image on a television screen in that both consist of many rows or lines of information placed one above the other. Unlike traditional microscopes, scanned-probe systems do not use lenses, so the size of the probe rather than diffraction effects generally limit their resolution.

The concept of resolution in AFM is different from radiation-based microscopies because AFM imaging is a three-dimensional imaging technique. There is an important distinction between images resolved by wave optics and those resolved by scanning probe techniques. The former is limited by diffraction, whereas the latter is limited primarily by apical probe geometry and sample geometry. Usually the width of a DNA molecule is loosely used as a measure of resolution, because it has a known diameter of 2.0 nm in its B form.

Many biological processes—DNA replication, protein synthesis, drug interactions, and others—are largely governed by intermolecular forces. AFM has the ability to measure these forces, some of which may be in the nanonewton range. This makes it possible to quantify molecular interactions in biological systems such as important ligand–receptor interactions. The dynamics of many biological systems depends on the electrical properties of the sample surface, and AFM is able to image and quantify electrical surface charges. In addition to measuring binding and electrostatic forces, the atomic force microscope can also probe the micromechanical properties of biological samples. Specifically, the AFM can observe the elasticity and, in fact, the viscosity of samples ranging from live cells and membranes to bone and cartilage.

One area of significant progress for AFM has beam the imaging of nucleic acids. The ability to generate nanometer-resolved images of unmodified nucleic acids has broad biological applications. Chromosome mapping, transcription, translation, and small-molecule–DNA interactions such as intercalating mutagens provide exciting topics for high-resolution studies. The first highly reproducible AFM images of DNA were obtained only in 1991. Four major advances that have enabled clear resolution of nucleic acids are (1) control of the local imaging environment including sample modification, (2) TappingMode™ scanning techniques, (3) improved AFM probes (such as standard silicon nitride probes modified by electron beam deposition and oxide-sharpened nanoprobes), and (4) compatible substrates (such as salinized mica and carbon coated mica).

There has been recent success in imaging individual proteins and other small molecules with the AFM. Smaller molecules that do not have a high affinity for common AFM substrates have been successfully imaged by employing selective affinity binding procedures. Thiol incorporation at both the 5′ and 3′ ends of short PCR (polymerase chain reaction, described in Section 2.3.5) products has been shown to confer a high affinity for ultraflat gold substrates and therefore improved AFM imaging.

Cell biologists have applied the AFM's unique capabilities to study the dynamic behavior of living and fixed cells such as red and white blood cells, bacteria, platelets, cardiac myocytes, living renal epithelial cells, and glial cells. AFM imaging of cells usually achieves a resolution of 20–50 nm, not sufficient for resolving membrane proteins but still suitable for imaging other surface features, such as rearrangements of plasma membrane or movement of submembrane filament bundles.

It is informative to compare AFM with other techniques. The scanning tunneling microscope (STM) is considered the predecessor technique to AFM. The STM may have better resolution than the AFM but can only be applied to conducting samples while AFM can be applied to both conductors and insulators. Compared with the scanning electron microscope (SEM), the AFM provides extraordinary topographic contrast, direct height measurements, and unobscured views of surface features (no coating is necessary). Compared with transmission electron microscopes, three-dimensional AFM images are obtained without expensive sample preparation and yield far more complete information than the two-dimensional profiles available from cross-sectioned samples. New approaches in AFM have provided a solid foundation from which research is expanding into more complex analyses. Higher-resolution imaging of a variety of small molecules is improving at a rapid pace.

3.7.2 Fast and Time-Resolved Methods

3.7.2.1 Stopped-Flow Kinetic Methods. Enzyme kinetics happen on very fast time scales; for instance, it is known that the rate of reaction for copper–zinc superoxide dismutase (CuZnSOD), $\sim 1 \times 10^9\,M^{-1}\,s^{-1}$, approaches the diffusion-controlled rate. Chemists use various methods to study fast reactions. One of the most frequently used rapid kinetic techniques is that of stopped-flow in which the reactants (enzyme and substrate) are rapidly mixed. The lower practical limit for mixing to take place is about 0.2 ms. The stopped-flow principle of operation allows small volumes of solutions to be driven from high-performance syringes to a high-efficiency mixer just before passing into a measurement flow cell. As the solutions flow through, a steady-state equilibrium is established and the resultant solution is only a few ms old as it passes through the cell. The mixed solution then passes into a stopping syringe, which then allows the flow to be instantaneously stopped. Some of the resultant solution will be trapped in the flow cell and as the reaction proceeds, the kinetics can be followed using the appropriate measurement technique. The most common method of following the kinetics is by absorbance or fluorescence spectrometry, and in these cases the measurement cell is an appropriate spectrometer flow cell. Many commercially available absorbance and fluorescence spectrometers may be modified to accept stopped-flow accessories.

In order to use the stopped-flow technique, the reaction under study must have a convenient absorbance or fluorescence that can be measured spectrophotometrically. Another method, called rapid quench or quench-flow, operates for enzymatic systems having no component (reactant or product) that can be spectrally monitored in real time. The quench-flow is a very finely tuned, computer-controlled machine that is designed to mix enzyme and reactants very rapidly to start the enzymatic reaction, and then quench it after a defined time. The time course of the reaction can then be analyzed by electrophoretic methods. The reaction time currently ranges from about 5 ms to several seconds.

Fee and Bull have studied the mechanism of CuZnSOD using a stopped-flow spectrophotometric system capable of forming aqueous solutions of O^{2-} having initial concentrations up to approximately 5 mM.[31] A detailed discussion of the enzyme superoxide dismutase is found in Section 5.2.3. By lowering the temperature to 5.5°C, it was possible to observe saturation of the CuZnSOD. At the lowered temperature and at pH 9.3, the Michaelis–Menten parameters extracted from the kinetic traces were turnover number (TN), approximately $1 \times 10^6\,s^{-1}$, and K_m, approximately 3.5×10^{-3} M. (Michaelis–Menten enzyme kinetic theory is discussed in Section 2.2.4.) Under the experimental conditions, the average rate at which O^{2-} binds to the active site, TN/K_m, is $0.26 \times 10^9\,M^{-1}\,s^{-1}$. The turnover number was decreased in the presence of D_2O, and a solvent isotope effect of approximately 3.6 was measured while TN/K_m was essentially unaffected by D_2O.

TN was increased by the presence of the general acid. These observations suggested that H_2O serves to donate the protons required to form product H_2O_2. Values of K_m and TN for the zinc-deficient enzyme were found to be approximately a factor of two less than those obtained for the holoenzyme under identical experimental conditions, whereas TN/K_m was largely unchanged. The authors concluded that the imidazolate bridge is thus not essential for catalytically competent extraction of a proton from the solvent by CuZnSOD.

3.7.2.2 Flash Photolysis. Time-resolved spectroscopy techniques are a powerful means of studying materials, giving information about the nature of the excitations, energy transfer, molecular motion, and molecular environment, information that is not available from steady-state measurements. Flash photolysis is a rapidly advancing field with applications in many areas of science and technology. The technique allows one to follow a reaction using fast (nanosecond to microsecond) laser excitation pulses to cause absorption in the species of interest. Following the excitation, one must use fast electronic devices to measure the light emission of absorption by the species of interest. For instance, one laboratory uses a Yag laser (266-, 355-, and 532-nm excitations) or excimer (308-nm excitation) sources with transmission (10-ns resolution) or diffuse reflectance (200-ns resolution) detection. A necessary criteria for the use of flash photolysis methods is that the molecule under study must show a detectable change upon laser excitation.

One research group used a flash-photometric method to show photochemical NO displacement by CO in myoglobin.[32] Previous investigations of thermal and photochemical NO displacement by CO suggest that the local heme pocket around the ligand, although significantly altered (according to circular dichroism investigations), imposed a barrier against the outward diffusion of ligand (NO or CO) into the solvent. (Find a complete discussion of ligand attachments to hemes in myoglobin and hemoglobin in Sections 4.2, 4.3, and 4.9.) The researchers found in this case that nanosecond and picosecond flash photolysis in proteins at low pH showed an extremely efficient geminate recombination of the ligand—that is, reattachment of the ligand before its leaving the heme pocket. The process involved a four-coordinated species within the heme and took place through a single-exponential process. This occurred to a significantly larger extent for the case of NO-“chelated” protoheme (where no distal barrier for ligand is present) than for CO ligated under the same circumstances. At neutral pH, when the proximal histidine–Fe bond is intact, the geminate recombination for NO takes longer and displays multiexponential kinetics. Altogether, these results suggested that even though heme distal ligand and protein environment effects play a role in NO or CO ligation and deligation from the iron heme center, proximal ligand and protein environment effects make an important contribution in modulating ligand–iron bond formation in hemes.

3.7.2.3 Time-Resolved Crystallography. Time-resolved crystallography (TC) uses an intense synchrotron X-ray source and Laue data collection techniques to greatly reduce crystallographic exposure times. Normal time resolution for X-ray crystallography has been in the range of seconds or tens of seconds. TC has the potential to take snapshots of protein structural changes on a nanosecond time scale. Consequently, multiple exposures may be taken that capture the evolution of the crystallographic unit cell as it reacts over time. Traditionally, crystallographers have applied several techniques to obtain detailed structural information on reaction intermediates. The most common approach has been to design a series of stable structures that mimic normally short-lived intermediates. However, these structures are stable precisely because they are not identical to the intermediates they seek to mimic, and key interactions are usually missing. Other experimental techniques and chemical intuition are called upon to supply the missing information, sometimes with only limited success. One successful attempt to understand how the attachment and release of carbon monoxide, and ultimately dioxygen, happens on a molecular scale is described in Section 4.9. In this case, Rodgers and Spiro studied the nanosecond dynamics of the R to T transition in hemoglobin.[33] Using pulse-probe Raman spectroscopy, with probe excitation at 230 nm, these workers were able to model the R–T interconversion of the hemoglobin molecule as it moved from the R state (HbCO) to the T state (Hb).

Time-resolved crystallography, TC, now has the potential to offer detailed structural information on short-lived intermediates in macromolecular reactions under near-physiological, crystalline conditions, and this aids elucidation of the underlying molecular mechanisms. Interpretation of TC data has been hindered, in part due to the difficulty in extracting structural information on intermediates from time-resolved electron density maps. Under certain assumptions, these maps are weighted averages of the electron density maps of the different structural species present at the experimental time points. That is, these time-dependent electron density maps are structurally heterogeneous. Various researchers, most notably Krebs and Moffat, have proposed techniques for interpreting these maps.

In their 1996 *Science* article "Photolysis of the Carbon Monoxide Complex of Myoglobin: Nanosecond Time-Resolved Crystallography,"[34] Moffat, Wulff, and co-workers described the nanosecond time resolution of structural changes that occur in the carbon monoxide complex of myoglobin (MbCO) at room temperature on CO photodissociation by a nanosecond laser pulse. The Fe–CO bond was broken with a 10-ns laser pulse, and X-ray data sets were collected at different time delays between the laser flash and the X-ray pulse (4 ns, 1 μs, 7.5 μs, 50 μs, and 1.9 ms). Although the difference maps clearly showed release of the CO molecule from the heme, they also suggested that CO recombination in the crystal form contains a fast, geminate phase with a recombination rate comparable with or greater than the maximum photolysis rate applied by the laser pulse of $10^9\ s^{-1}$. This result confirmed that it is much more difficult to photolyze MbCO molecules in the crystal form than in solution. A second prominent feature of the X-ray difference maps arose from the motion of the iron atom out of the heme plane and toward the proximal histidine. A third feature indicated a transient "docking site" for the photodissociated CO; however, well-populated docking sites indicating CO exit from the binding pocket were not identified. A number of small electron density features indicated structural rearrangements of aa residues surrounding the heme, especially the residues of the E and F helices implicated by other methods in heme and protein relaxation effects, and in iron ion displacement in or out of the heme plane. Their data suggested that complete iron displacement and heme relaxation occurred in < 4 ns, in agreement with other spectroscopic results.

More recently, the same group published the article "Protein Conformational Relaxation and Ligand Migration in Myoglobin: A Nanosecond to Millisecond Molecular Movie from Time-Resolved Laue X-ray Diffraction."[35] Using the same techniques in the same system (MbCO), the researchers photodissociated the CO ligand by a 7.5-ns laser pulse and also probed the subsequent structural changes by 150-ps or 1-μs X-ray pulses at 14 laser/X-ray delay times, ranging from 1 ns to 1.9 ms. Very fast heme and protein relaxation involving the E and F helices was evident from the data at a 1-ns time delay. The photodissociated CO molecules were detected at two locations: at a distal pocket docking site and at the Xe 1 binding site in the proximal pocket. The population by CO of the primary, distal site peaks at a 1-ns time delay and decays to half the peak value in 70 ns. The secondary, proximal docking site reaches its highest occupancy of 20% at approximately 100 ns and has a half-life of approximately 10 μs. At approximately 100 ns, all CO molecules are accounted for within the protein, in one of these two docking sites or bound to the heme. Thereafter, the CO molecules migrate to the solvent from which they rebind to deoxymyoglobin in a bimolecular process with a second-order rate coefficient of $4.5 \times 10^5\ M^{-1}\ s^{-1}$. The results demonstrate that structural changes as small as 0.2 Å and populations of CO docking sites of 10% can be detected by time-resolved X-ray diffraction.

The results discussed here indicate that time-resolved crystallography should continue to evolve and that its use will enhance understanding of more complex enzymatic systems in the future.

3.8 INTRODUCTION TO COMPUTER-BASED METHODS

Chemists use computers for many purposes. As the previous sections on instrumental methods have illustrated, every modern analytical instrument must include a computer interface. Chemical structure drawing, visualization, and modeling programs are important computer-supported applications required in academic, industrial, and governmental educational and research enterprises. Computational chemistry has allowed practicing chemists to predict molecular structures of known and theoretical compounds and to design and test new compounds on computers rather than at the laboratory bench.

3.9 COMPUTER HARDWARE

A basic review of computers and computing is given in reference 36. A short summary of this introductory material is presented here.

Table 3.5 Computing Units

Power of 10	Number and Unit	Number of Locations (bytes)
2^{10}	1024 = 1 kilobyte	1024
2^{20}	1024 k = 1 M (megabyte)	1,048,576 (1.045×10^6)
2^{30}	1024 M = 1 G (gigabyte)	1,073,741,824 (1.07×10^9)
2^{40}	1024 G = 1 T (terabyte)	($1.099511628 \times 10^{12}$)

The basic elements of a computer are:

1. The central processor unit (CPU) that does the work
2. Memory locations where programs are controlled and results are stored
3. Input and output (I/O) devices to communicate with other computers or devices
4. Buses that provide communication between various computer elements

Computers contain read-only memory whose contents are permanent (i.e., can only be read and not written to by the user) along with random access memory that can both be read from and written to by the user. The basic computing unit is a bit (b), which stands for binary digit; 8 bits comprise a byte (B). Table 3.5 illustrates calculation of computer memory bytes, i.e., the number of locations that can be addressed.

A small modern desktop or laptop computer might have 200 or more megabytes of ROM and 4 or 8 gigabytes of storage capability. Computer storage capacities are constantly increasing to keep pace with requirements for data storage and manipulation.

Microcomputers, familiar to all as PC and Macintosh desktop or laptop computers, are the basic interface for many instrumental methods, including all types of chromatography (LC, HPLC, GC), hyphenated methods (i.e., GC-MS, LC-IR), flame emission spectroscopy, and basic ultraviolet-visible and infrared spectroscopy. Microcomputers have operating systems (OS) such as (a) MS-DOS or Windows for the PC or (b) MacOS for Apple Macintosh computers. More recently, the Linux OS has become available for PCs and Macs. Computer and instrument users should become familiar with more than one OS and expect them to continually undergo change and upgrading. Microcomputers capable of performing the various analog-to-digital or digital-to-analog conversions (ADC or DAC, respectively), in addition to PIO (parallel input output) functions, are normally purchased as an integral part of the analytical instrument in question. Upgrades to software are to be expected during the instrument's lifetime, usually requiring hardware upgrades as well. Many molecular design, molecular modeling, visualization, and computational programs will run on high-end microcomputers as discussed below in the section on computer software for chemistry, Section 3.12.

Workstations, a step up in capability and cost from microcomputers, usually feature a multitasking operating system (OS) enabling the computer to run more than one operation simultaneously. Often the individual processes are graphically based and controlled through a graphical user interface (GUI). More sophisticated instrumentation such as high-field NMR require workstations for data acquisition, manipulation, and display of one-, two-, and three-dimensional NMR spectra. Low-end workstations may be powerful PCs and Macs, while high-end models are available from companies such as Silicon Graphics, Sun, Hewlett Packard, IBM, and so on. More sophisticated graphics, molecular modeling, molecular design, and analysis software such as SYBYL™, SYBYL CSCORE™, and UNITY™ components sold by Tripos, Inc. and discussed below in Section 3.12 require workstation capability.

Minicomputers are medium-sized machines usually having multiple users. They have one or more CPUs working together and large amounts of both RAM and storage. The OS will be some form of the Unix operating system. Most of the time the chemist will interface through a PC or workstation with the minicomputer acting as a server. Manufacturers of these machines are Sun, Silicon Graphics, or Hewlett Packard, among others.

Mainframes are large computers comprised of a cluster of tightly coupled machines or having multiple processors. These units will often be set up for specific applications as database servers, or for handling calculations such as those generated by quantum mechanics-based computational chemistry methods.

Supercomputers often have hundreds of linked high-speed processors and very large memories. Very large calculations are performed across many processors at once, and the so-called *parallel processing* is very efficient for manipulating large mathematical data arrays. Supercomputers often have a minicomputer front end providing the user interface, and only the largest computing jobs are actually run on the supercomputer. These machines are produced by Cray Research, Silicon Graphics, Fujitsu, NEC, and Hitachi companies, among others.

A relatively new computer architecture called cluster-type or parallel computers is comprised of groups of small machines hooked together. They can be assembled from off-the-shelf PCs, making their cost low. Chemists are particularly interested because they are useful for increasingly computing-intensive modeling and simulation projects. Chemists are engaged, alongside computing experts, in modifying chemistry software but also designing general tools to make clusters and their ilk run better.[37] The clusters contain a number of processors put together on a motherboard into a unit known as a node. The nodes are then hooked together with other boards via high-speed communications networks. Nodes can be hardwired into supercomputers produced by companies like IBM and Compaq or they can be composed of individual small PCs defining the low-end cluster. Next-generation powerful low-end and high-end computing systems may both be cluster-type or parallel species. Another developing computer cluster architecture might place different types of clusters together so that one part of the cluster does superfast calculations (needed for ab-initio computational chemistry methods) and another part carries out molecular dynamics simulations that require fast graphics engines. Either type of cluster needs an extremely fast communications network connecting the components. This, in turn, leads to the need for reformulated modeling software so that it is usable on the cluster-type computers. One test cluster used GAMESS (the General Atomic and Molecular Electronic Structure System), an open-source ab-initio quantum chemistry software package, in a performance run on a 512-processor Linux-based cluster in comparison to results obtained on a standard supercomputer. The cluster computers performed within a factor of two of the supercomputer, a good result considering that the cluster is about 10 times less expensive. GAMESS, a general ab-initio quantum chemistry package maintained by the members of Mark Gordon's research group at Iowa State University, maintains a web site at http://www.smps.ntu.edu.au/gamess/. In 2001 one could access the free ChemWeb computational chemistry toolkit, a fully functional GAMESS program and graphical user interface at http://toolkit.chemweb.com/gamess.

3.10 MOLECULAR MODELING AND MOLECULAR MECHANICS

3.10.1 Introduction to MM

In the summer 2001 issue of the ACS newspaper "Chemistry," computational chemistry or "cheminformatics" was described in the following manner:

> Computational chemistry and quantum chemistry have enlisted the computer and software in an entirely new kind of experimental methodology. Computational chemists, for example, don't study matter directly. In the past, chemists who wanted to determine molecular properties chose their instrumentation, prepared a sample, observed the reactions of the sample, and deduced the molecule's properties. Computational chemists now choose their computer and software packages and get their information by modeling and mathematical analyses.

This portion of Chapter 3 will provide an introduction to this fast-growing and changing area of chemistry.

Theoretical models include those based on classical (Newtonian) mechanical methods—force field methods known as molecular mechanical methods. These include MM2, MM3, Amber, Sybyl, UFF, and others described in the following paragraphs. These methods are based on Hook's law describing the parabolic potential for the stretching of a chemical bond, van der Waal's interactions, electrostatics, and other forces described more fully below. The combination assembled into the force field is parameterized based on fitting to experimental data. One can treat 1500–2500 atom systems by molecular mechanical methods. Only this method is treated in detail in this text. Other theoretical models are based on quantum mechanical methods. These include:

1. Hückel and extended Hückel treatments that consider many fundamental ideas of orbitals and bonding but neglect electron repulsions in molecules. Treats systems up to 200 atoms.

2. Semiempirical (CNDO, MNDO, ZINDO, AM1, PM3, PM3(tm) and others) methods based on the Hartree–Fock self-consistent field (HF–SCF) model, which treats valence electrons only and contains approximations to simplify (and shorten the time of) calculations. Semiempirical methods are parameterized to fit experimental results, and the PM3(tm) method treats transition metals. Treats systems of up to 200 atoms.
3. Ab-initio (nonempirical, from "first principles") methods also use the HF–SCF model but includes all electrons and uses minimal approximation. Basis sets of functions based on linear combinations of atomic orbitals (LCAO) increase in complexity from the simplest (STO-3G) to more complex (3-21G(*)) to extended basis sets (6-311 + G**) for the most accurate (and most time-consuming) results. Treat systems up to 50 atoms.
4. Density functional methods treat larger molecules more successfully than ab-initio methods and also make use of explicit atomic basis sets as described for ab-initio methods.

3.10.2 Molecular Modeling, Molecular Mechanics, and Molecular Dynamics

An excellent introduction to molecular modeling of inorganic compounds has been provided by Comba and Hambley in their 1995 book *Molecular Modeling of Inorganic Compounds*.[38] The book has been updated by a new enlarged edition containing an interactive tutorial on CD-ROM. The newer edition includes descriptions of calculations of stereoselective interactions of metal complexes with biomolecules.[39] Additionally, the authors have developed an inorganic compound-oriented molecular modeling system called MOMEC, which has been designed as an add-on to the HyperChem™ drawing and modeling program discussed below in Section 3.12. The reference 38 and 39 authors discuss the application of molecular mechanics to coordination and organometallic compounds, inorganic compounds involved in catalysis (including stereospecific catalysts), and design of new metal-based drugs, as well as to the interaction of metal ions with biological macromolecules such as proteins or DNA. In all cases, modeling of transition metal species is complicated by the number of different metals of interest as well as the variety of coordination numbers, geometries, and electronic states they may adopt. Other pitfalls abound. If a calculation is performed on the wrong electronic state of a molecule (singlet rather than triplet oxygen, for example), any energy minimization result will be incorrect as well. An incorrect starting geometry (the wrong conformer for instance) may yield the wrong minimum. The geometry optimization method defined may not produce an accurate geometric result.

In spite of its limitations, molecular mechanics (MM) is a technique that is widely used for the computation of molecular structures and relative stabilities. The advantage of MM over quantum mechanical methods is mainly based on the computational simplicity of empirical force field calculations, leading to a comparatively small computational effort for MM calculations. Therefore, even large molecular assemblies with hundreds of atoms are tractable by MM, and energy surfaces with many minima may be screened successfully. MM has been used routinely for many years in organic chemistry applications. Studies on inorganic systems have begun more recently because the effects of variable coordination numbers, geometries (square planar, tetrahedral, octahedral), oxidation and spin states, and electronic influences based on partly filled *d* subshells (such as Jahn–Teller effects), have been difficult to model with conventional MM approaches.

In references 38 and 39, Comba and Hambley introduce their topic in three parts: (1) basic concepts of molecular mechanics, (2) applications of the techniques and difficulties encountered, and (3) a guide to molecular modeling of a new system. Only the introductory section of reference 38 is summarized here.

Molecular modeling seeks to answer questions about molecular properties—stabilities, reactivities, electronic properties—as they are related to molecular structure. The visualization and analysis of such structures, as well as their molecular properties and molecular interactions, are based on some theoretical means for predicting the structures and properties of molecules and complexes. If an algorithm can be developed to calculate a structure with a given stoichiometry and connectivity, one can then attempt to compute properties based on calculated molecular structure and vice versa.

Molecular mechanics is defined as the calculation of the molecular structure and corresponding strain energy by minimization of the total energy. The energies of various minima are calculated using functions that relate internal coordinates to energy values. Molecular mechanics methods can provide excellent descriptions of equilibrium geometries and conformations but do not supply thermochemical information. They cannot yield acceptable results outside the range of their parameterization. This important point, repeated and amplified below, says that a force field parameter set assembled for one coordination complex or bioinorganic molecule may not be applied to other molecules unless some means of testing the parameter set (comparison of calculated bond distances and bond angles to those known experimentally from an X-ray crystallographic structure, for instance) is applied.

In molecular mechanics calculations, the arrangement of the electrons is assumed to be fixed and the positions of the nuclei are calculated. Bonded atoms are treated as if they are held together by mechanical springs (vibrational frequencies), and nonbonded interactions are assembled from van der Waals attractive and repulsive forces (gas compressibility data). Finding suitable values for van der Waals parameters presents one of the greatest problems in force field development. Fortunately, van der Waals forces have less influence on the final molecular geometry than do bond-stretch and angle-bend parameters. The typical equations and parameters simulating the various interactions that describe the potential energy surface of a molecule are assembled from (1) a function that quantifies the strain present in all bonds, (2) a bond angle function, (3) a function that calculates all the dihedral strain, and (4) a number of nonbonded terms. To use the equations to calculate the total strain in the molecule, one needs to know (1) the force constants (k) for all the bonds and bond angles in the molecule, (2) all the ideal bond lengths (r) and bond angles (q), (3) the periodicity of the dihedral angles (n) and the barriers to their rotation (V), (4) the van der Waals parameters A_{ij} and B_{ij} between the ith and jth atoms to simulate the nonbonded van der Waals interactions, and (5) the point charges q_i and q_j and the effective dielectric constant (ε) to model the electrostatic potential.[40] In inorganic molecular mechanics, these parameters are empirically derived usually by fitting a number of crystal structures and are derived for the specific force field used (see below). Parameters are transferable from one molecule to another (within limits described below) but are not transferable between force fields.

To optimize the geometry of a molecule, the total energy from all forces is minimized by computational methods. This so-called "strain energy" is related to the molecule's potential energy and stability. Because parameters such as bond length, bond angles, and torsional angles used to derive the strain energies are fitted quantities based on experimental data such as X-ray crystallographic structures, molecular mechanics methods are often referred to as empirical force field calculations or just force field calculations. A "force field" is defined as a collection of numbers that parameterize the potential energy functions. These functions include the force constants, ideal bond distances and angles, and parameters for van der Waals, electrostatic, and other terms. Because force field parameters are dependent on the potential energy functions, the entire set of functions and parameters are often referred to as "the force field." In essence, the molecular mechanics method interpolates the structure and strain energy of an unknown molecule from a series of similar molecules with known structure and properties. As mentioned above, molecular modeling of transition metal compounds is complicated by the partially filled d orbitals of the metal ions. However, the structure of a coordination compound, and thus its thermodynamics, reactivity, and electronic behavior, is strongly influenced by the ligand structure. Because empirical force field calculations have been very successful in modeling organic ligands, it follows that these techniques should be extendable to transition metal coordination complexes as well. The problem remains that molecular mechanics interpolates the structure of an unknown inorganic complex from a set of parameters derived from fitting a number of X-ray crystallographic structures. If the complex being investigated differs significantly from the structures used to determine the parameters, poor results will be obtained.

Specific molecular mechanics (MM) force fields have been assembled, and several of these are mentioned here. The MM2 molecular mechanics force field and its applications to inorganic chemistry is described in reference 41a,b and references therein. The MM3 force field, a third-generation refinement, is one used extensively in molecular mechanical computations.[42] For applications in inorganic chemistry see reference 43. Information on the MM2 and MM3 force fields is available on the website http://europa.chem.uga.edu/ccmsd/mm2mm3.html. The MMFF94 Merck molecular force field is described on the website http://www.tripos.com/fhome2.html and in reference 44. The MMX force field[45] has been used in many inorganic applications, one student exercise using MMX implemented in PCMODEL™ (Serena Software, Inc. in Section 3.12) is described in reference 46. CFF offers a molecular mechanics force field at website: http://struktur.kemi.dtu.dk/cff/cffhome.html. Dreiding offers a molecular mechanics generic force field described in reference 47 and at website: http://www.wag.caltech.edu/research.html#newthe. MOLBLD is a molecular mechanics force field program described in reference 48. The MOLMECH molecular mechanics program contains an extended MM2 force field.[49] SHAPES is a molecular mechanics program described in reference 50. UFF, a molecular mechanics generic force field, has been used in inorganic chemistry applications described in reference 51. MOMEC97 is a molecular mechanics program adapted to HyperChem™ as described below in Section 3.12. More information is available at the website http://www.uni-heidelberg.de/institute/fak12/AC/comba/. The current MOMEC force field is published on this website. The authors of reference 40 describe criteria that should be applied when describing molecules with these molecular mechanics programs. Some of these are as follows (1) Check the error file for interactions not in the parameter set because some programs will assign a force constant of zero to unrecognized atom types; (2) check all interactions generating $>5\,\mathrm{kJ/mol}$ of strain to determine, for instance, whether that bond or angle really is that strained or whether there is a parameterization or molecular structure problem; and (3) check the hybridization about the central atom because a commercial program not written for inorganic molecules may not recognize metal–ligand bonds. In summary, the significant advantage of molecular mechanics calculations is that they are relatively rapid; however, caution is needed in interpretation of molecular mechanical results produced by empirical force field calculations. A good analogy is to that of the neural network; that is, molecular mechanics is completely dependent on the facts it has been taught.

The types of input and output accepted by the computer program, as well as algorithms used to achieve energy minimization, are described here briefly. Generally, an energy minimization routine produces an optimized structure (conformer) most closely related to the input coordinates. That is, the routine falls into the *closest* energy minimum, which may or may not be the global energy minimum of a system. One carries out a conformational analysis by deriving the energy surface that results from changing a particular rotational value, often a set of dihedral angles. Conformations are then submitted to energy minimizing routines that include: (1) simplex (only the potential energy function is used), (2) gradient or first derivative methods (steepest descent), (3) conjugate-gradient methods where the history of the search can influence direction and step size (Fletcher–Reeves, Polak–Ribière), (4) second-derivative methods (Newton–Raphson and block-diagonal Newton–Raphson), and (5) least-squares methods (Marquardt). In applying molecular mechanics methods, particularly for those structures that are predominant in solution, it is important to find the lowest-energy structure, that is, the global energy minimum. Several methods exist for carrying out the minimization, but only the torsional Monte Carlo method will be discussed here. The Monte Carlo method is a stochastic search in which the variation from one starting conformation to another is limited in magnitude—for instance, by limiting the starting geometries to those that conform to some energy requirement (perhaps 15 kJ/mol from the energy minimum). Using the internal coordinate of torsional

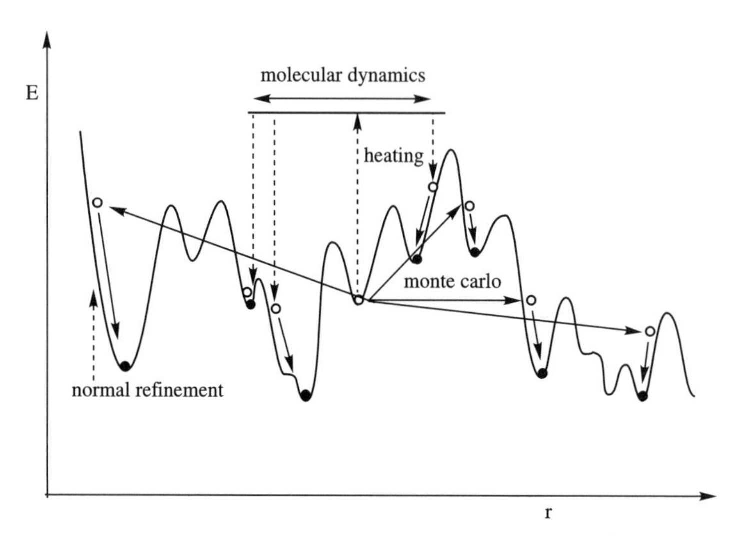

Figure 3.30 Conformational search procedures. (Adapted with permission from Figure 4.2 of Comba, P.; Hambley, T. W. *Molecular Modeling of Inorganic Compounds.* VCH Publishers, New York, 1995. Copyright 1995, VCH, Verlagsgesellschaft mbH, Weinheim, Federal Republic of Germany.)

angle causes significant differences in possible structure conformations—in contrast to changes in bond distances or bond angles, which do not. In each Monte Carlo step a random number of torsional angles are varied by a random amount generating a new starting geometry that can be minimized.

Molecular dynamics involves the calculation of the time-dependent movement of each atom of a molecule, achieved by solving and applying Newton's equations of motion. Structures for starting geometries are sampled as a function of time or geometry during a molecular dynamics run of a few nanoseconds. Usually, an elevated temperature is used to favor faster and more complete molecular dynamics searching. A good conformational search systematically (deterministically) scans the entire potential energy surface, generates starting geometries, and then minimizes the geometries. A combination of the above-described methods—deterministic, stochastic, and molecular dynamics—illustrated in Figure 3.30 as adapted from reference 38, screens the entire potential energy surface and produces the best possible results.

3.10.3 Biomolecule Modeling

The modeling of biomolecules is discussed in reference 52, which provides good background on protein folding, secondary protein structure prediction, sequence alignment for finding comparable natural or synthetic proteins, and modeling by homology to produce three-dimensional protein structures in advance of their experimental elucidation. The last topic makes use of the Ramachandran map, which places protein structure into allowed conformational regions for amino acids.

Fitting of the modeled protein into allowed regions—α-helical (α), β-strand (β), and left-handed helical (L)—indicates that the protein conforms to known protein secondary structural constraints. Reference 52 includes a short discussion of theoretical investigations of enzyme mechanism, useful for understanding the molecular basis of enzyme activity. Karplus and Petsko have provided a review of the application of molecular dynamics to biological problems.[53] Bioinorganic applications and examples of molecular modeling studies discussed in reference 38 often are limited by the lack of suitable small molecule analogs to the metal ion environments found in metalloproteins. Two examples are the grossly distorted tetrahedral type I Cu(II) sites in blue copper proteins and three-coordinate iron sites found in the enzyme nitrogenase. Models for metalloporphyrin active sites have been derived by modeling small molecules and extending the AMBER[54] and MM2[55] force fields parameterization schemes. A molecular dynamics simulation of hydrogen peroxide binding to the heme iron in cytochrome c peroxidase has been modeled using a modified AMBER force field.[56] The CHARMM force field was used in a study of the temperature dependence of both the structure and the internal dynamics of (carbonmonoxy)myoglobin.[57] The geometry about the iron center was assumed to be octahedral. Analysis of the movements of the iron center with respect to the heme group indicated that the largest-amplitude motions were perpendicular to the heme plane. In an article describing the computer-aided design (CAD) of superoxide dismutase synzymes (functional mimics), Mn(II) complexes of carbon substituted macrocycles were studied using molecular mechanics.[58] The molecular mechanics methods were described in detail as being based on an MM2 method with extensions provided by the software maker, CAChe™ (described in Section 3.12). The structural parameters used were modified based on six-coordinate crystal structures of Mn(II). In this study the researchers were attempting to discover better Mn(II) superoxide dismutase mimics for pharmaceutical applications. Their goal was to find complexes with greater kinetic stability at $pH = 5$ for treatment of ischemia reperfusion injury (ischemia is defined as localized tissue anemia due to obstruction of the inflow of arterial blood). The success of the adopted modeling and mechanics method to predict the structure of complexes having the desired characteristics was subsequently tested by synthesizing molecules and testing them for the desired activity. The authors found that they were able to predict the expected rate constants (and thus the desired medicinal characteristics) for computer-modeled Mn(II) complexes. This was accomplished by testing hundreds of structures by computer-based molecular mechanics methods and choosing the most promising molecules as determined by their MM methods. Only these most promising species were selected for submission to laboratory analysis—that is, complicated syntheses of the macrocyclic ligands and their metal complexes. The results are further discussed in Chapter 7, Section 7.2.1.

3.10.4 Molecular Modeling Descriptive Examples

Up until the time of publication of reference 38, there had been few modeling studies of metal–amino acid, metal–protein, metal–nucleotide, or metal–nucleic

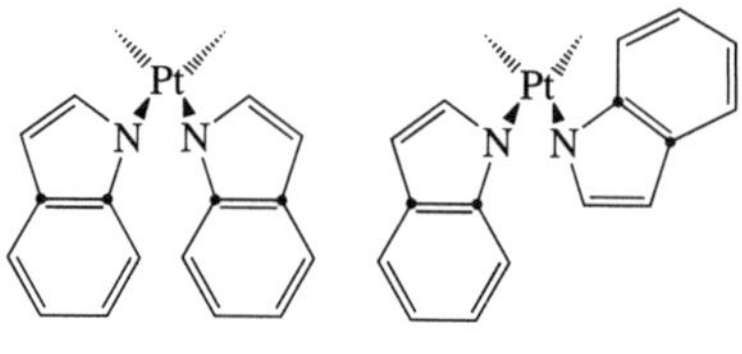

Figure 3.31 Head-to-head and head-to-tail conformers of bis(purine)–platinum(II) units. (Adapted from Figure 12.4 of reference 38.)

acid interactions. Hambley[59] studied *bis(*purine)diamineplatinum(II) complexes demonstrating how steric factors influenced the barriers to rotation about the Pt(II)–N(purine) coordinate bonds in interconversions of the head-to-head (HTH) to head-to-tail (HTT) isomers (see Figure 3.31). During the study, force field parameters for Pt(II)–nucleotide interactions were developed.

Numerous studies of the interaction of the anticancer drug cisplatin, *cis*-$[Pt(NH_3)_2Cl_2]$, and other platinum-containing drugs with DNA have been reported. (Section 7.2.3 contains a detailed discussion of platinum antitumor drugs.) The majority of platinum agent–DNA interactions have concerned the adduct formed between the platinum(II) ion with *cis*-diammine ligands and two adjacent guanine bases on one strand of DNA. The studies have emphasized analysis of the effect of platinum–DNA adduct formation on the local and global conformation of DNA.[60] Models have been produced to study the variation in hydrogen bonding found in the adducts. All studies indicate direct or indirect hydrogen bonds between the ammine ligands and the DNA molecule. It is known that replacement of all ammine hydrogens usually (but not always as indicated below) leads to a loss of anticancer activity in platinum antitumor compounds.[61] It is now known that molecules with *trans*-conformation of amine and chloro ligands about the platinum center and even complexes without hydrogen bonding capability have anticancer activity.

The molecular modeling study of Bierbach and Farrell[62] investigated *trans*-$[Pt_2(NH_3)_2Cl_2]$ and *trans*-$[Pt(NH_3)(quinoline)Cl_2]$. Reaction with the modified base 9-ethylguanine (9-EtGua) was used to model the first DNA-binding step followed by reaction with the simple nucleotide 5′-guanosine monophosphate (5′-GMP) to simulate the second step, that is, DNA–DNA crosslink formation. As an example, strain energies (energy minimization) for the compound *trans*-$[Pt(NH_3)(quinoline)(9\text{-}EtGua)Cl]^+$ shown in Figure 3.32 were calculated with the all-atom AMBER force field[63] employing the module of HyperChem^{TM} version 4.5 (1995). Calculations were carried out on a PC (120 MHz, 16 MB) or on an SGI workstation. Point minimizations employing the Polak–Ribiere minimizer were used to optimize structures. A Δ rms gradient of 0.001 kcal/(mol Å^2) was chosen as the convergence criterion. The systematic conformational search was carried out with a module in ChemPlus 1.0a (1994), a HyperChem extension. Calculations were based on a usage-directed scheme of the Monte Carlo multiple-minimum

trans-[Pt(NH_3)(quinoline)(9-EtGua)Cl]$^+$

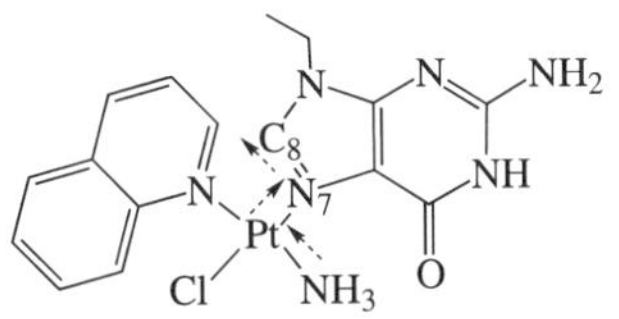

----→ = H_3N-Pt-N7-C8 dihedral angle

Figure 3.32 Structure of *trans*-[Pt(NH_3)(quinoline)(9-EtGua)Cl]$^+$ minimized by molecular mechanics calculations in reference 62.

(MMCM) method.[64] For each planar base bound to platinum, one torsional angle that describes its dihedral angle with the platinum coordination plane (for example, H_3N–Pt–N7–C8) was included in the random variations (angles between 60° and 180°). Optimizations were carried out as described above. Figures 8 and 9 of reference 62 show stereoviews of energy minimized conformations of *trans*-[Pt(NH_3)(quinoline)(9-EtGua)Cl]$^+$ and [Pt(NH_3)(quinoline)(9-EtGua)$_2$]$^{2+}$. The authors' results suggested that replacement of the NH_3 ligand with a planar amine (quinoline) may lead to significant changes in the rate of formation and structure of DNA adducts. These changes may be significant for understanding the anticancer activity of the so-called nonclassical *trans*-platinum antitumor complexes. More information on the *trans* complexes is also found in Chapter 7, Section 7.2.3.5.

A major problem in modeling of large biomolecules arises from the flexibility of proteins and DNA with corresponding numbers of adoptable geometries. The experimental system modeled is therefore only one possible representation of the many possible geometries. A second difficulty is that the accuracy of molecular mechanics models of biomolecules is substantially lower than that of small molecules. Large numbers of independent parameters are needed, but there are few experimentally known structures and these are often of low precision. Electrostatic considerations and solvent effects cause further limitations. The real value in molecular modeling of macromolecular systems, as stated by the reference 38 authors, emerges when the models make predictions that can be tested experimentally. Qualitatively, the models can be used to visualize molecules whose structures are not accessible by any other means.

In Part 3 of reference 38, the authors give much needed advice on developing a force field, taking into account bond length deformation, valence angle deformation, torsion angle deformation, out-of-plane deformation, van der Waals and electrostatic interactions, and hydrogen bonding interactions. All of these parameters are interrelated and modification of one must lead to further testing of all. They reiterate that the force field parameter set must model the molecule under consideration as accurately as possible and that results of the calculation should be compared to experimental data whenever this is available. The authors then discuss carrying out the calculation, first listing the important considerations. These include: having an adequate starting model, choosing an appropriate energy minimization method, and considering the probability that there are many possible energy minima.

3.11 QUANTUM MECHANICS-BASED COMPUTATIONAL METHODS

3.11.1 Introduction

Many other approaches for finding a correct structural model are possible. A short description of ab-initio, density functional, and semiempirical methods are included here. This information has been summarized from the paperback book *Chemistry with Computation: An Introduction to Spartan*. The Spartan program is described in the Computer Software section below.[65] Another description of computational chemistry including more mathematical treatments of quantum mechanical, molecular mechanical, and statistical mechanical methods is found in the Oxford Chemistry Primers volume: *Computational Chemistry*.[52]

3.11.2 Ab-Initio Methods

Ab-initio calculations are based on first principles using molecular orbital (MO) calculations based on Gaussian functions. Combinations of Gaussian functions yield Slater-type orbitals (STOs), also called Slater determinants. STOs are mathematical functions closely related to exact solutions for the hydrogen atom. In their ultimate applications, ab-initio methods would use Gaussian-type wave functions rather than STOs. The ab-initio method assumes that from the point of view of the electrons the nuclei are stationary, whereas molecular mechanical methods assume that the motions of the nuclei of a molecule are independent of the motions of surrounding electrons. In ab-initio calculations, the Born–Oppenheimer approximation—separation of the movement of atom nuclei and electrons is possible because electrons move much more rapidly than nuclei—is used to solve the Schrödinger equation ($H\Psi = E\Psi$) with a large but finite basis set of atomic orbitals. In practice, most ab-initio methods use the Hartree–Fock approximation, which represents the many-electron wavefunction as a sum of products of one-electron wavefunctions, termed molecular orbitals. Many different basis sets of orbitals have been generated for use with ab-initio calculations. A practical minimal basis set (such as the popular STO-3G representation; STO = Slater-type orbitals or Slater-type basis functions) for lithium and beryllium would contain 1*s*, 2*s*, and 2*p* occupied atomic orbitals supplemented by a set of unoccupied (in the atom) but energetically low-lying *p*-type functions. Minimal basis sets do not adequately describe nonspherical (anisotropic) electron distributions in molecules. One finds this basis set in computer modeling programs, but it should not be expected to yield realistic results for any inorganic complexes. To remedy this, one "splits" the valence description into "inner" and "outer" components. The result is the split-valence basis set, and the valence manifolds of main-group elements for instance are represented by two complete sets of *s*- and *p*-type functions. A simple split-valence basis set is called 3-21G. For heavier main-group elements unoccupied (in the atom) but energetically low-lying *d*-type functions are added to create the 3-21G(*) basis set. Polarization basis sets account for the displacement in molecular orbitals resulting from hybridization (an *sp* hybrid would be an example). Two popular polarization basis sets are the 6-31G* and 6-31G** representations. Ab-initio methods provide excellent accounts of equilibrium and transition state geometries and conformations as well as reaction thermochemistry. They are computationally very intensive and are usually limited to molecules containing 50 atoms or less.

3.11.3 Density Function Theory

Density functional models attempt to describe the total energy of a molecular system also from the standpoint of Hartree–Fock theory. These methods accurately describe equilibrium and transition-state geometries as well as thermochemistry. For density functional models, one must calculate the kinetic energy (KE) of the individual electrons (nuclear KE is zero in the Born–Oppenheimer approximation), the attractive potential energy between nuclei and electrons, and the repulsive potential energy between electrons (the Coulomb electron repulsive term in Hartree–Fock theory adjusted by the exchange term which takes into account a repulsive overestimation). The exchange and correlation terms ($E_{\text{correlation}} = E_{\text{true}} - E_{\text{Hartree-Fock}} - E_{\text{relativistic}}$) in so-called "local" density functional models originate from the exact solution of an idealized many-electron problem, mainly an electron gas of constant total electron density. In practice, one establishes functional relationships between the exchange and correlation energies for the "idealized gas" and the total electron density. However, such relationships are not unique and do not as yet lead to systematic progression to a low-energy structure. Practical density functional calculations make use of explicit atomic basis sets as described for ab-initio methods. Additionally, numerical integration steps are necessary. These may lead to loss of precision and slower calculations compared to ab-initio methods, especially for small molecules. Costwise density function methods are particularly useful for large systems when compared to other Hartree–Fock methods. In Section 3.12, density functional methods are offered for high-end programs, usually run on workstation-level computers.

Application of density function methods to bioinorganic systems was the subject in the American Chemical Society Inorganic Section at the August 2000 national meeting. In introductory remarks, David A. Dixon of the Pacific Northwest National Laboratory stated that density functional theory has been shown to be an effective way—in accuracy and computational cost—to predict structures and vibrational spectra of inorganic compounds in comparison to other methods. He stated that DFT can be used to predict other properties including reaction energies, electron densities, excited-state spectra, and NMR chemical shifts. Another speaker in this section, M. B. Hall, reported density functional calculations on models for [Fe]-hydrogenase, and he related these to structures and vibrational frequencies of the observed redox forms of the enzyme and the reaction mechanism at the enzyme's diiron active center.[66] R. A. Friesner spoke on large-scale ab-initio quantum chemical calculations on biological systems.[67] These authors have applied localized perturbation approaches and density functional theory to systems with hundreds of atoms, allowing accurate calculations including electron correlation to be carried out. Application of these methods to large-scale modeling of biological systems was discussed at the same session.

3.11.4 Semiempirical Methods

Semiempirical molecular orbital (MO) methods follow from the Hartree–Fock models with inclusion of additional approximations and the introduction of empirical parameters. These methods provide acceptable equilibrium and transition–state geometries but fail to account reliably for known thermochemistry. The single additional approximation used in semiempirical methods, termed NDDO, eliminates overlap of atomic basis functions on different atoms, a severe limitation. However, it leads to a great reduction in computation effort. Most currently-used semiempirical models are restricted to a minimal valence basis set of atomic functions. Because of this, calculations involving a heavy main-group element (e.g., gallium) take place in the same time frame as those for a first-row element (e.g., boron). Slater-type basis functions (STOs closely related to exact solutions for the hydrogen atom) are used in place of Gaussian functions (bell-shaped wavefunctions) employed for ab-initio calculations. The adjustable parameters introduced to reproduce experimental data in semiempirical calculations can improve the model obtained only if the correct method is chosen for the molecule in question. Three of the commonly used methods are called AM1 (limited data set of parameters mostly applicable to organic molecules and ligands),[68] PM3 (larger and more diverse "training set" of parameters),[69] and MNDO (the oldest method using fewer parameters and slightly different functional forms).[70] A newer MNDO/d method includes *d*–type functions for second-row and heavier main-group elements [much like the 3-21G(*) basis set for second-row and heavier main-group elements versus 3-21G for first-row elements].[71] A newer PM3(tm)[72] describes transition metals in terms of a minimal valence basis set [nd, $(n+1)s$, $(n+1)p$, where n is the principal quantum number]. PM3(tm) is normally used in conjunction with PM3. Semiempirical calculations on molecules containing up to 200 atoms are practical.

One example of the use of semiempirical methodology is provided in an article detailing a molecular-dynamics simulation of the beta domain of metallothionein with a semiempirical treatment of the metal core.[73] The beta domain of rat liver metallothionein-2 contains three-metal centers. In this study, three molecular variants with different metal contents—(1) three cadmium ions, (2) three zinc ions, and (3) one cadmium ion and two zinc ions—were investigated using a conventional molecular dynamics simulation, as well as a simulation with a semiempirical quantum chemical description (MNDO and MNDO/d) of the metal core embedded in a classical environment. For the purely classical simulations, the standard GROMOS96 force-field parameters were used, and parameters were estimated for cadmium. The results of both kinds of simulations were compared to each other and to the corresponding experimental X-ray crystallographic and NMR solution data. The purely classical simulations were found to produce a too compact metal cluster with partially incorrect geometries, which affected the enfolding protein backbone structure. The inclusion of MNDO/d for the treatment of the metal cluster improved the results to give correct cluster geometries and an overall protein structure in agreement with the experiment. The metal cluster and the cysteine residues bound to it were found to be structurally stable, while the irregular polypeptide backbone loops between the cysteines exhibited a considerable flexibility. MNDO without extension to *d* orbitals failed to maintain the structure of the metal core.

3.12 COMPUTER SOFTWARE FOR CHEMISTRY

This section provides a snapshot of chemistry software and websites, arranged alphabetically, available in 2002. Any software listing will necessarily be incomplete and out-of-date before this text's publication. The emphasis in the following listing is on software that contains algorithms suitable for use with transition metals and biomolecules. More sophisticated software programs allowing quantum mechanical (ab-initio or density functional theory calculations) may require workstation-level computer hardware, although requirements change frequently as desktop and laptop machines become more powerful. At the current time, most software suppliers offer very capable programs suitable for use with desktop PCs and Macintosh computers. For instance, Wavefunction, Inc. provides Spartan 6.0™ and Linux Spartan™ requiring Pentium II or higher hardware with appropriate Linux programs. PC Spartan Plus™ requires a 486 or Pentium/Pentium Pro desktop computer running Windows, and MacSpartan Plus™ requires a Power Macintosh with Mac OS 8.6 or higher. This author suggests that prospective users evaluate as many programs as possible by visiting websites or vendor booths at professional society meetings. Many vendors offer useful software demonstrations at professional meetings and provide on-line demonstration software or evaluation CDs for further evaluation. Many other programs not discussed here are available for drawing, visualization, computation, conformational analysis, combinatorial chemistry analysis, quantitative structure activity relationship (QSAR) searching, and other database searching. Not all capabilities of the software programs discussed here are mentioned. The idiosyncratic nature of the information below arises from its accessibility on various software company websites. Some product description pages were well-organized and very accessible, while others were difficult and required unnecessarily long download times for complex visualizations better left to demonstration downloads or CDs. In the listing below, each software supplier or program mentioned will list a website that should be consulted for the most accurate and up-to-date information.

CambridgeSoft at http://www.cambridgesoft.com/products offers a variety of tools including the software suite, ChemOffice™. The combination package consists of ChemDraw™, Chem3D™, and ChemFinder™. Every undergraduate chemistry student should be familiar with the chemical structure drawing program ChemDraw. Many undergraduate organic chemistry texts bundle ChemOffice, as a course resource. Chem3D Pro™ allows viewing of molecular surfaces, orbitals, electrostatic potentials, charge densities, and spin densities. Extended Hückel computations of partial atomic charges are possible, and MM2 calculations are performed for energy minimizations and molecular dynamics simulations on molecules of interest. Chem3D Ultra™ includes MOPAC 97 (see description under Serena Software, Inc.) for calculation of transition-state geometries and physical properties using PM3, AM1, MNDO, MINDO/3, and MNDO/d. A *CS Gaussian* interface (see description under Gaussian, Inc.) within Chem3D allows the user to create, run jobs, and visualize Gaussian calculation results.

ChemDraw Ultra 7.0, Chem3D Ultra 7.0, and ChemFinder Pro 7.0 became available in 2002. E-Notebook Ultra 7.0, BioAssay Pro 7.0, MOPAC, Gaussian & GAMESS interfaces, ChemSAR Server Excel, CLogP, Purchasing for Excel, CombiChem/Excel, as well as the full set of ChemInfo databases, including ChemACX & ChemACX-SC, The Merck Index and ChemMSDX have been added to ChemOffice Pro.

Fujitsu offers the CAChe™ software systems at http://www.cache.fujitsu.com. A variety of programs are available, depending on the user's needs and pocketbook. Personal CAChe™ is available in Macintosh or PC Windows-based versions. This program's editor contains a fragment library of amino acids, nucleic acids, transition states, organometallic, organics, inorganics, and drugs. All elements up to lawrencium are accessible. Translators import and export from a variety of external file formats including viewing and editing of crystallographic structure files, MDL, Protein Data Bank (PDB), and Cambridge crystallographic database (CCSD) files. Energy minimization routines are molecular mechanics or extended Hückel computations. Both methods may be applied to all elements and organic and inorganic molecules. However, the user should observe cautions described above in Section 3.10. Available force fields are augmented MM2 and MM3; mechanics parameters are adjustable and may be customized.

Quantum CAChe™ has the above-described capabilities plus the other features listed here. Molecular dynamics calculations are possible for all elements using augmented MM2[42] for organics and inorganics. MOPAC (see description under Serena Software, Inc.) calculations can be applied to main group elements, organics, and inorganics. Adjustable parameters are possible for semiempirical methods AM1,[68] PM3,[69] MINDO/3, MNDO,[70] and COSMO solvent model calculation models. ESR, IR spectra, heat of formation, free energy, transition-state searching, intrinsic reaction coordinates, dynamic reaction coordinates, activation energy, polarizabilities, reactivity surfaces, dipole moments, atom partial charges, bond orders, potential energy surfaces, and geometry optimization are additional capabilities for MOPAC calculations. ZINDO calculations include *d* orbitals. INDO and CNDO methods for organics and inorganics include capabilities for atom partial charges, bond orders, UV-visible spectra, SCRF (self-consistent reaction field) solvent modeling, reactivity surfaces, and augmented Hessian (a matrix of second derivatives of energy with respect to geometric distortions).

Ab-initio CAChe™ features all of the above plus ab-initio and density functional methods. This program requires a workstation (Windows NT minimum or SGI and IBM unix-based machines) and can be used to build and visualize results from ab-initio programs (e.g., Gaussian, see description under Gaussian, Inc.). Also, CAChe directly interfaces to Dgauss™, a computational chemistry package that uses density functional theory to predict molecular structures, properties, and energetics.

CAChe 5.0, available in 2002, includes a new, more powerful, semiempirical method that uses the PM5 Hamiltonian, a MOPAC 2002 offering, modeling of molecules with up to 20,000 atoms, the inclusion of all main group elements in one semiempirical method, and using MOPAC AM1-d, supports the transition metals Pt, Fe, Cu, Ag, Mo, V, and Pd. Researchers can now import and display, in 3D, proteins from the Protein Data Bank (PDB), optimize proteins, dock ligands, and model reactions on protein molecules.

Gaussian, Inc. programs, available at http://www.gaussian.com/product.htm, offer electronic structure modeling using ab-initio quantum mechanical methods to solve the electronic Schrödinger equation. The more rigorous ab-initio methods provide greater accuracy of bond lengths and angles when optimizing a molecule. The disadvantage for ab-initio methods will be a prohibitively long calculation time on a PC computer for any but the smallest and simplest molecules. Gaussian 98™, requiring a unix-capable workstation computing environment, and Gaussian 98W™, requiring a PC with a Pentium processor running Windows, are designed to model a broad range of molecular systems under a variety of conditions. GaussView™ is a Windows-based graphical user interface (GUI) for use with Gaussian 98. Gaussian 98 can predict energies, molecular structures, vibrational frequencies—along with the numerous molecular properties that are derived from these three basic computation types—for systems in the gas phase and in solution.

Gaussian 98M™, available in 2002, is an implementation of the Gaussian 98 electronic structure modeling program for the Mac OS X environment. It models a broad range of molecular systems under a variety of conditions, and performs its computations starting from the basic laws of quantum mechanics.

Hypercube, Inc. at http://www.hyper.com offers molecular modeling packages under the HyperChem™ name. In 2001 the newest versions were HyperChem 6™ and HyperChem Lite™. HyperChemLite was evaluated from a demonstration disk on a PC running Windows. After an organic structure had been sketched, the model builder feature easily and quickly turned the rough sketch into 3D structures for further manipulation. Orbitals and electron densities were displayed as wire-mesh or shaded solid surfaces or as contour plots. Molecular mechanics and molecular orbital calculations were integrated features of the software. The MM+ force field is used and includes geometry optimization to find stable structures. The extended Hückel method is used to calculate electronic properties and orbital energies. A molecular dynamic playback feature analyzes chemical reactions and the trajectories of colliding molecules generated in HyperChem Lite®. One example in HyperChem Lite® illustrated building a highly coordinated organometallic iron complex. However, the geometry constraint system for building simple square-planar, tetrahedral, or octahedral metal complexes did not function well, especially when square-planar geometry was specified.

HyperChem 6™ is a more advanced modeling package that also runs on Windows-based PC desktop computers. Visualizations in 3D and animations are possible. Computational methods include ab-initio, semiempirical, and molecular mechanics. Quantum mechanical calculations can include IR and UV-visible spectra. Biological macromolecules such as proteins and polynucleotides can be modeled and displayed. This version can be used to create HTML files for placement on websites, allowing chemists to publish structures and results on the web. Inorganic coordination complexes can be drawn; however, this user continued to have problems displaying square-planar structures as they reverted to tetrahedral geometry upon selecting the Build command. Square-planar structures were achieved by creating octahedral geometry and removing two ligands.

HyperChem Release 7™, available in 2002, is a full 32-bit application, developed for the Windows 95, 98, NT, ME, 2000 and XP operating systems. Density Functional Theory (DFT) has been added to complement Molecular Mechanics, Semi-Empirical Quantum Mechanics and *Ab Initio* Quantum Mechanics already available. The HyperNMR package has been integrated into the core of HyperChem, allowing for the simulation of NMR spectra. A full database capability is integrated into HyperChem 7. Many other features are updated and improved.

MOMEC, available at http://www.uni-heidelberg.de/institute/fak12/AC/comba/ is a program for strain energy minimization designed to be used in conjugation with HyperChem in a Windows environment. Because the memory space of PCs is limited, the present PC version of MOMEC is limited to molecules with less than 150 atoms and to force fields with less than 150 atom types and less than 100 stretching, 350 bending, 50 torsional bending, and 30 out-of-plane functions. Optimization of geometries is carried out in MOMEC; model building and visualization are developed in HyperChem. The program was developed for modeling transition metal compounds; however, it may be used in any area of inorganic and organic chemistry. Minimization is achieved by either the conjugate gradients (first derivatives) or the full-matrix Newton–Raphson refinement, or by a combination using both minimizers. Most energy minimization programs rely on variants of what are called "first-derivative" techniques to achieve energy minimization. The most commonly used technique is called "conjugate gradients." The advantage of these techniques is their modest memory requirement and ability to cope with crude starting models. A disadvantage is that the convergence criterion is based on the rate of change of the strain energy. It is not possible to verify whether a true minimum has been reached, and convergence can occur some distance away from the minimum. "Second-derivative" methods such as the Newton–Raphson method yield mathematically verifiable minima and generally do so after far fewer iterations. Also, it is possible with second-derivative methods to impose mathematically precise constraints; in first-derivative methods, only restraints are available. The disadvantages of second-derivative methods are their large memory requirements and the need to have a good starting model. In some cases, second-derivative methods will converge at a saddle point, where one of the second derivatives is positive rather than negative. The conjugate gradient Fletcher–Reeves method is superior if structures far away from the energy minimum are minimized. Therefore, it is suggested that this algorithm be used when the structure has been produced by HyperChem. The second derivative Newton–Raphson method has to be used when mathematically precise constraints are defined. Also, the precise definition of the energy minimum requires second derivatives. Therefore, it is advisable to switch to the Newton–Raphson minimizer in the final cycles of the structure optimization.

In order to facilitate modeling, MOMEC has been designed to have maximum flexibility. All force-field parameters are external to the program and can be readily modified with the force-field editor from the MOMEC window. In MOMEC the structures of Jahn–Teller active hexacoordinate molecules may be refined with the Jahn–Teller module. The approach used is based on a first-order harmonic model that minimizes the sum of steric strain and electronic stabilization. The strain energy calculation uses the Energy module with the metal–donor distances following a Jahn–Teller mode (i.e., elongation of the metal–ligand distances on the z-axis by $2d$ and compression of the in-plane metal–ligand bonds by d). The electronic term depends on the ideal metal–donor distance, the type of ligand (σ or π bonding), and the ligand field strength. The following references provide molecular modeling and computation examples using MOMEC and force-field parameters for use with MOMEC and HyperChem. In reference 74, Bernhardt gives details on molecular mechanics calculations of transition metal complexes. Reference 75 speaks to the directional nature of *d* orbitals and molecular mechanics calculations of octahedral transition metal compounds. Reference 76 discusses the molecular mechanics modeling of the organic backbone of metal-free and coordinated ligands. Reference 77 provides information on molecular mechanics calculations and the metal ion selective extraction of lanthanide ions. Reference 78 details a simple new structural force field for the computation of linear metallocenes.

Schrödinger, Inc. at http://www.schrödinger.com/Products/products.html offers a number of modeling and calculation packages. Jaguar™, operating in a workstation environment, contains an ab-initio software package and other quantum mechanical methods. The Jaguar program uses the self-consistent reaction field (SCRF) model to obtain geometries and solvation energies of molecules in solution. Reference 79 uses Jaguar to study spin states and charge transfer states of a bimetallic iron–chromium organometallic compound. The MacroModel™ program, also requiring workstation architecture, features molecular graphics, molecular mechanics, and a large selection of force fields and advanced methods for conformational analysis, molecular dynamics, and free energy calculations. See reference 80 for applications of Macromodel to problems inorganic chemistry. Both Jaguar and Macromodel may use Maestro™, Schrödinger's graphical user interface (GUI). The Titan™ program contains Schrödinger's Jaguar and Wavefunction's Spartan™ programs in an integrated package. Wavefunction, Inc. programs are separately described below.

SemiChem products are available at http://www.semichem.com/prods.html. AMPAC™, available as a stand-alone product with Windows-based and workstation-level interfaces, is a semiempirical quantum mechanical program featuring SAM1, AM1, MNDO, MNDO/d, PM3, MNDO/C, and MINDO/3 semiempirical methods. SAM1, introduced in 1993 as a new semiempirical method, replaces a key part of the AM1/PM3/MNDO theoretical model with a new approach. This key component is the two-electron repulsion integrals (TERIs), which is one of the basic terms in Hartree–Fock theory. Previously, this quantity was computed using a multipole expansion, but SAM1 uses a minimal Gaussian basis set (with appropriate semiempirical scaling) to compute these terms directly. SAM1 is currently parameterized for some first-row transition metals with work on others continuing. AMPAC™ also includes a graphical user interface (GUI) that builds molecules and offers full visualization of results. The SYBYL/Base™ program offered by Tripos, Inc (products described below) provides an interface with interactive graphing and structural display tools that can be used to access AMPAC's calculation tools.

Serena Software, Inc. at http://www.serenasoft.com provides PCMODEL™, GMMX™, MOPAC™, and other computational and visualization programs. PCMODEL offers basic molecular modeling with a molecule builder and a small molecule force field. It is available in desktop PC running Windows, Macintosh OS, and Unix-based workstation versions. In 2001, PCMODEL's Version 7 included support for Amber, MMX, MM3 and MMFF94 force fields. Graphical polypeptide backbone ribbons and Ramachandran plots provide the tools to make PCMODEL useful to biochemists and educators in a large number of applications. PCMODEL also reads and writes files for many other types of molecular calculation programs including MM2, MM3, MOPAC, Gaussian (Gaussian, Inc.), Macromodel (Schrödinger, Inc.), Alchemy (Tripos, Inc.), SYBYL (Tripos, Inc.), AMPAC (SemiChem), Chem-3D (Fujitsu), and the PDB (Protein Data Bank). Transition-metal complexes can be built with explicit sigma bonding, lone-pair coordination, and pi-system coordination. Parameters are available for all transition metals. Reference 46 offers a modeling exercise for undergraduate students that utilizes the PCMODEL program.

PCMODEL Version 8 became available in 2002. New features in version 8 include support for different and improved force fields along with the MMX, MM3, MMFF94, Amber, and Oplsaa force fields currently supported. The atom limit has been increased to 2500 atoms, and support for reading and writing PDB and SDF files has been added. Transition–metal complexes can be built with explicit sigma bonding, lone-pair coordination, and pi-system coordination. Parameters are available for all transition metals.

Global-MMX™ (GMMX) is a steric energy minimization program that uses the MMX force field and operates in batch mode. Its main purpose is to search conformational space and to list the lowest-energy unique conformations found. It is available in desktop PC running Windows, Macintosh OS, and Unix-based workstation versions.

MOPAC™ is a general-purpose semiempirical molecular orbital program for the study of chemical structures and reactions. It is available in desktop PC running Windows, Macintosh OS, and Unix-based workstation versions. It uses semiempirical quantum mechanical methods that are based on Hartree–Fock (HF) theory with some parameterized functions and empirically determined parameters replacing some sections of the complete HF treatment. The approximations in semiempirical theory result in more rapid single-energy calculations, which allow much larger structures to be studied. MOPAC® can use the semiempirical Hamiltonians MNDO, MINDO/3, AM1, and PM3 to obtain molecular orbitals, heats of formation, and their derivatives with respect to molecular geometry. Using these results, MOPAC can calculate vibrational spectra, thermodynamic quantities, dipole moments, molecular orbitals, and electron densities. MOPAC's greatest disadvantage for the bioinorganic chemist is that parameters are not available for many of the *d*-block elements. This situation is changing rapidly as additional parameters for all heavier elements become available and usable. When usable, MOPAC is a quantum chemistry program recognizing electron reorganization and can be used to study reaction mechanisms. This is in contrast to molecular mechanics programs that can only deal with a particular valence representation of a molecule.

Tripos offers computational and modeling software programs available at http://www.tripos.com/software/index.html. Those of most interest to bioinorganic chemists are described briefly here. The SYBYL/Base™ program includes a comprehensive set of molecular modeling tools for structure building, optimization, comparison, visualization of structures and associated data, annotation, hard-copy and screen capture capabilities, and a wide range of quantum mechanical methods and force fields. Its use requires a workstation environment. Reference 81 discusses several applications in inorganic chemistry. The SYBYL/Base program includes implementations of the Amber united-atom and all-atom force fields as well as MMFF94 and AMPAC (SemiChem). Other calculation software available through Tripos include MM2(91) and MM3(2000). Alchemy 2000™, for PC desktop computers, is a molecular modeling and visualization program capable of energy calculations and conformational searching. New versions of Tripos software that became available in 2002 included SYBYL 6.8 with updated SYBYL/Base features.

Wavefunction, Inc. at http://www.wavefun.com offers molecular modeling software. PC Spartan Plus™ and MacSpartan Plus™ use desktop PCs and Macintosh computers. Its graphical user interface (GUI) is user friendly. One can build organic molecules from included atomic fragments (sp^3, sp^2, sp carbons, oxygens, and nitrogens) and peptides using included amino acid residues. Inorganic molecules, transition-metal coordination complexes, and organometallic structures may be built from on-screen selection modules allowing any central atom geometry. The program contains a small ligand library, a peptide library, and allows modifiable hybridization as well as adjustable bond lengths and strengths. The user may import and export files from the Protein Data Bank (PDB) and other drawing and modeling programs. For molecular mechanics energy minimizations, the programs use the SYBYL force field. Semiempirical methods included are AM1 and AM1-SM2 for solvation, along with PM3 and PM3(tm). Possible Hartree–Fock ab-initio calculations use the basis sets 3-21G, 3-21G(*) and 6-31G*. In addition to the previous features, PC Spartan Pro and MacSpartan Pro programs additionally may use the MMFF94 force field for molecular mechanics minimizations. Up to 1000 atoms may be treated. Semiempirical methods include MNDO/d. More basis sets for

Hartree–Fock and MP2 methods are included. Density function theory methods include local (VWN) and nonlocal Becke–Perdew (BP86 and pBP86) for faster minimizations of medium to large molecules. One may simulate chemical reactions, manually and graphically compare molecules, and carry out conformational searches using Monte Carlo methods. Linux- and Unix-based Spartan programs require workstation computer architecture but have additional import–export and extended and enhanced computational capability over their lower-priced alternatives. Determination of equilibrium geometries and transition-state geometries are possible as well as calculation of normal-mode vibrational frequencies. Conformational space may be searched to establish whether or not a geometry corresponds to an energy minimum or to a transition state. Scanning of geometrical coordinates to locate the transition state along a reaction coordinate is possible.

The newest software version, Spartan '02 for Windows™, became available in 2002. In this update, the software added a polynucleotide–builder from included nucleotide fragments, as well as a substituent–builder. A transition–state library is available. Import/export choices are expanded. More DFT (density functional theory) basis sets are included.

Some of the many other drawing, modeling, and computation programs are listed below. Some of these have been already mentioned or referenced in previous material in this section. Most of the information listed here was taken from a larger list at the website http://ep.11n1.gov/msds/dvc/viewrs.html. AMBER carries out molecular simulations, particularly of large biomolecules.[82] Website: http://www.amber.ucsf.edu/amber/amber.html. CHARM carries out molecular simulations, particularly of large biomolecules.[83] Website: http://yuri.harvard.edu/. The Discover program carries out molecular simulations.[84] Website: http://www.msi.com/life/products/insight/index.html. The GROMOS program carries out molecular simulations.[85] Website: http://igc.ethz.ch/gromos/. MOBY is a molecular modeling package for the PC. For applications in inorganic chemistry see reference 86.

3.12.1 Mathematical Software

SPSS offers the useful graphing programs SigmaPlot at http://www.spssscience.com/SigmaPlot/ (for Windows-based PCs and Macintosh operating systems) and DeltaGraph (for Macintosh or Windows) at http://www.spss.com/deltagraph/. SigmaPlot will display multiple graphs, bar charts, 3D mesh, and line plots and ternary plots. It contains more powerful curve-fitting programs than does DeltaGraph. Both programs import data entered into Microsoft Office Excel™ files. A web-based version of SigmaPlot became available in 2001.

Design Science, Inc. provides MathType (www.mathtype.com), the professional version of the Microsoft Equation Editor. It is an interactive tool for Windows and Macintosh that lets users create mathematical notation for word processing and desktop publishing documents, web pages, and presentations.

Wolfram Research at www.wolfram.com supplies Mathematica. This program does everything mathematical from simple calculator operations to large-scale programming and interactive document preparation. Mathematica combines interactive calculation (both numeric and symbolic), visualization tools, and a complete programming environment, and it runs on Windows-based PCs and workstations.

3.13 WORLD WIDE WEB ONLINE RESOURCES

Any listing of World Wide Web online resources becomes obsolete and incomplete as soon as it is written down. Sites come and go daily, probably hourly or by the millisecond. In assembling the listings below, this author used many sources including a Chemical Society Reviews article that appeared in 1997.[87] The article not only lists World Wide Web sites but gives some information and advice on terminology, searching, chemical standards and guidelines, chemical markup language, applications to chemical teleconferencing, and innovations in chemical electronic journals. A useful glossary of terms is included. The article is still available online (good news) at http://www.rsc.org/is/journals/current/chsocrev/csr398.htm; however, this address represents a change from the original website address listed in the printed article (bad news).

3.13.1 Nomenclature and Visualization Resources

Nomenclature resources help the user to give correct names to chemical structures.

MDL Information Services, Inc. offers free software downloads at http://www.mdli.com/cgi/dynamic/downloadsect.html?uid=&key=&id=1. These include AutoNom Standard (**auto**matic **nom**enclature), which generates IUPAC chemical names directly from graphical structures created in ISIS/Draw or registered in ISIS/Base.

ACD/Name by Advanced Chemistry Development working in Windows generates accurate systematic names according to IUPAC (International Union of Pure and Applied Chemistry) and IUMBM recommendations on nomenclature of organic chemistry and selected classes of natural products, biochemical, organometallic, and inorganic compounds. A free download of the web version is available at http://www.acdlabs.com/products/java/sda. The ACD Structure Drawing Applet (ACD/SDA) is a complete structure drawing, editing, and visualization tool written in Java that can be incorporated into HTML documents. The applet can be used for composing substructure queries to databases and visualizing results. The only requirement is a Java-compatible browser such as Netscape 2.0 or later or MS Internet Explorer 3.0 or later.

Visualization software allows the user to display molecular structures imported from databases or other software programs. Chime, RasMol, and Protein Explorer programs are available at the websites listed below for Windows operating PCs and Macintosh PowerPC computers.

One important technique that users should acquire is the ability to transform two-dimensional stereoviews into three dimensions. The stereoviews may be found in visualization software such as Chime and RasMol. They are also found in journal articles in *Science* or *Nature* and others, especially those describing new biomolecule X-ray crystallographic or NMR solution structures. The following hints for learning to transform the stereoviews come from the extremely helpful website http://www.usm.maine.edu/~rhodes/0Help/StereoView.html.

Computer and projected stereo images require convergent (cross-eye) viewing—that is, looking at the left-hand image with the right eye and the right-hand image with the left eye. Gaze at a projected stereo pair with your head level. Cross your eyes slightly and slowly so that the two center images (of the four you see) come together. When they fuse, you will see them as a single 3D image. Ignore the other images at the periphery of your vision. Another approach: With your head level and about 2.5 feet from the computer screen, hold up a finger, with its tip about 6 inches in from of your face, centered between the stereo pair on the screen. Focus on your finger tip. If you see four images, move your finger slowly back and forth until the middle images converge, then change your focus to the screen. Try removing the finger. Texts and journal articles require divergent viewing—that is, left-hand image with the left eye and right-hand image with the right eye. Put your nose on the page between the two views. With both eyes open, the two images should be superimposed, but blurred. Slowly move the paper away from your face, trying to keep the images superimposed until you can focus on them. At this point, the middle image should appear three-dimensional and the two peripheral ones should be ignored. Another technique suggests that you tape a divergent stereo pair to a mirror, just below eye level. Look at your eyes in the mirror above the image and then bend your knees so that your view passes through the stereo pair on the way to looking at your eyes below the image. Rise, then repeat. At some point the images should fuse. If not, consider the practice a good leg-strengthening exercise. Practice with any or all of the techniques helps.

The Research Collaboratory for Structural Bioinformatics' Protein Data Bank (RCSB-PDB at http://www.rcsb.org/pdb/) is the online source for X-ray and NMR structural data. Many software programs mentioned in Section 3.12 include the facility to visualize imported data; however, two free software programs operate well in this regard. These are RasMol and Chime, described below.

Chime is a chemical structure visualization plug-in for Internet Explorer and Netscape Communicator. This web browser plug-in was developed in 1996 from a stand-alone program called RasMol (written by Roger A. Sayle in the early 1990s at Imperial College). Chime supports a wide variety of molecule coordinate formats, including PDB (Protein Data Bank), Molfile (from ISIS/Draw), MOP (MOPAC input files), and GAU (Gaussian Input files). It also supports the RasMol scripting language developed by Roger Sayle, which allows complex "molecular style sheets" and animations to be developed. Other Chime features include the display of NMR, MS, IR, and UV spectra in JCAMP-DX format, along with the rendering of 3D volume information in the form of Gaussian Cube files (i.e., molecular orbitals, surfaces, etc.).

The free RasMol visualization program is available at http://www.umass.edu/microbio/rasmol/ or http://www.ch.ic.ac.uk/. The latter is the Department of Chemistry site at Imperial College of Science, Technology and Medicine, London.

RasMol 2.7, an open source code free version, became available in 2001. This molecular graphics program displays macromolecules in three dimensions. RasMol renders each atom in a macromolecule in its specified 3D position according to the data in an atomic coordinate file, adding the appropriate covalent bonds. The program also interprets data about atom identities, secondary structure, disulfide bridges, hydrogen bonding, polypeptide chains, and more, if present in the coordinate file. RasMol is unequaled in its ability to rotate smoothly a space-filled macromolecule in three dimensions. Simple combinations of mouse and keyboard allow the user to rotate, move, and zoom the molecule in 3D. Drop-down menus allow changes to color and display modes. RasMol also accepts a large vocabulary of commands, providing the savvy user with almost unlimited control over the display. One RasMol drawback is that the user must learn a specialized command language to tap the majority of the program's power. Another limitation is that RasMol, while excellent for self-directed exploration by a knowledgeable user, is quite limited as a presentation tool and can be troublesome for beginners. Despite these limitations, RasMol remains a popular and very powerful molecular visualization tool.

ISIS/Draw is a chemically intelligent drawing package that enables the drawing of chemical structures using the same intuitive signs and symbols used for paper sketches. It is offered free for academic and personal home use only.

François Savary in the group of Professor J. Weber, Department of Physical Chemistry, University of Geneva offers a website with a great introduction to various molecular modeling and rendering techniques at http://scsg9.unige.ch/fln/eng/toc.html. The National Institutes of Health (NIH) displays a molecular modeling page at http://cmm.info.nih.gov/modeling/. Its software list is available at http://cmm.info.nih.gov/modeling/software.html. This site offers a free Chime download at http://www.mdli.com/cgi/dynamic/product.html?uid=$uid&key=$key&id=6. or http://www.mdlchime.com/chime/.

Protein Explorer, a free RasMol and Chime derivative with extended capabilities, is available at http://www.umass.edu/microbio/chime/explorer/. The program allows the user to visualize macromolecular structure in relation to function.

3.13.2 Online Societies, Literature, Materials, Equipment Web Servers

The American Chemical Society maintains the chemistry.org website at http://chemistry.org/portal/Chemistry. One can access Chemical Abstracts (CAS) from this site. The site lists meetings and publications and includes a careers and jobs site and an online store. American Chemical Society (ACS) publications at http://pubs.acs.org provides its members information products and services. Currently, over 30 magazines and peer-reviewed journals are published or co-published by the Publications Division.

The website http://www.annualreviews.org/ offers online searching capability to the entire *Annual Reviews* series. The *Annual Reviews of Biochemistry* and *Annual Reviews of Biophysics and Biomolecular Structure* are of most pertinence to bioinorganic chemists. This nonprofit by-subscription scientific publisher provides free searching of the site and no-cost abstract retrieval.

The Library of Congress site at http://www.loc.gov/catalog/ provides free searching access to the Library of Congress' huge collection of books and journals. Some journal articles are available online.

Chemical Abstracts, a subscription service at http://www.cas.org/ offers access to CAS, a large and comprehensive database of chemical information.

STN International at http://www.cas.org/stn.html is a full-featured online service that offers information for a broad range of scientific fields, including chemistry, engineering, life sciences, pharmaceutics, biotechnology, regulatory compliance, patents, and business.

SciFinder at http://www.cas.org/SCIFINDER/scicover2.html is an easy-to-use desktop research tool that allows the user to explore research topics, browse scientific journals, and access information on the most recent scientific developments. SciFinder Scholar is a desktop research tool designed especially for use by students and faculty to easily access the information in the CAS databases. With either tool, one can search Chemical Abstracts and the CA Registry by author name, research topic, substance identifier, chemical structure, or chemical reaction.

Scirus at http://www.scirus.com is a comprehensive search engine designed specifically for finding relevant scientific information everywhere on the World Wide Web. Currently, it indexes more than 60 million science-related pages.

The chemistry societies network at http://www.chemsoc.org/ provides information made available by about 30 national chemistry societies worldwide. The site includes the following: a Careers and Job Center; a Chembytes Infozone section providing science, industry, and product news; a guide to funding sources; publications updates; and an online magazine. The Conferences and Events section allows the user to search for events, advertise meetings, and view online abstracts from major conferences. The Learning Resources center presents (a) Web tutorials and videotapes for teaching and learning, (b) tutorial chemistry texts, and (c) the Visual Elements periodic table.

The National Cancer Institute (NCI) research resources database lists scientific information, tools, reagents, and services for cancer researchers on its website at http://resresources.nci.nih.gov.

The ISI web of science at http://www.isinet.com is a multidisciplinary database that provides web access to current and retrospective journal literature. It includes three citation databases: Science Citation Index Expanded, Social Sciences Citation Index, and Arts and Humanities Citation Index. The ISI® Science Citation Index (SCI®) provides access to current and retrospective bibliographic information, author abstracts, and cited references found in 3500 science and technical journals covering more than 150 disciplines. The Science Citation Index Expanded format available through the ISI Web of Science™ and the online version, SciSearch, cover more than 5700 journals.

WWW chemicals at http://www.chem.com enables chemists to search catalogs and directories of suppliers, distributors, manufacturers to find chemicals and equipment. Its Structures database offers access to 127,000 3-D structures from the NCI (National Cancer Institute) database.

The Scientific World at http://www.thescientificworld.com/ offers literature searching through SciBase, a collection of databases of scientific, technical, and medical research literature. SciBase currently covers more than 19 million documents published since 1965 in more than 30,000 journals. SciBase content is derived from databases created by the National Library of Medicine (MEDLINE), the British Library, BIOSIS, and PASCAL, as well as CAB ABSTRACTS. Abstracts are sometimes available free and individual articles are available for purchase.

BioMedNet at http://www.bmn.com provides access to biomedical databases, journal searching through MedLine, web links to 3500 sites, news, and access to conference reports. It includes a books and lab equipment site and a jobs site.

Access to Medline and PubMed is provided through the National Library of Medicine at http://www.nlm.nih.gov. MEDLINE contains bibliographic citations and author abstracts from more than 4000 biomedical journals published in the United States and 70 other countries. The service provides titles and abstracts for over 130 chemistry journals.

3.14 SUMMARY AND CONCLUSIONS

This chapter has provided an introduction to some instrumental methods used in bioinorganic chemistry with emphasis on methods referred to in later chapters. The structural methods of X-ray crystallography (solid state) and NMR (in solution) have proved indispensable to researchers studying complex bioinorganic systems. Mössbauer spectroscopy is an invaluable aid for researchers analyzing bioinorganic systems containing iron. Electron paramagnetic resonance methods provide information on the many bioinorganic systems containing unpaired electrons. The introduction to computer hardware and software provides a snapshot of some systems available in 2001–2002. A brief survey of some online resources should prove useful to students and researchers. Of all the fields one could address in this short text, those of computer hardware, software, and online resources will be the one most quickly outdated. Fortunately, updates are readily available through a search of the World Wide Web.

REFERENCES

1. Que, L., ed. *Physical Methods in Bioinorganic Chemistry: Spectroscopy and Magnetism*, University Science Books, Sausalito, CA, 2000.
2. Skoog, D. A.; Holler, F. J.; Nieman, T. A. *Principles of Instrumental Analysis*, 5th ed., Saunders Publishing, Philadelphia, PA, 1998.
3. Cowan, J. A. *Inorganic Biochemistry, An Introduction*, 2nd ed., Wiley-VCH, New York, 1997.

4. Scott, R. A., in Que, L., ed., *Physical Methods in Bioinorganic Chemistry: Spectroscopy and Magnetism*, University Science Books Sausalito, CA, 2000, 465–503.
5. Kau, L.-S.; Spira-Solomon, D. J.; Penner-Hahn, J. E.; Hodgson, K. O.; Solomon, E. I. *J. Am. Chem. Soc.*, 1987, **88**, 595–598.
6. Christiansen, J.; Tittsworth, R. C.; Hales, B. J.; Cramer, S. P. *J. Am. Chem. Soc.*, 1995, **117**, 10017–10024.
7. Peters, J. W.; Stowell, M. H. B.; Soltis, S. M.; Finnegan, M. G.; Johnson, M. K.; Rees, D. C. *Biochemistry*, 1997, **36**, 1181–1187.
8. Nobbs, C. L.; Watson, H. C.; Kendrew, J. C. *Nature*, 1966, **209**, 339.
9. Fermi, G.; Perutz, M. F.; Shaanan, B.; Fourme, R. *J. Mol. Biol.*, 1984, **159**, 175.
10. Drenth, J. *Principles of Protein X-Ray Crystallography*, 2nd ed., Springer-Verlag, New York, 1999.
11. (a) Gilliland, G. L.; Ladner, J. E. *Curr. Opin. Struct. Biol.*, 1996, **6**, 595–603. (b) Gilliland, G. L. *Methods Enzymol.*, 1997, **277**, 546–556.
12. Hahn, T., ed., *International Tables for Crystallography*, Vol. A, D. Reidel, Dordrecht, 1993.
13. Matthews, B. W. *J. Mol. Biol.*, 1968, **33**, 491–497.
14. Kitajima, N.; Fujisawa, K; Fujimoto, C.; Moro-oka, Y.; Hashimoto, S.; Kitagawa, T.; Toriumi, K.; Tatsumi, K.; Nakamura, A. *J. Am. Chem. Soc.*, 1992, **114**, 1277–1291.
15. (a) Laskowski, R. A.; MacArthur, M. W.; Moss, D. S.; Thorton, J. M. *J. Appl. Crystallogr.*, 1993, **26**, 283–291. (b) Laskowski, R. A.; Moss, D. S.; Thorton, J. M. *J. Mol. Biol.*, 1993, **231**, 1049–1067. (c) MacArthur, M. W.; Laskowski, R.A.; Thorton, J. M. *Curr. Opin. Struct. Biol.*, 1994, **4**, 731–737. (d) Otwinowski, A. in Sawyer, L.; Issacs, N.; Bailey, S., eds., *Data Collection and Processing*, SERC Daresbury Laboratory, Daresbury, U.K. 1993, pp. 56–62. (e) Brünger, A. R.; Kuriyan, J.; Karplus, M. *Science*, 1987, **235**, 458–460.
16. Palmer, G., in Que, L., ed. *Physical Methods in Bioinorganic Chemistry: Spectroscopy and Magnetism*, University Science Books, Sausalito, CA, 2000, 121–185.
17. Valentine, J. S.; DeFreitas, D. M. *J. Chem. Ed.*, 1985, **62**(11), 990–997.
18. Solomon, E. I.; Baldwin, M. J.; Lowery, M. D. *Chem. Rev.*, 1992, **92**, 521–542.
19. Akitt, J. W. *NMR and Chemistry: An Introduction to Modern NMR Spectroscopy*, 3rd ed., Chapman and Hall, London, 1992.
20. Ming, L.-J., in Que, L., ed. *Physical Methods in Bioinorganic Chemistry: Spectroscopy and Magnetism*. University Science Books, Sausalito, CA, 2000, pp. 375–464.
21. Bertini, I.; Luchinat, C.; Ming, L.-J.; Piccioli, M.; Sola, M.; Valentine, J. S. *Inorg. Chem.* 1992, **31**, 4433–4435.
22. Clore, G. M.; Gronenborn, A. M. *Proc. Natl. Acad. Sci.* 1998, **95**(11), 5891–5898.
23. Drago, R. S. *Physical Methods in Chemistry*, W. B. Saunders, Philadelphia, PA, 1977.
24. Münck, E., in Que, L., ed. *Physical Methods in Bioinorganic Chemistry: Spectroscopy and Magnetism*, University Science Books, Sausalito, CA, 2000, pp. 287–319.
25. Suslick, K. S.; Reinert, T. J. *J. Chem. Ed.*, 1985, **62**, 974–983.
26. 26. Beinert, H.; Holm, R.H.; Münck, E. *Science*, 1997, **277**, 653–659.
27. Venters, R. A.; Nelson, M. J.; McLean, P. A.; True, A. E.; Levy, M. A.; Hoffman, B. M.; Orme-Johnson, W. H. *J. Am. Chem. Soc.*, 1986, **108**, 3487–3498.

28. True, A. E.; Nelson, M. J.; Venters, R. A.; Orme-Johnson, W. H.; Hoffman, B. M. *J. Am. Chem. Soc.*, 1988, **110**, 1935–1943.
29. Chasteen, N. D.; Snetsinger, P. A., in Que, L., ed. *Physical Methods in Bioinorganic Chemistry: Spectroscopy and Magnetism*, University Science Books, Sausalito, CA, 2000, pp. 187–231.
30. Yoo, S. J.; Angove, H. C.; Papaefthymiou, V.; Burgess, B. K.; Münck, E. *J. Am. Chem. Soc.*, 2000, **122**, 4926–4936.
31. Fee, J. A.; Bull, C. *J. Biol. Chem.*, 1986, **261**, 13000–13005.
32. Duprat, A. F.; Traylor, T. G.; Wu, G. Z.; Coletta, M.; Sharma,V. S.; Walda, K. N.; Magde, D. *Biochemistry*, 1995, **34**, 2634–2644.
33. Rodgers, K. R.; Spiro, T. G. *Science*, 1994, **265**, 1697–1699.
34. Srajer, V.; Teng, T.-y.; Ursby, T.; Pradervand, D.; Ren, Z.; Adachi, S.-i.; Schildkamp, W.; Bourgeois, D.; Wulff, M.; Moffat, K. *Science*, 1996, **274**, 1726–1729.
35. Srajer, V.; Ren, Z.; Teng T-.y.; Schmidt M.; Ursby T.; Bourgeois D.; Pradervand C.; Schildkamp, W. *Biochemistry*, 2001, **40**, 13802–13815.
36. Biggs, P. *Computers in Chemistry*, Oxford University Press, New York, 1999.
37. Wilson, E. K. *C&EN*, 2001, April 9, 46.
38. Comba, P.; Hambley, T. W. *Molecular Modeling of Inorganic Compounds*, VCH Publishers, New York, 1995.
39. Comba, P.; Hambley, T. W. *Molecular Modeling of Inorganic Compounds*, John Wiley & Sons, New York, 2001.
40. Comba, P; Zimmer, M. *J. Chem. Ed.*, 1996, **73**, 108.
41. (a) Hay, B. P. *Coord. Chem. Rev.*, 1993, **126**, 177. (b) Yates, P. C.; Marsden, A. K. *Comput. Chem.*, 1994, **18**, 89.
42. (a) Allinger, N. L.; Yuh, Y. H.; Li, J.-H. *J. Am. Chem. Soc.*, 1989, **111**, 8551–8566. (b) Bowen, J. P.; Allinger, N. L. *Rev. Comput. Chem.*, 1991, **2**, 81.
43. (a) Albinati, A.; Lianza, F.; Berger, H.; Pregosin, P. S.; Rüegger, H.; Kunz, R. W. *Inorg. Chem.* 1993, **32**, 478. (b) Hay, B. P.; Rustad, J. R. *J. Am. Chem. Soc.*, 1994, **116**, 6313.
44. (a) Halgren, T. A. *J. Am. Chem. Soc.*, 1992, **114**, 7827. (b) Halgren, T. A. *J. Comp. Chem.*, 1996, **17**, 490, 520, 553.
45. Gajewski, J. J.; Gilbert, K. E.; McKelvey, J., in *Advances in Molecular Modeling*, vol. 2. Liotta, D., ed., JAI Press, Greenwich, CT, 1990.
46. Lipkowitz, K. B.; Pearl, G. M.; Robertson, D. H.; Schultz, F. A. *J. Chem. Ed.*, 1996, **73**(2), 105.
47. Mayo, S. L.; Olafson, B. D.; Goddard, W. A., III. *J. Phys. Chem.*, 1990, **94**, 8897.
48. Hancock, R. D.; Dobson, S. M.; Evers, A.; Wade, P. W.; Ngwenya, M. P.; Boeyens, J. C. A.; Wainwright, K. P. *J. Am. Chem. Soc.*, 1988, **110**, 2794.
49. Adam, K. R.; Antolovich, M.; Brigden, L. G.; Lindoy, L. F. *J. Am. Chem. Soc.*, 1991, **113**, 3346.
50. Allured, V. S.; Kelly, C. M.; Landis, C. R. *J. Am. Chem. Soc.*, 1991, **113**, 1.
51. (a) Rappé, A. K.; Casewit, C. J.; Colwell, K. S.; Goddard III, W. A.; Skiff, W. M. *J. Am. Chem. Soc.*, 1992, **114**, 10024. (b) Casewit, C. J.; Colwell, K. S.; Rappé, A. K. *J. Am. Chem. Soc.*, 1992, **114**, 10046. (c) Rappé, A. K.; Colwell, K. S.; Casewit, C. J. *Inorg. Chem.*, 1993, **32**, 3438.

52. Grant, G. H.; Richards, W. G. *Computational Chemistry*, Oxford University Press, New York, 1996.
53. Karplus, M.; Petsko, G. A. *Nature*, 1990, **347**, 631–639.
54. Lopez, M. A.; Kollman, P. A. *J. Am. Chem. Soc.*, 1989, **111**, 6212.
55. Charles, R.; Ganly-Cunningham, M.; Warren, R.; Zimmer, M. *J. Mol. Struct.*, 1992, **265**, 385.
56. Collins, J. R.; Du, P.; Loew, G. H. *Biochemistry*, 1992, **31**, 11166.
57. Kuczera, K.; Kuriqan, J.; Karplus, M. *J. Mol. Biol.*, 1990, **213**, 351.
58. Riley, D. P.; Henke, S. L.; Lennon, P. J.; Aston, K. *Inorg. Chem.*, 1999, **38**, 1908–1917.
59. Hambley, T. W. *Inorg. Chem.*, 1988, **27**, 1073.
60. (a) Kozelka, J.; Petsko, G. A.; Lippard, S. J.; Quigley, G. J. *J. Am. Chem. Soc.*, 1985, **107**, 4079. (b) Kozelka, J.; Petsko, G. A.; Quigley, G. J.; Lippard, S. J. *Inorg. Chem.* 1986, **25**, 1075. (c) Kozelka, J.; Archer, S.; Petsko, G. A.; Lippard, S. J.; Quigley, G. J. *Biopolymers* 1987, **26**, 1245.
61. McCarthy, S. L.; Hinde, R. J.; Miller, K. J.; Anderson, J. S.; Basch, H.; Krauss, M. *Biopolymers*, 1990, **29**, 785. (b) Herman, F.; Kozelka, J.; Stoven, V.; Guittet, E.; Girault, J.-P.; Huynh-Dinh, T.; Igolen, J. Lallemand, J.-Y.; Chottard, J.-C. *Eur. J. Biochem.*, 1990, **194**, 119.
62. Bierbach, U.; Farrell, N. *Inorg. Chem.*, 1997, **36**, 3657–3665.
63. Weiner, S. J.; Kollman, P. A.; Case, D. A.; Singh, U. C.; Ghil, C.; Alagona, G.; Profeta, S., Jr.; Weiner, P. *J. Am. Chem. Soc.*, 1984, **106**, 765.
64. Chang, G.; Guida, W. C.; Still, W. C. *J. Am. Chem. Soc.*, 1989, **111**, 4379.
65. Hehre, W. J.; Huang, W. W. *Chemistry with Computation: An Introduction to Spartan*, Wavefunction, Inc., Irvine, CA, 1995.
66. (a) Fan, H. J.; Hall, M. B. *J. Am. Chem. Soc.*, 2001, **123**, 3828–3829. (b) Cao, Z.; Hall, M.B. *J. Am. Chem. Soc.*, 2001, **123**, 3734–3742.
67. Friesner, R. A.; Dunietz, B. D. *Acc. Chem. Res.*, 2001, **34**, 351–358.
68. Dewar, M. J. S.; Zoebisch, E. G.; Healy, E. F.; Stewart, J. J. P. *J. Am. Chem. Soc.*, 1985, **107**, 3902.
69. Stewart, J. J. P. *J. Comput. Chem.*, 1989, **10**, 209.
70. Dewar, M. J. S.; Thiel, W. J. *J. Am. Chem. Soc.*, 1977, **99**, 4899.
71. (a) Thiel, W. *Adv. Chem. Phys.*, 1996, **93**, 703. (b) Thiel, W.; Voityuk, A. A. *J. Phys. Chem.*, 1996, **100**, 616. (c) Thiel, W.; Voityuk, A. *Theor. Chem. Acta*, 1992, **81**, 391. (d) Thiel, W.; Voityuk, A. *Int. J. Quantum Chem.*, 1985, **44**, 807.
72. Yu, J.; Hehre, W. J. *Polym. Mater. Sci. Eng.* (*Washington*), 1996, **74**, 439.
73. Berweger, C. D.; Thiel, W.; van Gensteren, W. F. *Proteins*, 2000, **41**(3), 299–315.
74. Bernhardt, P. V.; Comba, P. *Inorg. Chem.*, 1992, **31**, 2638.
75. Comba, P.; Hambley, T. W.; Ströhle, M. *Helv. Chim. Acta,* 1995, **78**, 2042.
76. Bol, J. E.; Buning, C.; Comba, P.; Reedijk, J.; Ströhle, M. *J. Comput. Chem.*, 1998, **19**, 512.
77. Comba, P.; Gloe, K.; Inoue, K.; Krueger, T.; Stephan, H.; Yoshizuka, K. *Inorg. Chem.*, 1998, **37**, 3310.
78. Comba, P.; Gyr, T. *Eur. J. Inorg. Chem.*, 1999, 1787–1792.
79. Vacek, P. L. *Chem. Phys. Lett.*, 1999, **310**, 189.

80. Zimmer, M.; Crabtree, R. H. *J. Am. Chem. Soc.*, 1990, **112**, 1062.
81. Hancock, R. D.; Hegetschweiler, K. *J. Chem. Soc., Dalton Trans.*, 1993, 2137.
82. Cornell, W. D.; Cieplak, P.; Bayly, C. I.; Gould, I. R.; Merz, K. M., Jr.; Ferguson, D. M.; Spellmeyer, D. C.; Fox, T.; Caldwell, J. W.; Kollman, P. A. *J. Am. Chem. Soc.*, 1995, **117**, 5179.
83. (a) Brooks, B. R.; Bruccoleri, R. E.; Olafson, B. D.; States, D. J.; Swaminathan, S.; Karplus, M. *J. Comp. Chem.*, 1983, **4**, 187. (b) Momany, F. A.; Rone, R. *J. Comp. Chem.*, 1992, **13**, 888.
84. Hwang, M.-J.; Stockfisch, T. P.; Hagler, A. T. *J. Am. Chem. Soc.*, 1994, **116**, 2515.
85. van Gunsteren, W. F.; Daura, X.; Mark, A. E. "The GROMOS force field" in *The Encyclopaedia of Computational Chemistry*, Schleyer, P. v. R.; Allinger, N. L.; Clark, T.; Gasteiger, J.; Kollman, P. A.; Schaefer, H. F., III; Schreiner, P. R., eds., John Wiley & Sons, Chichester, 1998.
86. Wiesemann, F.; Teipel, S.; Krebs, B.; Höweler, U. *Inorg. Chem.*, 1994, **33**, 1891.
87. Murray-Rust, P.; Rzepa, H. S.; Whitaker, B. J. *Chem. Soc. Rev.*, 1997, 1–10.

4

IRON-CONTAINING OXYGEN CARRIERS AND THEIR SYNTHETIC MODELS

4.1 INTRODUCTION

Reversible coordination of dioxygen (O_2), along with its transport through the blood stream of vertebrates (and many invertebrates) by the iron-containing metalloproteins myoglobin (Mb) and hemoglobin (Hb), is critical to the maintenance of biological function. While dioxygen's solubility in water is quite low (6.6 ml/liter, or 3×10^{-4} M), myoglobin and hemoglobin increase O_2's solubility in blood approximately 30 times—that is, to 200 ml/liter, or 9×10^{-3} M.[1] Myoglobin and hemoglobin complex iron through use of a prosthetic heme group, a planar four-coordinate porphyrin ligand such as that shown in Figure 4.1. Protein side-chain ε-nitrogen atoms of histidine complete iron's ligand coordination sphere. See Section 2.2.1 for amino acid structures.

Hemerythrins, used for dioxygen transport by certain marine invertebrates, ligate iron through protein side-chain–ligand atoms only. Hemocyanins, copper-containing oxygen transport metalloproteins found in arthropods and mollusks, coordinate copper through sulfur and nitrogen amino acid side-chain ligands of the surrounding protein. Neither hemerythrins nor hemocyanins contain the porphyrin ligand heme system of Mb or Hb. Hemocyanins will be discussed in Chapter 5. Properties of oxygen transport proteins are summarized in Table 4.1 as adapted from references 5 and 7.

Hemoglobin's dioxygen binding is regulated by local concentrations of H^+ (known as the Bohr effect), CO_2 concentration, and organic phosphates such as diphosphoglycerate (DPG), whose structure is shown in Figure 4.2.[17]

Decreasing pH and increasing CO_2 concentration lower hemoglobin's O_2 affinity so that conditions of low pH and high carbon dioxide concentration promote

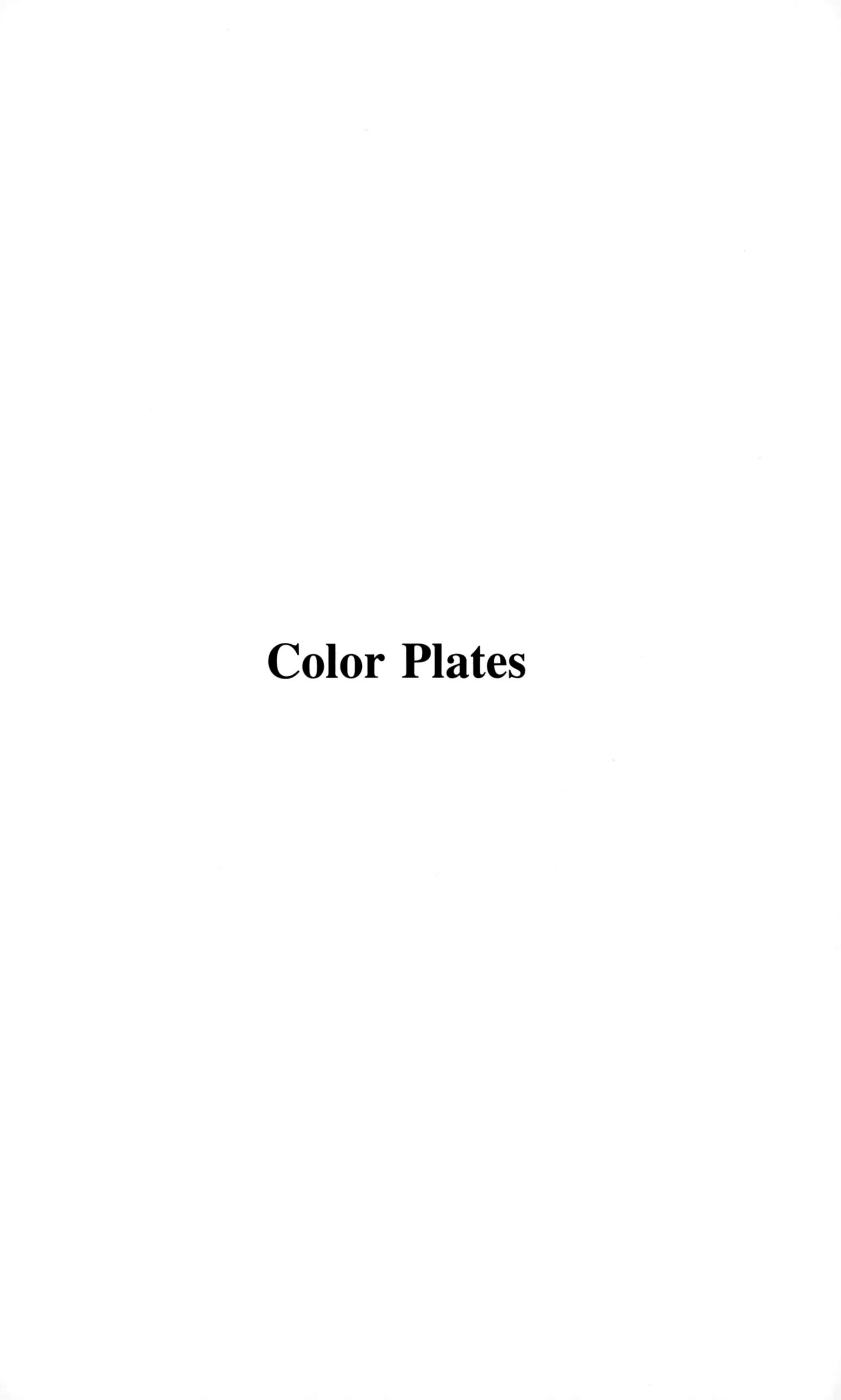

Color Plates

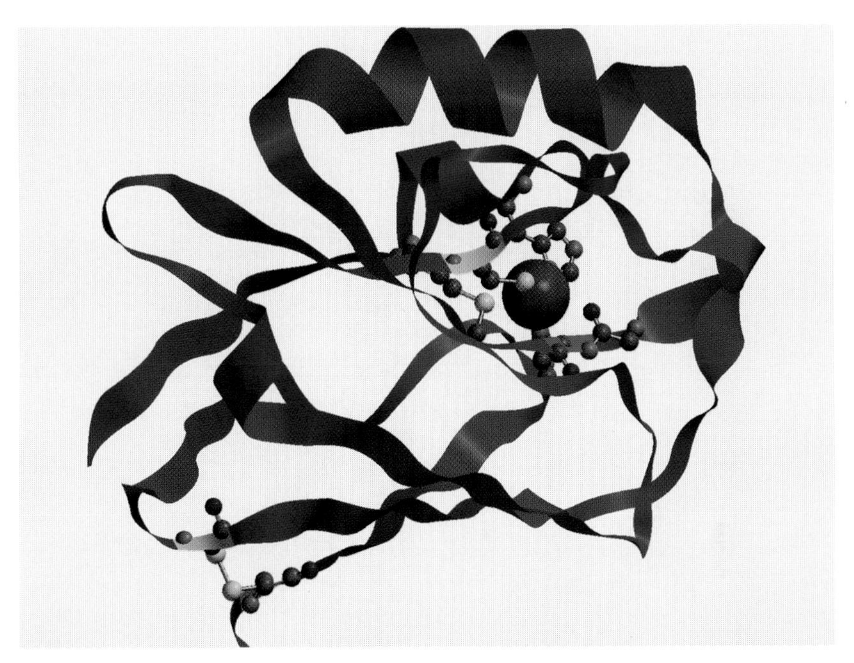

Figure 2.10 Secondary and tertiary structure of the copper enzyme azurin visualized using Wavefunction, Inc. Spartan '02 for Windows^{TM} from PDB data deposited as 1JOI. See text for visualization details. Printed with permission of Wavefunction, Inc., Irvine, CA.

Figure 2.21 Zinc-finger protein from the yeast transcription factors SWI as visualized using Wavefunction, Inc. Spartan '02 for Windows^{TM} from PDB data deposited as 1NCS. See text for visualization details. Printed with permission of Wavefunction, Inc., Irvine, CA.

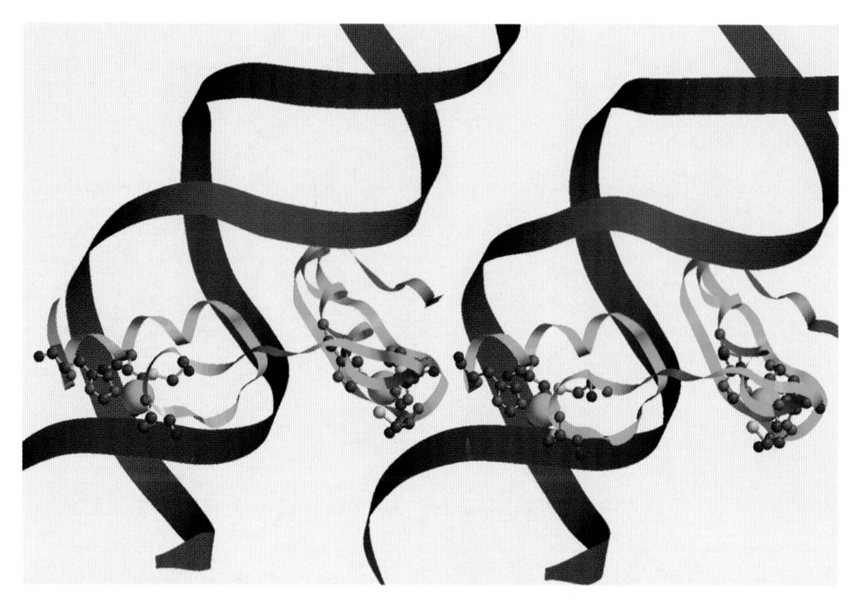

Figure 2.23 Wavefunction, Inc. Spartan '02 for Windows^{TM} visualization (side-on) of the PDB structural data (2DRP) for zinc-finger–dsDNA contacts as described in reference 32. See text for visualization details. Printed with permission of Wavefunction, Inc., Irvine, CA.

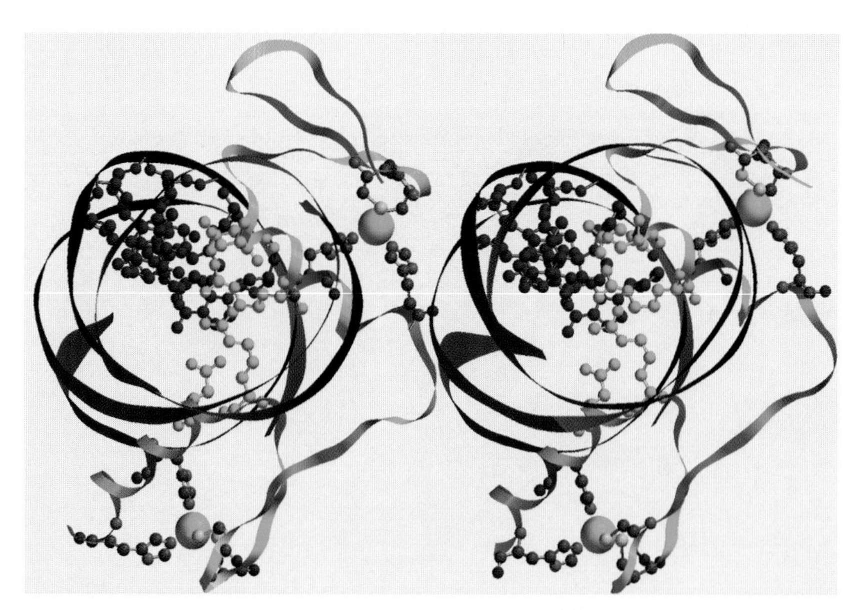

Figure 2.24 Wavefunction, Inc. Spartan '02 for Windows^{TM} visualization (end-on) of the PDB structural data (2DRP) for zinc-finger–dsDNA contacts as described in reference 32. See text for visualization details. Printed with permission of Wavefunction, Inc., Irvine, CA.

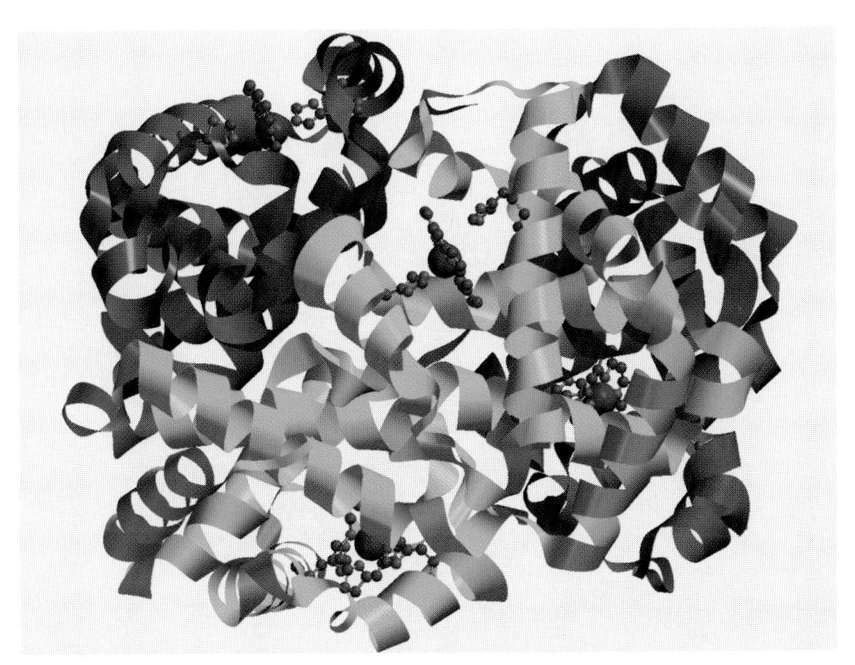

Figure 4.4 Quaternary structure of deoxyhemoglobin tetramer visualized using Wavefunction, Inc. Spartan '02 for Windows™ from PDB data deposited as 4HHB. See text for visualization details. Printed with permission of Wavefunction, Inc., Irvine, CA.

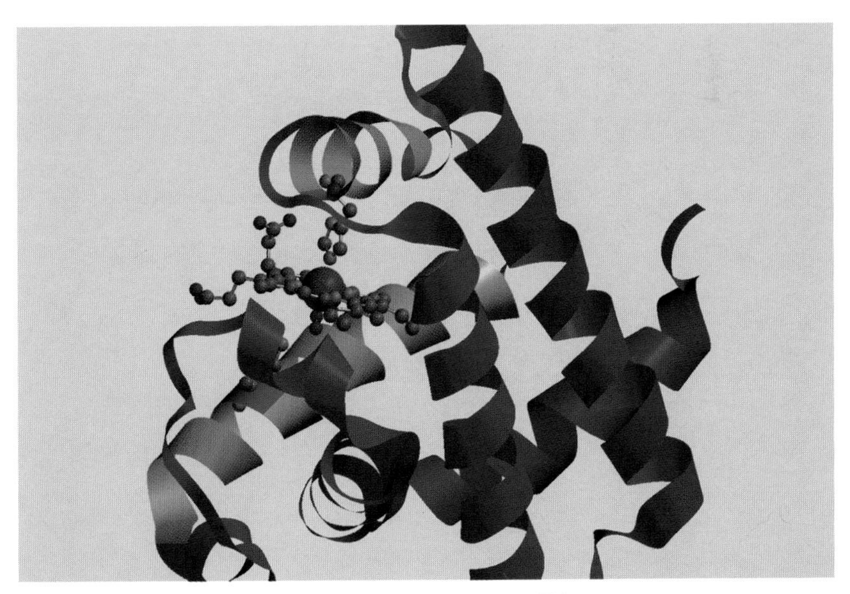

Figure 4.5 Wavefunction, Inc. Spartan '02 for Windows™ visualization of oxymyoglobin from PDB data deposited as 1MBO.[6] See text for visualization details. Printed with permission of Wavefunction, Inc., Irvine, CA.

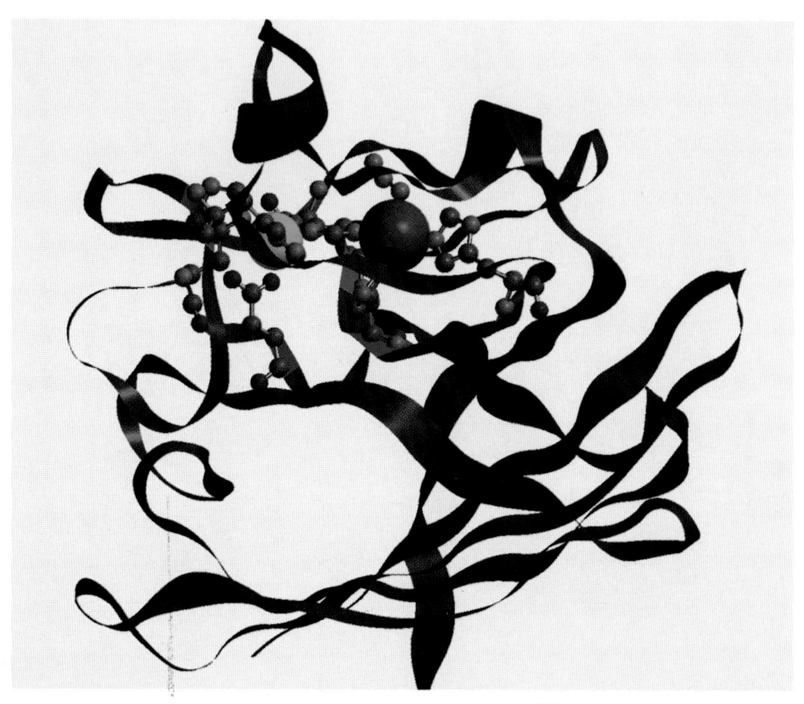

Figure 5.3 Wavefunction, Inc. Spartan '02 for Windows^{TM} visualization of superoxide dismutase with azide ion from PDB data deposited as 1 YAZ from reference 25. See text for visualization details. Printed with permission of Wavefunction, Inc., Irvine, CA.

Figure 5.8 Wavefunction, Inc. Spartan '02 for Windows^{TM} visualization of oxygenated hemocyanin from PDB data deposited as 1 OXY from reference 37a. See text for visualization details. Printed with permission of Wavefunction, Inc., Irvine, CA.

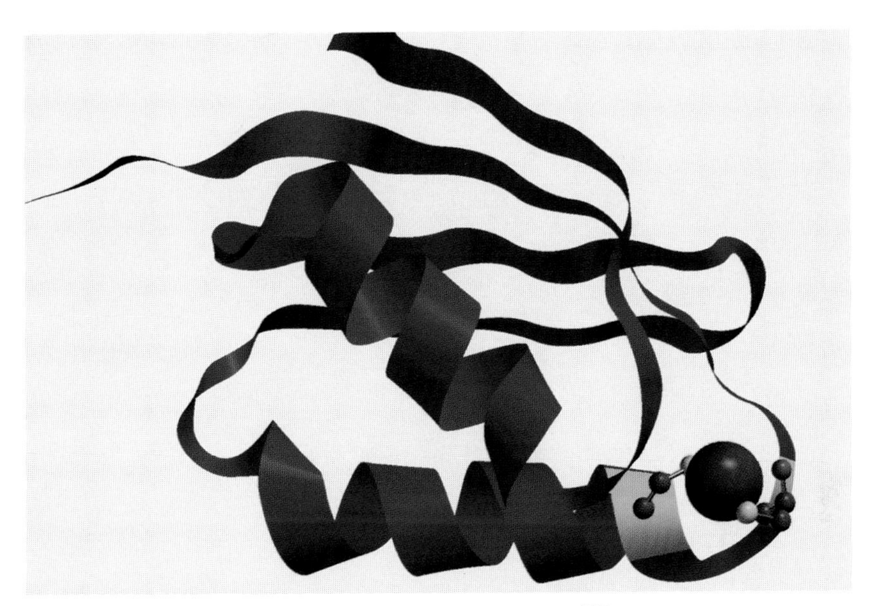

Figure 7.28 Wavefunction, Inc. Spartan '02 for Windows^{TM} representation of the Atx1 copper chaperone protein (PDB code: 1FD8) with data from reference 116. See text for visualization details. Printed with permission of Wavefunction, Inc., Irvine, CA.

Figure 7.29 Wavefunction, Inc. Spartan '02 for Windows^{TM} representation of the Hah1 metallochaperone from reference 118. X-ray data deposited as PDB code 1FEE. See text for visualization details. Printed with permission of Wavefunction, Inc., Irvine, CA.

Figure 7.32 Apo form of metal binding domain 4 of Menkes copper-transporting ATPase described in reference 133 (PDB: 1AW0). Visualized using Wavefunction, Inc. Spartan '02 for Windows^{TM}. See text for visualization details. Printed with permission of Wavefunction, Inc., Irvine, CA.

Figure 7.33 Metallated form of metal-binding domain 4 of Menkes copper-transporting ATPase described in reference 133 (PDB: 1AW0). Visualized using Wavefunction, Inc. Spartan '02 for Windows^{TM}. See text for visualization details. Printed with permission of Wavefunction, Inc., Irvine, CA.

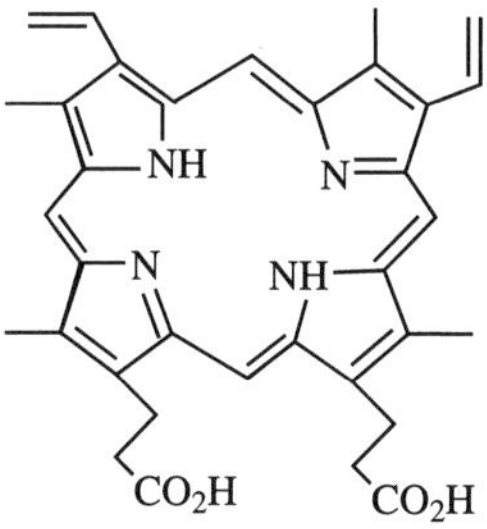

Figure 4.1 Protoporphyrin IX as found in Hb and Mb.

dioxygen release from hemoglobin. DPG binds to hemoglobin at a site remote from the heme groups. Nevertheless, the equilibrium shown in equation 4.1 indicates that high DPG concentration lowers hemoglobin's O_2 affinity and that low DPG concentration inhibits release of dioxygen from hemoglobin.

$$\text{Hb—DPG} + 4\text{O}_2 \Leftrightarrow \text{Hb(O}_2)_4 + \text{DPG} \quad (4.1)$$

Schematic drawings of the tertiary and quaternary structures of Mb and Hb are shown in Figure 4.3 as reprinted from Figure 4.2 of reference 7.

The structure of hemoglobin and myoglobin were among the first to be solved by X-ray crystallography: for deoxymyoglobin by J. C. Kendrew and H. C. Watson beginning in 1966,[2] and for deoxyhemoglobin by G. Fermi beginning in 1975.[3] These first deoxyhemoglobin structures have been declared obsolete in the Research Collaboratory for Structural Bioinformatics (RCSB) Protein Data Bank (PDB). More information may be found on the PDB at the website http://www.rcsb.org/pdb/ and in Chapter 3, Section 3.13.1. In 1984 Fermi and co-workers published the deoxyhemoglobin quaternary structure visualized in Figure 4.4.[4] (See also Colour Plate.) These X-ray crystallographic data, deposited in the PDB under codes 2HHB, 3HHB, and 4HHB, have been updated by many later structures of deoxy- and oxyhemoglobins, many having been modified by site-directed mutagenesis of important aa residues that affect the position of the iron heme as well as dioxygen binding and release. (Chapter 2 Section 2.3.4 discusses the practice and utility of site-directed mutagenesis.) In late 2001, 186 different data depositions of

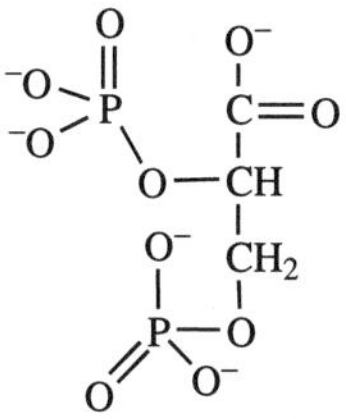

Figure 4.2 Diphosphoglycerate (DPG), regulator of heme dioxygen binding.

Table 4.1 Some Properties of Oxygen Transport Proteins

O_2 carrier:	Myoglobin	Hemoglobin	Hemerythrin	Hemocyanin
Source:	Higher animals, some invertebrates	Higher animals, some invertebrates	invertebrates	Arthropods, mollusks
Metal:	Fe	Fe	Fe	Cu
Metal:bound O_2 stoichiometry (ligands):	$Fe:O_2$ (heme, histidine)	$Fe:O_2$ (heme, histidine)	2 $Fe:O_2$ (nonheme, protein side chains)	2 $Cu:O_2$ (nonheme, protein side chains)
Metal ox state in deoxy form/*d* electrons (color):	II/d^6 (red-purple, violet)	II/d^6 (red-purple, violet)	II/d^6 (colorless)	I/d^{10} (colorless)
Metal ox state in oxy form/ *d* electrons (color):	II/d^6–O_2 or III/d^5–O_2^- (red)	II/d^6–O_2 or III/d^5–O_2^- (red)	III/d^5 (burgundy)	II/d^9 (blue)
Approximate molecular weight (kDa):	17	65	108	400 to 2×10^4
Number of subunits:	1	4 (some species have up to 10)	8	Many

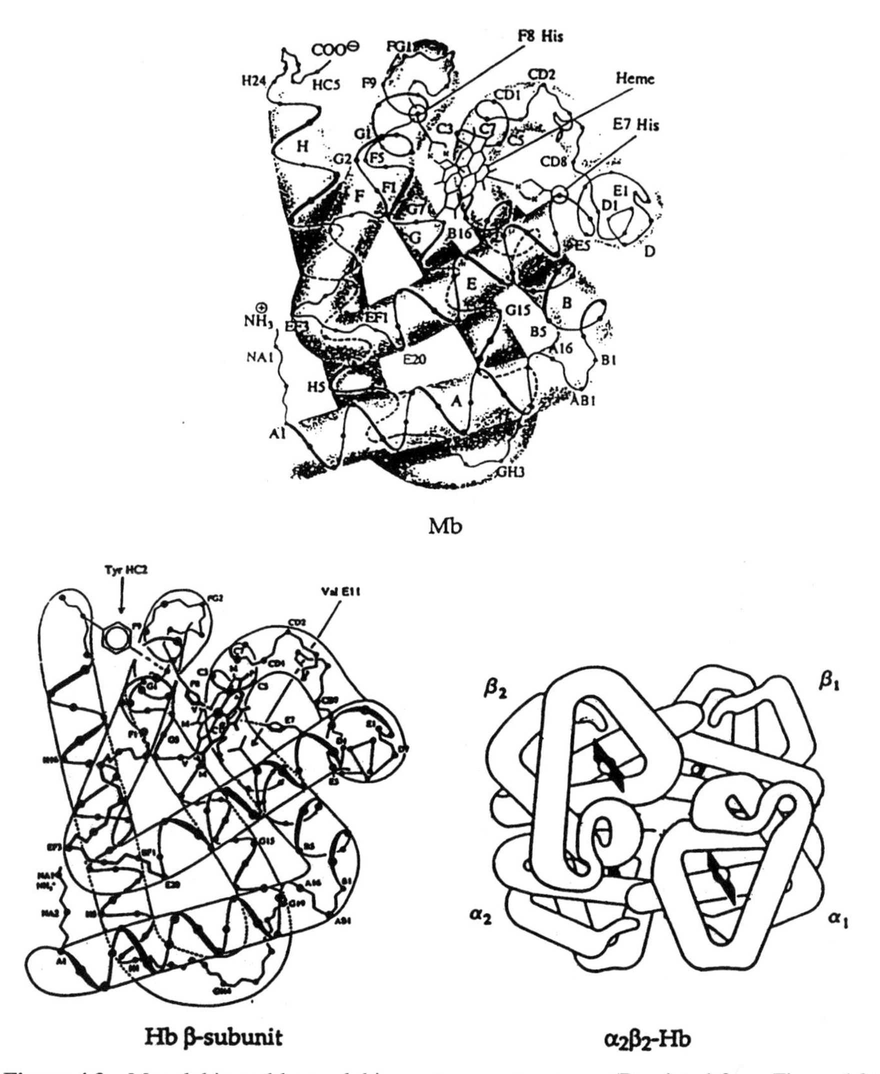

Figure 4.3 Myoglobin and hemoglobin quaternary structures. (Reprinted from Figure 4.2 of Cowan, J. A. *Inorganic Biochemistry, An Introduction*, 2nd ed., Wiley-VCH, New York, 1997. Copyright 1997, Wiley-VCH.)

X-ray and NMR solution studies for deoxy, oxy, carbonmonoxy, nitroso, and cyano variants of hemoglobin and its mutant congeners were available in the PDB. Figure 4.4/Color Plate 5 is a Wavefunction, Inc. Spartan '02 for Windows™ visualization of 4HHB data. In this view, α-helical protein secondary structures have been displayed in blue ribbon form (α-chains with proximal his87 and distal his58 iron ligands), or brown ribbon form (β-chains with proximal his92 and distal his63 iron ligands). The planar hemes are shown in ball-and-spoke form, and

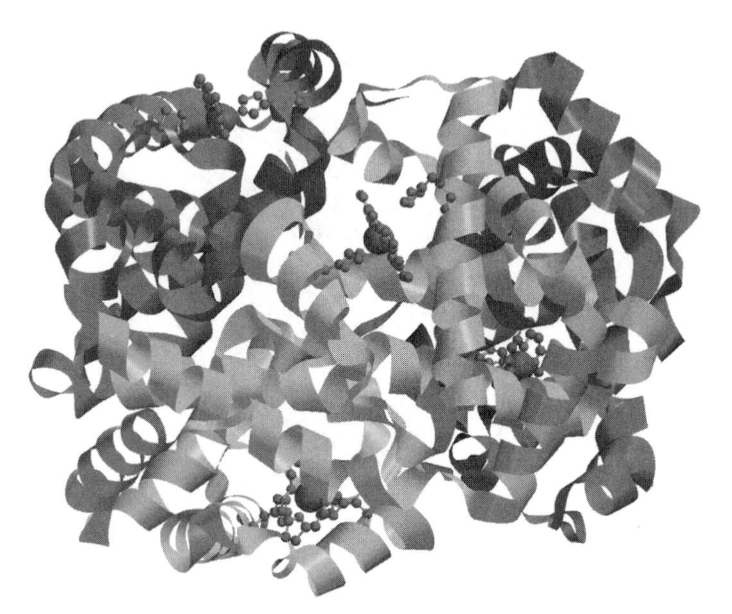

Figure 4.4 Quaternary structure of deoxyhemoglobin tetramer visualized using Wavefunction, Inc. Spartan '02 for Windows^{TM} from PDB data deposited as 4HHB. See text for visualization details. Printed with permission of Wavefunction, Inc., Irvine, CA. (See color plate.)

the iron atoms have been displayed in space-fill form. The proximal his87 coordinating the iron atom in Figure 4.4's upper center appears within bonding distance of the iron atom, whereas the distal his58 to the upper right of this same iron atom does not.

The Figure 4.4 visualization of human deoxyhemoglobin was refined at 1.74-Å resolution using data collected from a synchrotron X-ray source. The crystallographic R-factor is 16.0% or 0.16. The estimated error in atomic positions is 0.1 Å overall, 0.14 Å for main-chain atoms of internal segments, and 0.05 Å for the iron atoms. The geometry of the iron–nitrogen complex closely resembles that of the deoxymyoglobin structures and of the five-coordinate model compounds to be discussed below in Section 4.8.2. In deoxyhemoglobin, the distances of the iron from the mean plane of N(porphyrin) are 0.40(5) Å and 0.36(5) Å, respectively, at the alpha and beta heme centers, in contrast to the corresponding distance of +0.12(8) Å and −0.11(8) Å in oxyhemoglobin (see Tables 4.3 and 4.4 and reference 30). The Fe–N_{Im}–(proximal histidine Nε2) bond length is 2.12(4) Å and the Fe–N_p (porphyrin nitrogen) bond length is 2.06(2) Å, in good agreement with data presented in Table 4.3 for other deoxyhemoglobin X-ray structures.[4]

Chemists began working on model compounds for myoglobin and hemoglobin in the early 1970s.[5] Because of the study of myoglobin and hemoglobin structures and functions and through the design of model compounds to mimic these, scientists now know more about the active site and the mechanism for the behavior of myoglobin and hemoglobin than for any other metalloprotein. The discussion here will focus on: (1) structure of the active heme site of myoglobin and hemoglobin as determined by NMR, X-ray crystallography, and other methods; (2) analytical techniques used to study Mb and Hb; (3) the mechanism for reversible binding of dioxygen and cooperativity of oxygen binding; (4) structure of the active site in Mb and Hb compared to that in model compounds; and (5) the selectivity of Mb and Hb for O_2 versus CO binding.

4.2 MYOGLOBIN AND HEMOGLOBIN BASICS

Myoglobin (Mb) is a globular monomeric protein containing a single polypeptide chain of 160 amino acid residues (MW 17.8 kDa) made up of seven α-helical segments (A–G) and six nonhelical segments. An example of oxymyoglobin's tertiary structure is found in the work of Phillips, who refined X-ray crystallographic data collected for oxymyoglobin at 1.6-Å resolution (deposited under PDB code: 1MBO).[6] The structure has been visualized using Wavefunction, Inc. Spartan '02 for Windows™ in Figure 4.5. (See Section 3.13.1 for information on visualization programs.) The protoporphyrin IX heme is visualized in ball-and-spoke form, while the iron atom is shown in tan space-fill form. The attached dioxygen (or superoxide anion), pointing down, is shown in its bent format within hydrogen bonding distance of the distal histidine, his64, (E7), shown in ball-and-spoke form. The E helix is displayed in blue/green ribbon form, with the E7 residue in deep blue, to distinguish it from the majority of myoglobin's protein

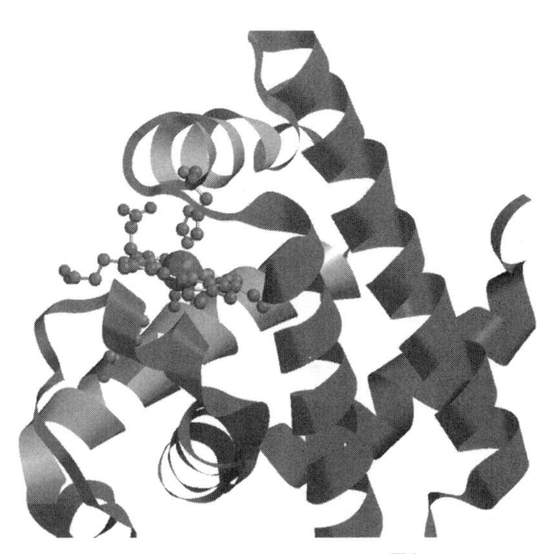

Figure 4.5 Wavefunction, Inc. Spartan '02 for Windows™ visualization of oxymyoglobin from PDB data deposited as 1MBO.[6] See text for visualization details. Printed with permission of Wavefunction, Inc., Irvine, CA. (See color plate.)

chain, visualized in purple ribbon form. The proximal histidine bonded to the heme iron ion, his93 (F8), is also shown in ball-and-spoke form within the orange F helix.

The heme prosthetic center contains an iron ion complexed by a porphyrin known as protoporphyrin IX (see Figure 4.1). The Fe(II) protoporphyrin cofactor is held in place in the protein principally by noncovalent hydrophobic interactions of some 80 or so residues—principally leucine, isoleucine, valine, and phenylalanine aa residues—and one covalent linkage at the proximal His F8 (his93) residue. In the myoglobin structure shown in Figure 4.5, the terminology His F8 refers to the eighth residue of the F α-helical region of the protein's tertiary structure. In newer publications this histidine will usually be referred to as his93, counting aa residues sequentially beginning at the N-terminal end. The so-called distal histidine, described more fully below, is identified as His E7 or his64. Mb stores oxygen in muscle and other cellular tissue binding one oxygen molecule per protein subunit. Hemoglobin (Hb), a tetramer of four globular protein subunits, each of which is nearly identical to a Mb unit (see Figure 4.4), transports oxygen through the blood plasma. Hb's four subunits are comprised of two α chains of 141 residues and two β chains of 146 residues with a total MW of 64.5 kDa. In hemoglobin, α and β chains differ slightly, especially in the manner in which the porphyrin is held within the protein. Bond lengths and angles reported in Tables 4.3 and 4.4 illustrate these differences.

Hemoglobin binds dioxygen in a cooperative manner; that is, once one O_2 molecule is bound to the enzyme, the second, third, and fourth attach themselves more readily. This behavior leads to the oxygen-binding curves shown in Figures 4.9 and 4.10. Both Mb and Hb bind dioxygen only when the iron ion is in its reduced state as iron(II). The terminology *oxy*- and *deoxy*-Mb and Hb refer to the enzyme in its oxygenated or deoxygenated forms, respectively, both with iron(II) metal centers, while *met*- describes oxidized heme proteins containing iron(III) centers. Comparison of reduction potentials in equations 4.2 and 4.3 indicates that dioxygen should oxidize iron(II) under most expected concentration conditions.[7]

$$O_2 + 4H^+ + 4e^- \rightarrow 2H_2O \qquad E^0 = +0.82\,V \tag{4.2}$$

$$Hb(Fe^{3+}) + e^- \rightarrow Hb(Fe^{2+}) \qquad E^0 = +0.17\,V \tag{4.3}$$

Therefore the stability of biological heme–O_2 complexes must arise from kinetic rather than thermodynamic considerations. Some circumstances favoring heme–O_2 stability include: (1) placement of the heme in a hydrophobic pocket within the enzyme that is inaccessible to water molecules and protons; (2) a bent binding mode for dioxygen favored by the prosthetic group's pocket shape that prevents μ-oxo dimer formation (see Figures 4.14 and 4.16 and Section 4.8.2; (3) σ-bonding donation from an sp^2-rehybridized superoxide ion to an empty d_z^2 Fe(II) orbital facilitated by a bent orientation of the bound dioxygen (see Figures 4.6, 4.7, and 4.8); and (4) the formation of π-back-bonds through interaction of a half-filled d_{xz} orbital of Fe(II) with a half-filled π^* orbital of the superoxide ion (see Figure 4.6).[8]

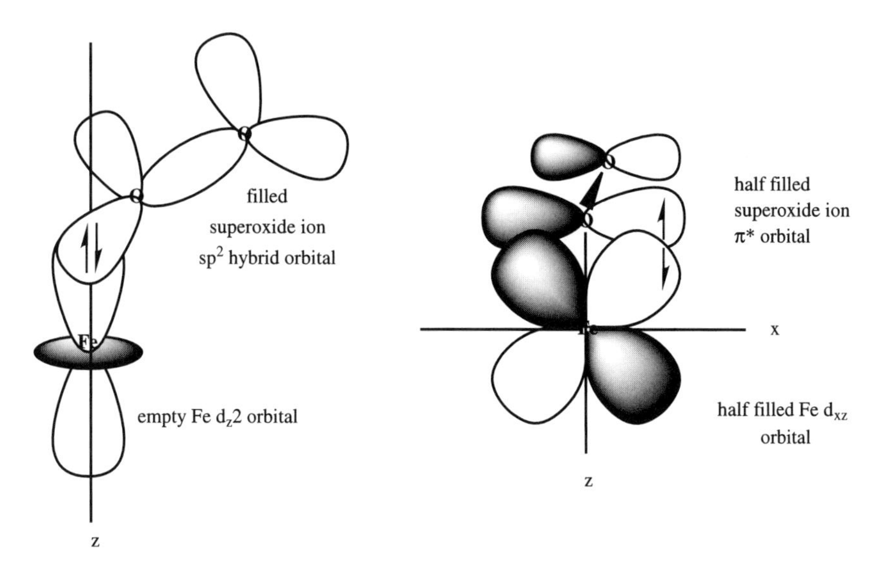

Figure 4.6 Bonding of the O_2^- ion to Fe(III) in Mb or Hb. (Adapted with permission from Figure 3 of Momenteau, M.; Reed, C. A. *Chem. Rev.*, 1994, **94**, 659–698. Copyright 1994, American Chemical Society.)

4.3 STRUCTURE OF THE PROSTHETIC GROUP

Deoxymyoglobin and deoxyhemoglobin contain pentacoordinate iron(II) centers in which the metal ion lies out of the plane of the porphyrin's four pyrrole–nitrogen donor ligands.[9] Perutz has called this state the T- or tense state.[10] The T-state, a term describing the quaternary structure of the Hb tetramer, is one of low oxygen affinity in which the protein is restrained by binding of the so-called proximal histidine. In the T-state the Fe(II) center is held approximately 0.55 Å outside of the porphyrin plane and none of the four Hb subunits possess dioxygen ligands. The

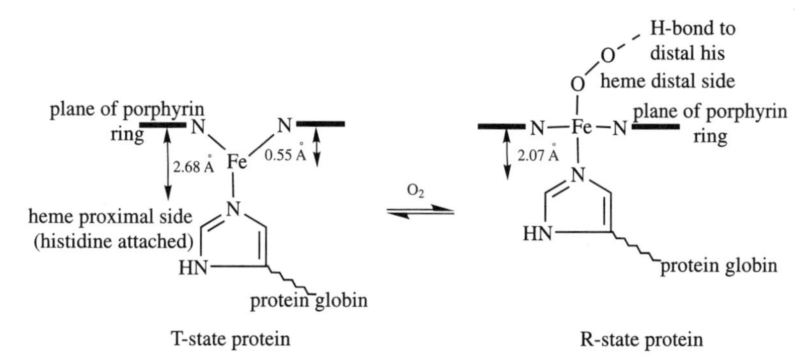

Figure 4.7 T- and R–states for iron hemes. (Adapted with permission from Figure 4 of Collman, J. P. *Inorg. Chem.*, 1997, **36**, 5145–5154. Copyright 1997, American Chemical Society.)

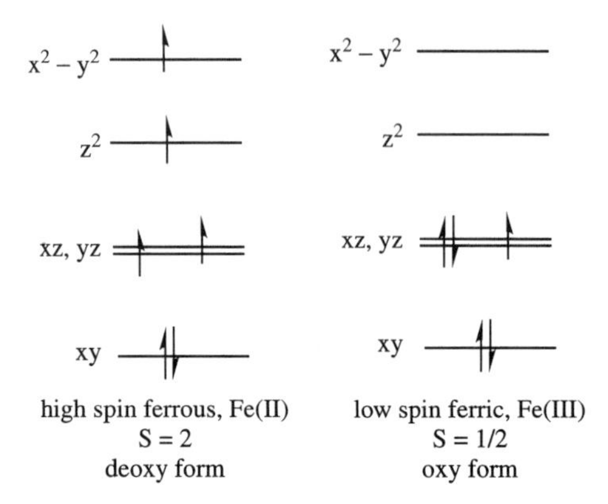

Figure 4.8 Change from the high, Fe(II), to low, Fe(III), spin states allows iron ion to enter porphyrin plane.

porphyrin ring also is anchored at the active site by iron's coordination to the proximal histidine's imidazole nitrogen (see Figures 4.7 and 4.8).[11] In the R-state or relaxed quaternary state, dioxygen is bound to iron on the so-called distal side of the porphyrin ring. Switching from the T-state to the R-state in the Hb tetramer takes place during or after binding of approximately two dioxygen molecules. Upon binding of dioxygen, constraints within the surrounding protein matrix are relaxed, allowing the iron ion to move toward and nearly into the porphyrin plane.

Binding of the sixth ligand (O_2) and consequent movement of the iron atom into the porphyrin plane induces a tertiary structure change as the proximal histidine changes its bond angle with the iron atom. The F helix containing the proximal histidine also changes position. In hemoglobin, these factors in turn change the quaternary structure of the Hb tetramer and influence the affinity of the four hemes for dioxygen as discussed in Section 4.5. The metal ion's movement is accompanied by a change from a high- ($S = 2$) to a low-spin state ($S = 1/2$) and from the Fe(II) to Fe(III) oxidation states. In the change from the high- to low- spin states and from Fe(II) to Fe(III), the Fe ion becomes smaller and is thus able to fit better into the porphyrin ring's cavity (see Figures 4.7 and 4.8). The dioxygen molecule is guided by the protein pocket surrounding it to attach in a bent structure with an Fe–O–O bond angle of 115°. As Fe(II) is oxidized to Fe(III), dioxygen becomes the superoxide ion (O_2^-). Evidence for this behavior is found in the Fe–O, Fe–O–O, and O–O distances and bond angles determined by X-ray crystallography and from O–O bond orders determined by infrared and resonance Raman spectroscopy. Additional evidence for the Fe(III)–O_2^- moiety is its experimental spin state $S = 0$ indicating magnetic coupling of the Fe(III) ion's unpaired electron with that of the superoxide ion. This evidence is presented below in Section 4.7.

4.4 ANALYTICAL TECHNIQUES

Many analytical techniques have been used to study deoxy- and oxymyoglobin and hemoglobin and their model compounds. A summary of these techniques, adapted from reference 20, is broken down into two categories: (1) analytical techniques discussed in Chapter 3 and (2) those not described in this text in any detail.

Category 1

1. Extended X-ray absorption fine structure (EXAFS) and X-ray absorption near-edge structure (XANES) methods yield information on the number, type, and radial distance of ligand donor atoms bonded to the metal.[43] XANES may yield geometric information as well.
2. X-ray single-crystal diffraction yields precise three-dimensional structure, bond distances, and angles for small molecules with the same information generated at lower resolution and precision for proteins.[3–5,22,24,29–32]
3. Electron paramagnetic resonance (EPR) yields the location of unpaired electron density from hyperfine splitting by metals or atoms with nuclear spin.[21] The $S = 0$ Fe(III)–O_2^- state of oxy-Mb or Hb would be indicated by the absence of an EPR signal, although other results such as the IR or resonance Raman absorption of the O_2^- moiety would be needed for positive confirmation.
4. Nuclear magnetic resonance may yield identification of histidine by deuterium exchange (N–H versus N–D) at or near the metal, especially if paramagnetic.[25a] The resonances are moved away from the 0–10 ppm region as a result of a paramagnetic center in proximity.
5. Mössbauer spectroscopy gives information on oxidation and spin states of iron-containing species and may diagnose antiferromagnetic coupling.

Category 2

1. Magnetic susceptibility measures the strength of interaction of a sample with a magnetic field. Information obtained includes identification of spin state and spin coupling—ferromagnetic or antiferromagnetic. An example would be identification of the Fe(III)–O–Fe(III) μ-oxo dimer moiety.
2. Infrared spectroscopy (IR) measures vibrational modes involving changes in dipole moments. One may classify O–O moieties—superoxo versus peroxo—or identify ν(M–O) and ν(M–O–M) modes.[35]
3. Raman (R) and resonance Raman (RR) spectroscopy detects vibrational modes involving a change in polarizability. For RR, enhancement of modes is coupled with electronic transition excited by a laser light source. This technique is complementary to IR and is used for detection of ν(O–O) and ν(M–O), especially in metalloproteins. In porphyrins, one may identify oxidation and spin states.
4. UV-visible (UV-vis) spectroscopy detects valence electron transitions. One may detect the electronic state of metal ions from *d–d* transitions. Identification of unusual ligands—that is, Cu(II)–SR, Fe(III)–OPh, Fe(III)–O–Fe(III)—may be possible. UV-vis spectroscopy on single crystals using polarized light may yield geometric information.
5. X-ray photoelectron (XPS-ESCA) spectroscopy uses inner-shell electron transitions to determine the oxidation state of metals in bioinorganic species.

Table 4.2 Parameters for Oxygen Molecules and Ions

Compound	Bond Type	O–O Bond Length (Å)	O–O Bond Order	IR Stretching Frequency (cm^{-1})
O_2^+	$(O{=}O)^+$ radical	1.12	2.5	~1860
O_2	O=O	1.21	2.0	~1556
KO_2	Superoxide (O_2^-)	1.28–1.30	1.5	1075–1200
H_2O_2	Peroxide (O_2^{2-})	1.49 (1.40–1.50)	1.0	740–930

6. Neutron diffraction involves bombarding a solid sample with neutrons to reveal proton and deuteron locations unavailable from X-ray crystallographic analyses. One can place H and D atoms within a structure—for instance, to identify hydrogen bonding of distal histidines to bound O_2 in Mb and Hb.[23]

Important characteristics of the dioxygen, superoxo, and peroxo ions are summarized in Table 4.2. These become important data for determining dioxygen's manner of coordination and oxidation state within biological oxygen carriers.

As discussed in Section 4.8, iron(II) centers in the absence of heme and protein surroundings are readily and irreversibly oxidized to iron(III) species with formation of the μ-oxo dimer of Figure 4.14. See also the reaction sequence 4.14–4.17. The Fe(III)–O–Fe(III) fragment, also known as μ-oxodiiron(III), appears as a motif in many systems besides hemoglobin—for example in hemerythrin,[12] the hydrolase purple acid phosphatase,[13] the oxidoreductases ribonucleotide reductase[14] and methane monoxygenase,[15] an iron sulfur protein rubrerythrin,[16] and the iron-transport protein ferritin.[17] The μ-oxodiiron(III) complex has distinctive spectroscopic fingerprints that allow it to be detected in proteins. Its magnetic susceptibility lies in the range 1.5 to 2.0 Bohr magnetons per μ-oxodiiron(III) group regardless of the number, geometry, or type of ligand around the iron center. The Bohr magneton value translates to the equivalent of one unpaired electron per μ-oxodiiron(III) group; therefore the high spin ($S = 5/2$) iron(III) centers must be strongly antiferromagnetically coupled. The μ-oxodiiron(III) group is also identified by its asymmetric Fe–O stretch ν_{as}(Fe–O), which lies in the range 730–880 cm^{-1}. The symmetric vibration ν_s(Fe–O), forbidden in the IR for linear, symmetric Fe–O–Fe groups, occurs in the range 360–545 cm^{-1}.[20]

4.5 MECHANISM FOR REVERSIBLE BINDING OF DIOXYGEN AND COOPERATIVITY OF OXYGEN BINDING

One can measure the fraction of oxygenated sites in Hb and Mb as a function of oxygen partial pressure. As expected, a hyperbolic curve, as shown Figure 4.9, is found for Mb—one dioxygen molecule binding to one iron–porphyrin heme. In contrast, Hb's curve is sigmoidal, indicating that the binding of dioxygen is

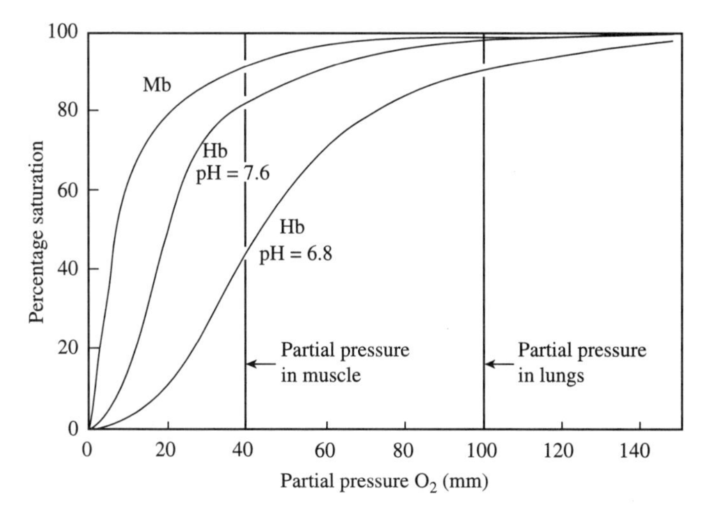

Figure 4.9 Dioxygen binding curves for myoglobin and hemoglobin. (Reprinted with permission from Figure 4.4 of Cowan, J. A. *Inorganic Biochemistry, An Introduction*, 2nd ed., Wiley-VCH, New York, 1997. Copyright 1997, Wiley-VCH.)

"cooperative"; that is, the binding of a second O_2 to a second heme in the Hb tetramer is assisted by the binding of the first and so on for binding of all four oxygen molecules.[18] Allosterism—rearrangement of the protein tertiary and quaternary structure caused by binding of a small molecule remote from the catalytic site—is responsible for this behavior as the iron atoms themselves are 25–40 Å apart in the assembled Hb tetramer. The binding of the first dioxygen to a subunit of tetrameric Hb induces the tertiary structural changes described in Section 4.3. Tertiary structural changes within one subunit in turn alters the stereochemical contacts at the heme and between other subunits, changes the electrostatic attractions of charged subunits, and thereby affects the O_2 affinity of other subunits.[18] Because molecules of the same type (i.e., O_2) affect the binding of successive dioxygen molecules, this behavior is called a homotropic allosteric interaction.[20]

Physiologically, it is important that Mb, the muscle oxygen storage heme protein, have a greater affinity for dioxygen at lower partial pressures, whereas the transport protein Hb should be nearly 100% oxygen saturated at the higher partial pressures in the lung. Both these behaviors are evident from Figure 4.9. It is also true that Mb has higher affinity for dioxygen than Hb at any dioxygen partial pressure. The result is that once Hb reaches the muscle tissue, its oxygen will be transferred to Mb.

In terms of the equilibrium expressions for Mb and Hb oxygenation, one can write equations 4.4 and 4.5 where $P(O_2)$ is the partial pressure of dioxygen:

$$\mathrm{Mb} + \mathrm{O_2} \Leftrightarrow \mathrm{MbO_2} \tag{4.4}$$

$$K = \frac{[\mathrm{MbO_2}]}{[\mathrm{Mb}][\mathrm{O_2}]} \quad \text{or} \quad K_p = \frac{[\mathrm{MbO_2}]}{[\mathrm{Mb}]P(\mathrm{O_2})} \tag{4.5}$$

When Mb is half-saturated with oxygen, $[MbO_2] = [Mb]$ and one uses the term $P_{1/2}$ or P_{50} (50% saturation of Mb with oxygen). For this situation, equation 4.6 holds.

$$K_p = \frac{1}{P_{1/2}(O_2)} \quad \text{or} \quad P_{1/2}(O_2) = \frac{1}{K_p} \tag{4.6}$$

Then, oxygen saturation (Y) can be written as

$$Y = \frac{[MbO_2]}{[Mb] + [MbO_2]} \tag{4.7}$$

Substitution of the equilibrium statement and rearrangement leads to

$$Y = \frac{[O_2]}{K + [O_2]} \tag{4.8}$$

One can also state $[O_2]$ in terms of pressure described by the hyperbolic function of equation 4.9 giving the noncooperative, hyperbolic Mb curve in Figure 4.9:

$$Y = \frac{P(O_2)}{P_{1/2}(O_2) + P(O_2)} \quad \text{or} \quad Y = \frac{K_p(O_2)}{1 + K_p P(O_2)} \tag{4.9}$$

For hemoglobin, similar equations are written as

$$Hb + 4O_2 \Leftrightarrow Hb(O_2)_4 \tag{4.10}$$

$$K = \frac{[Hb(O_2)_4]}{[Hb][O_2]^4} \tag{4.11}$$

In terms of pressure and oxygen saturation, the equation for Hb is written as

$$Y = \frac{(P(O_2))^4}{P_{1/2}(O_2)^4 + P(O_2)^4} \tag{4.12}$$

This behavior leads to the sigmoidal curves for Hb in Figure 4.9 as shown for two different pH values. Further analysis leads to a straight-line Hill plot arising from

$$\log\left(\frac{Y}{1 - Y}\right) = n \log P(O_2) - n \log P_{1/2}(O_2) \tag{4.13}$$

where

Y = oxygen saturation

$P(O_2)$ = partial pressure of dioxygen

$P_{1/2}(O_2)$ = partial pressure at which half the iron will be oxygenated

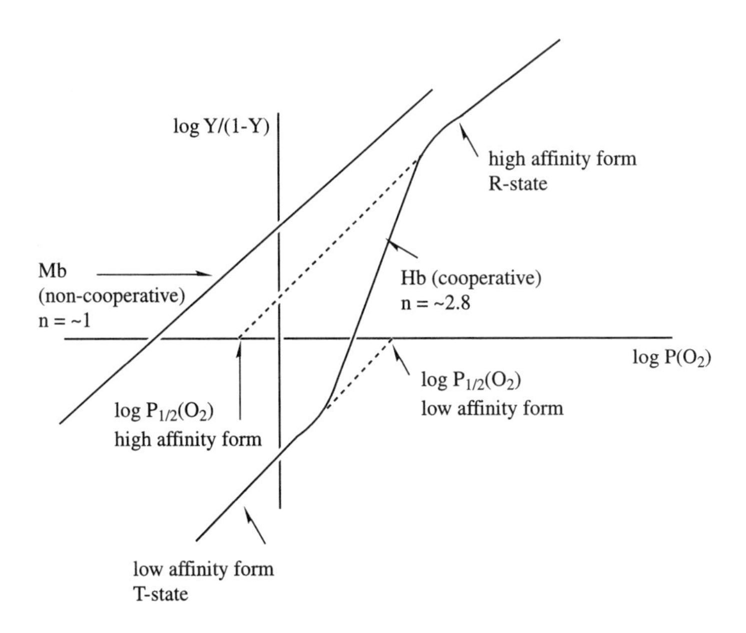

Figure 4.10 Hill plot O_2 binding curves for Mb and Hb. (Adapted with permission from Figure 6 of Momenteau, M.; Reed, C. A. *Chem. Rev.*, 1994, **94**, 659–698. Copyright 1994, American Chemical Society.)

In Figure 4.10 the so-called Hill *n* term, equal to the slope of the line, measures cooperativity. It is close to 1.0 for Mb and 2.8 for Hb, indicating noncooperative behavior for Mb and a positive cooperativity for Hb.[18,19]

In the fully oxygenated R-state in Figure 4.10, Hb's dioxygen affinity is still less than that for Mb. Once Hb transfers one O_2 to Mb, the transfer of Hb's remaining dioxygen molecules is facilitated, leading to the deoxygenated T-state of Hb in Figure 4.10. The effects on Hb's dioxygen binding due to diphosphoglycerate (DPG) concentration and pH (the Bohr effect) are called heterotropic allosteric interactions because the molecule affecting dioxygen binding is different from the molecule bound. Figure 4.9 shows hemoglobin's O_2 binding for two different pH values. A more complete discussion of dioxygen binding, both noncooperative and cooperative, is given in Chapter 4 of reference 20.

4.6 BEHAVIOR OF DIOXYGEN BOUND TO METALS

As can be seen from its ground-state molecular orbital diagram in Figure 4.11, dioxygen has a paramagnetic ground state. It is the only stable homonuclear diatomic molecule with this property.

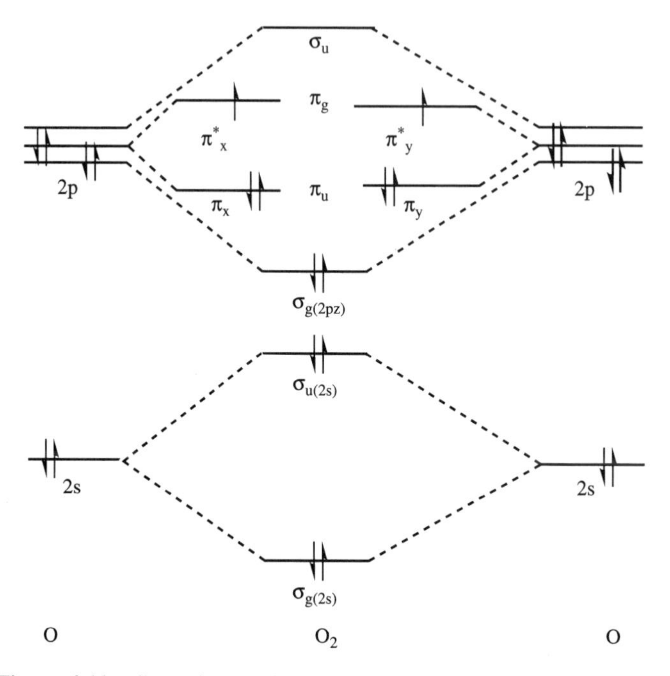

Figure 4.11 Ground-state electron configuration for the O_2 molecule.

The ground state (0 kJ/mol) for the O_2 molecule is represented by the term symbol $^3\Sigma_g^+$. The first excited state (92 kJ/mol above the ground state) is a $^1\Delta_g$ singlet (electrons spin paired with both electrons in either the π^*_x or the π^*_y level). The $^1\Sigma_g^+$ state with paired spin electrons, one each in the π^*_x and π^*_y levels, is the next excited level 155 kJ/mol above the ground state. Reduction of O_2 by one electron yields the superoxide ion (O_2^-), a radical anion. Reduction by two electrons yields the peroxide ion, (O_2^{2-}). Bond lengths and bond orders for these are given in Table 4.2. As noted in equation 4.2, the reduction potential for O_2 in the presence of protons is thermodynamically favorable. Therefore, reversible binding of O_2 to a metal can only be achieved if competition with protons and further reduction to superoxide and peroxide are both controlled.[8]

4.7 STRUCTURE OF THE ACTIVE SITE IN MYOGLOBIN AND HEMOGLOBIN: COMPARISON TO MODEL COMPOUNDS

X-ray crystallographic structures of myoglobin and hemoglobin were first completed in 1966[2] and 1975[3], respectively. Since then, many other X-ray crystallographic studies of deoxy- and oxy- as well as met-myoglobin and hemoglobin have been carried out.[22,24] Additionally, researchers have studied the carbon monoxide bound moieties MbCO and HbCO as well as MbNO. Site–directed mutagenesis of residues near the active sites of Mb and Hb have yielded information on the exact nature of O_2, CO, and NO binding and the small molecule's orientation at the heme site. With information from these studies confirmed by many other instrumental and analytical techniques, a clear picture of the metalloprotein's active site has emerged.

The active site consists of an iron(II) protoporphyrin IX (the ligand shown in Figure 4.1) encapsulated in a water-resistant pocket and bound to the protein through a single coordinate bond between the imidazole nitrogen of the proximal histidine residue (his93, F8 for myoglobin) and the iron(II) (see Figure 4.5). The proximal Fe–N bond vector has an approximate 10° tilt off the heme normal.

Additionally, other protein residues such as leucine, isoleucine, valine, and phenylalanine interact with the heme, holding it in place through hydrophobic interactions. The five-coordinate Fe(II) can add dioxygen in its sixth, vacant, coordination site and a variety of other small ligands (CO, NO, RCN) may also bind there. Other amino acid residues that control the immediate environment with respect to polar, hydrophobic, or steric interactions surround the distal, vacant coordination site of the deoxy form. When O_2 is bound, it is stabilized by hydrogen-bond interactions through the distal histidine (his64, E7). See Figures 4.5, 4.7, and 4.12. The H-bond interactions may affect O_2 affinity and inhibit pathways leading to further oxidation and μ-oxo dimer formation.[8]

Direct evidence for hydrogen-bonding interactions for heme-bound O_2 were first shown in an EPR study on cobalt-substituted hemoglobin.[21] These studies are further described in Sections 4.8.1 and 4.8.2. Subsequently, X-ray structure analysis[22] and neutron diffraction studies[23] on MbO_2 provided direct evidence for hydrogen bonding of iron-bound oxygen with distal histidine, yielding an N(H)–O distance of 2.97 Å. Hydrogen bonding of the same sort for HbO_2 with N(H)–O = 2.7 Å and 3.2–3.4 Å for α and β subunits respectively was shown by X-ray crystallography.[24] The distal histidine hydrogen-bonding structure for hemoprotein and for a model "amide basket handle" heme is illustrated in Figure 4.12. Modern NMR techniques allow one to assign most of the heme and distal amino acid proton resonances, but these cannot be interpreted in terms of H bonding to dioxygen.[25] However, MbO_2 models derived from picket-fence porphyrins having some suitably situated substituents with differing H-bonding capabilities do show differing dioxygen affinities.[26]

Despite the many model compounds that have been prepared, picket-fence porphyrin[27] remains the only structure yielding crystallographic data comparable to Mb oxyheme stereochemistry (see Table 4.4). In fact the oxygenated picket–fence porphyrin model's X-ray structure was known before that of the oxyhemoproteins and anticipated the correct bent geometry of the dioxygen molecule. As described above, the bent geometry facilitates metal *dπ* to O_2 (π^*) bonding (see Figure 4.6). The proximal 2-MeIm ligand of the model compounds is sterically more demanding, causing lengthening of the Fe–O bond, and models the T-state (lower dioxygen affinity), whereas the proximal 1-MeIm ligand models the R-state (see Figure 4.18).

The information of Tables 4.3 and 4.4 indicates that a water molecule is found in the binding cavity of α chains of hemoglobin A (HbA) even though this cavity has been called hydrophobic. Indeed, although many hydrophobic groups such as

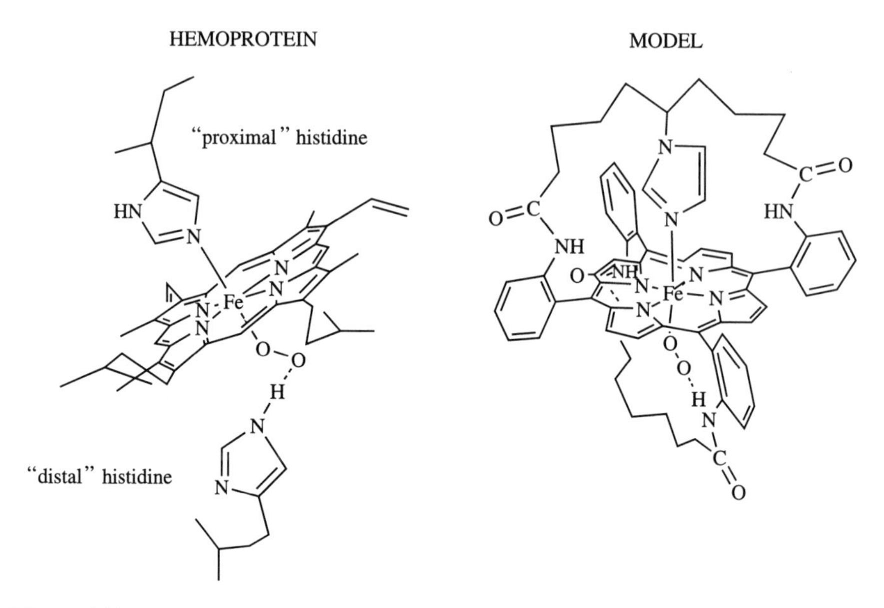

Figure 4.12 Distal histidine hydrogen bonding structure for hemoglobin (left) and a heme model (right). (Reprinted with permission from Figure 12 of Momenteau, M.; Reed, C. A. *Chem. Rev.*, 1994, **94**, 659–698. Copyright 1994, American Chemical Society.)

Table 4.3 Deoxy–Heme Stereochemistry for Mb, Hb, and Model Compounds

	Mb[28]	HbA (α-H_2O)[a] (β)[29]	Fe Model 1[b] (deoxy)[30]
Fe–N_p, Å[c]	2.03(10)	2.08(3) 2.05(3)	2.072(5)
Fe–N_{Im}, Å[c]	2.22	2.16(6) 2.09(6)	2.095(6)
Fe–Porph, Å	0.42	0.40(5)	0.43
ϕ, deg[d]	19	18(1)	22.8
tilt, deg[d]	11	12(2)	9.6

[a] α and β refer to the particular protein chains in myoglobin or hemoglobin. One notes from the table that the bond distances and angles are slightly different for the two moieties. In each table cell, data are reported in order α, then β.

[b] Fe model 1 (deoxy form) Fe(T_{piv})PP-(2-MeIm). The ligand is (T_{piv})PP-(2-MeIm).

[c] Fe–N_p is the bond distance to porphyrin nitrogen ligand atoms. Fe–N_{Im} is the bond distance to histidine (or imidazole in model compounds) nitrogen ligand atoms.

[d] ϕ is defined as the angle between the plane of the axial base (imidazole or histidine) and the plane defined by the metal, the N ligands of the porphyrin ring, and the axial base. Tilt is defined as the angle the axial base moves away from the normal to the metal–porphyrin plane and is affected by methyl substituents such as those on Fe(T_{piv})PP-(2-MeIm).

Table 4.4 Oxy–Heme Stereochemistry for Mb, Hb, and Model Compounds

	MbO_2[22]	HbO_2 (α-H_2O)[a] (β)[31]	Fe Model 2[b] (oxy)[32]	Fe Model 3[c] (oxy)[33]
Fe–N_p, Å[d]	1.95(6)	1.99(5) 1.96(6)	1.98(1)	1.996(4)
Fe–N_{Im}, Å[d]	2.07(6)	1.94(9) 2.07(9)	2.07(2)	2.107(4)
Fe–O, Å	1.83(6)	1.66(8) 1.87(3)	1.75(2)	1.898(7)
O–O, Å	1.22(6)		1.24	1.22(2)
Fe–O–O, deg	115(5)	153(7) 159(12)	131	129(2)
Fe–Porph, Å	0.18(3)	0.12(8) −0.11(8)	−0.03	0.09
ϕ, deg[e]	1	11 27	20	22
tilt, deg[e]	4	3 5	0	7

[a] α and β refer to the particular protein chains in myoglobin or hemoglobin. One notes from the table that the bond distances and angles are slightly different for the two moieties. In each table cell, data are reported in order α, then β.

[b] Fe model 2 (oxy form) Fe(T_{piv})PP-(1-MeIm)O_2. The ligand is Fe(T_{piv})PP or picket-fence porphyrin, meso-*tetrakis*($\alpha,\alpha,\alpha,\alpha$-*o*-pivalamidephenyl)porphyrin (see H_2PF in Figure 4.23 of reference 20, Figure 16 of reference 8). In this chapter see Figures 4.17 and 14.18.

[c] Fe model 3 (oxy form) Fe(T_{piv})PP-(2-MeIm)O_2. The ligand is (T_{piv})PP-(2-MeIm).

[d] Fe–N_p is the bond distance to porphyrin nitrogen ligand atoms. Fe–N_{Im} is the bond distance to histidine (or imidazole in model compounds) nitrogen ligand atoms.

[e] ϕ is defined as the angle between the plane of the axial base (imidazole or histidine) and the plane defined by the metal, the N ligands of the porphyrin ring, and the axial base. Tilt is defined as the angle the axial base moves away from the normal to the metal–porphyrin plane and is affected by methyl substituents such as those on Fe(T_{piv})PP-(2-MeIm).

valine, leucine, isoleucine, and phenylalanine are positioned over the porphyrin, the immediate vicinity of the binding site is in fact polar, containing distal histidine, the heme itself, and associated water molecules. Model 1 (deoxy) in Table 4.3 should be a good match for deoxy Hb and be considered as a T–state model compound (see Figures 4.7, left, and 4.18A). Model 2 (oxy) in Table 4.4 should be a good match with HbO_2 and be considered an R–state model compound (see Figures 4.7, right, and 4.18B). The orientation of the proximal, axial base (i.e., the angle ϕ in Table 4.3) and the distance of Fe out of the porphyrin plane for deoxy-HbA, as well as Mb, and the model compound Fe(T_{piv})PP-(2-MeIm), model 1, are similar. In Table 4.4, model 2, Fe(T_{piv})PP-(1–MeIm)O_2, works better as a model for HbO_2 and MbO_2 than does model 3, Fe(T_{piv})PP–(2-MeIm)O_2, although the ϕ and tilt values are not good matches with MbO_2.

4.8 MODEL COMPOUNDS

Kenneth Suslick[18] has written several pertinent statements about the study of metalloproteins and their model compounds. In Chapters 4 through 7, this paradigm will guide the study of the varied bioinorganic systems described. Additions to Suslick's statement in point 2 stress the importance of modern structural analysis (X-ray crystallography, EXAFS, EPR, and two-and three–dimensional high-field NMR) of the purified metalloproteins and their model compounds. As paraphrased from reference 18, the following statements outline the steps involved in the study of metalloproteins:

1. *Isolation* and *purification* of the metalloprotein. For instance, Hb was first crystallized in 1849, its physiological purpose of oxygen transport was recognized by 1864, and its molecular weight and primary amino acid sequence was known by 1930.
2. *Measurement* of physical and spectroscopic properties of the active site. X-ray crystallography has been extensively used to characterize metalloproteins, especially with regard to their metal cofactors. Interpretation of the X-ray data must be consistent with spectroscopic data from EXAFS, EPR, and NMR studies. Currently, X-ray crystallographic data are deposited with the Protein Data Bank (PDB) for met-, deoxy-, and oxyhemoglobin and met-, deoxy-, and oxymyoglobin from a number of species. More recently, X-ray studies of myoglobin and hemoglobin modified by site-directed mutagenesis of aa residues near and at the active site have led to more detailed information on structure–function relationships.
3. *Characterization* of structural, spectroscopic, and reactivity properties of model compounds—that is, metal cofactor small molecule analogs.
4. *Comparisons* between the protein and the analogs to reveal new structure–function relationships.

Chemists find many advantages in studying small molecule analogs of metalloproteins. For hemoglobin, as for other systems, one can systematically change one variable at a time. These variables might include modifications of axial ligation, binding site polarity, steric restraint, and solvent effects, among others. The disadvantage for all small molecule analogs is that the protective environment of the protein itself is lost. In the heme example under discussion, one must synthesize a complex porphyrin ligand system, keep iron in a five-coordinate state until dioxygen is added, prevent irreversible oxidation of the iron center, and produce an environment in which O_2 ligation is favored over CO binding.

4.8.1 Cobalt-Containing Model Compounds

Early in the search for myoglobin or hemoglobin model compounds for study, it was learned that cobaltoheme substituted into apo–myoglobin or -hemoglobin

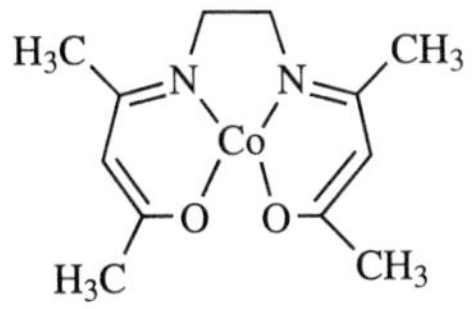

Figure 4.13 Co(acacen), an early model compound exhibiting reversible dioxygen binding at low temperatures.

exhibited dioxygen-binding and cooperativity.[34] The first small molecule cobalt-containing model systems for hemoglobins involved a $1:1$ $Co:O_2$ complex and utilized Schiff-base ligands. It was first prepared in the late 1960s.[35] The example shown in Figure 4.13, Co(acacen), acacen = acetylacetonate ethylenediamine, was found to slowly uptake oxygen over a period of days at room temperature, oxidizing the organic ligand. The reaction took place in a coordinating solvent such as dimethylformamide or a noncoordinating solvent such as toluene with an added base (pyridine or imidazole). However, at temperatures of 0°C or below, a rapid and reversible uptake of dioxygen was seen. The solid complexes exhibited a strong IR band near 1140 cm^{-1}, which disappeared when O_2 was removed. This is in the region expected for the O_2^- superoxo ligand as noted in Table 4.2.

Because Co(II) (d^7) complexes are paramagnetic, one may study these model compounds using electron paramagnetic resonance (EPR) spectroscopy. The EPR spectra in both liquid and frozen solutions exhibit eight-line hyperfine splitting indicating interaction of one electron with a single ^{59}Co nucleus ($I = 7/2$). A 90% transfer of spin density from Co(II) to O_2 indicates that the $1:1$ $Co:O_2$ complex should be described as Co(III)–O_2^-.[35,36] X-ray crystallography of the complex established a Co–O–O bond angle of 125° and an O–O bond distance of 1.2 Å, close to that of superoxide (1.32 Å) and much shorter than that of the peroxide ion (1.49 Å).

Many other cobalt complexes including Co(porphyrin) complexes were and are useful for understanding the dioxygen affinity of cobalt and, by extrapolation, that of iron transition metal complexes. Cobalt complexes having tetradentate square-planar chelating ligands are simpler to study than their iron counterparts for two important reasons: (1) Reactions 4.14 and 4.17 leading to the formation of μ-oxo dimer are less likely to take place with Co(II); and (2) the equilibrium constant for the formation of six-coordinate CoL_4B_2 from five-coordinate CoL_4B is much smaller than that of CoL_4B's formation, so that the likelihood of the disproportionation reaction to CoL_4 and CoL_4B_2 from two CoL_4B species is low.

4.8.2 Iron-Containing Model Compounds

In the case of iron-containing small molecule analogs of Mb and Hb a much rockier road to successful model compounds was encountered. Even though the syntheses of iron porphyrin complexes were carried out in analogous manner to the cobalt species described above, their irreversible oxidation to the μ-oxo dimer upon

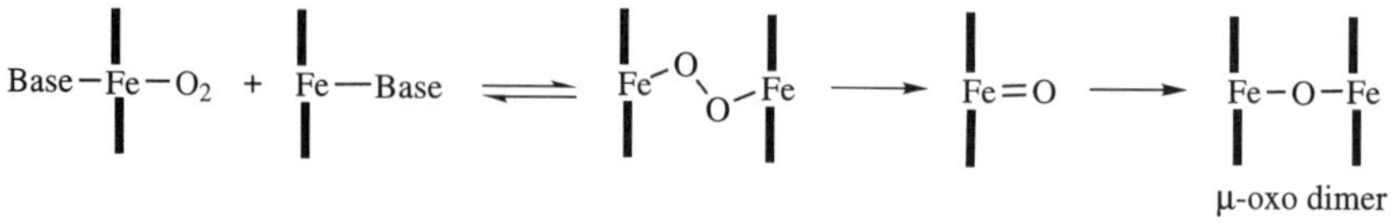

Base—Fe—O_2 + Fe—Base ⇌ Fe–O–O–Fe → Fe=O → Fe–O–Fe

μ-oxo dimer

Figure 4.14 Formation of μ-oxo dimer. (Adapted with permission from Figure 9 of Suslick, K. S.; Reinert, T. J. *J. Chem. Ed.*, 1985, **62**(11), 974–983. Copyright 1985, Division of Chemical Education, Inc.)

addition of O_2 remained a stumbling block to their study as small molecule analogs of Mb and Hb. Addition of dioxygen to simple, undecorated iron porphyrins led to irreversible oxidation and into a thermodynamic Fe(III) pit with formation of the μ-oxo dimer. This behavior is illustrated in the following reactions adapted from reference 20 and the scheme adapted from Suslick's article as shown in Figure 4.14.[18] Initially attempts were made to solve this problem for solution studies through the use of low temperatures and aprotic solvents although the experimental conditions were far from those of physiological systems.

$$\mathrm{Fe(II)} + \mathrm{O_2} \Leftrightarrow \mathrm{Fe(III)}\text{—}\mathrm{O_2^-} \tag{4.14}$$

$$\mathrm{Fe(III)}\text{—}\mathrm{O_2^-} + \mathrm{Fe(II)} \Leftrightarrow \mathrm{Fe(III)}\text{—}\mathrm{O_2^{2-}}\text{—}\mathrm{Fe(III)} \tag{4.15}$$

$$\mathrm{Fe(III)}\text{—}\mathrm{O_2^{2-}}\text{—}\mathrm{Fe(III)} \rightarrow \mathrm{Fe(IV)}{=}\mathrm{O} \tag{4.16}$$

$$\mathrm{Fe(IV)}{=}\mathrm{O} + \mathrm{Fe(II)} \rightarrow \underset{\mu\text{-oxo dimer}}{\mathrm{Fe(III)}\text{—}\mathrm{O}\text{—}\mathrm{Fe(III)}} \tag{4.17}$$

A simple molecular orbital scheme showing why μ-oxo dimer formation is more likely to happen for Fe(III), d^5, than for Co(III), d^6, is shown in Figure 4.15. For d^6 Co(III), the last electron goes into a π* antibonding orbital, making reactions 4.16 and 4.17 less likely.

Successful iron-containing Mb and Hb models have simulated the protein surroundings of the heme porphyrin through introduction of steric hindrance on the distal side of the porphyrin (Figures 4.16 and 4.17). The porphyrin's decoration is necessary to prevent irreversible oxidation of Fe(II) by dioxygen and the formation of the μ-oxo dimer shown in Figure 4.14. As stated previously, one can make the irreversible oxidation slower by lowering the temperature. Other methods include attaching the Fe(II) complex to a solid support or making the observation time faster by using stopped flow kinetics.[18]

In order to prepare and isolate solid-state, crystalline, oxygenated iron–heme model complexes, chemists learned to synthesize (by self-assembly methods) and oxygenate many types of hindered porphyrins. For instance, "capped" porphyrins were synthesized by direct condensation of a suitable tetraaldehyde with four pyrrole molecules.[37] "Picket-fence" porphyrins such as [Fe(TPP)(*N*–MeIm)] (where TPP = *meso*-tetraphenylporphyrin and *N*–MeIm = *N*–methylimidazole)

A π orbital scheme for M-O with M = Fe(III)

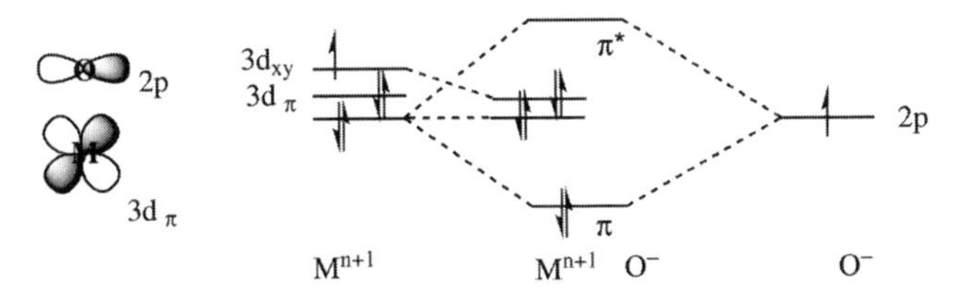

B π orbital scheme for M-O with M = Co(III)

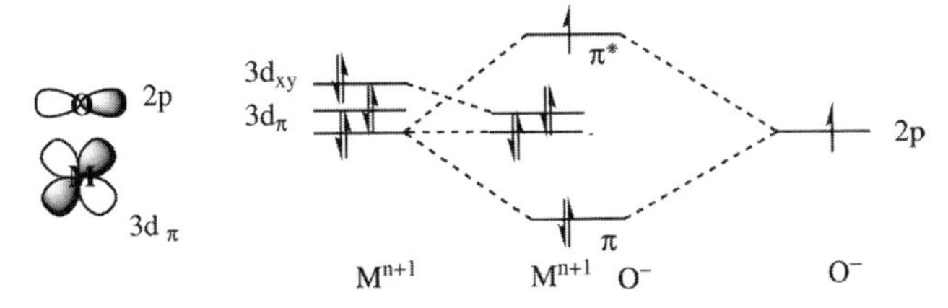

Figure 4.15 Molecular orbital scheme for (A) Fe(III)—O_2^- and (B) Co(III)—O_2^- (Adapted with permission from Figure 4.15 of Jameson, G. B.; Ibers, J. A., in Bertini, I.; Gray, H.; Lippard, S. J.; Valentine, J. S. *Bioinorganic Chemistry*, University Science Books, Sausalito, CA, 1994.)

Figure 4.16 Prevention of μ-oxo dimer formation by distal-side porphyrin modifications. (Adapted with permission from Figure 10 of Suslick, K. S.; Reinert, T. J. *J. Chem. Ed.*, 1985, **62**(11), 974–983. Copyright 1985, Division of Chemical Education, Inc.)

Figure 4.17 Picket-fence porphyrin Fe(T_{piv})PP, meso-*tetrakis*(α,α,α,α-*o*-pivalamidephenyl)porphyrin described in references 5b, 30, 32, and 33.

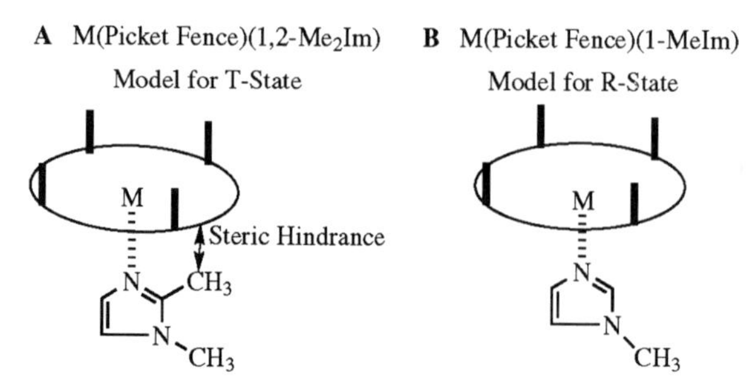

Figure 4.18 Picket-fence porphyrin models. (A) Model for T-state. (B) Model for R-state. (Adapted with permission from Figure 5 of Collman, J. P. *Inorg. Chem.*, 1997, **36**, 5145–5154. Copyright 1997, American Chemical Society.)

were synthesized and purified by separation of isomeric mixtures of *tetra*(*O*–substituted phenyl)porphin.[5a] Another picket-fence porphyrin, Fe(T_{piv})PP, meso–*tetrakis*(α, α, α, α-*o*-pivalamidephenyl)porphyrin, synthesized and studied by Collman's group in the 1970s, is shown in Figure 4.17.[5b] Structural and geometric data on deoxy and oxy forms of this model porphyrin are listed in Tables 4.3 and 4.4. Many other representative examples of hemoglobin model compounds, some of which became known as the "basket" and "pocket" porphyrins, are shown in Figures 8 and 9 of reference 8.

All successful myoglobin and hemoglobin model compounds provide steric bulk on the distal side of the porphyrin ring with a hydrophobic pocket for complexation of dioxygen as well as a bulky alkyl imidazole proximal ligand (not shown in Figure 4.17) occupying the other, proximal side of the porphyrin ring. Design of such systems have been many and varied and have led, for example, to models for the T– and R–states as shown in Figure 4.18.[11] The T–state model (deoxy form in Figure 4.7) uses steric hindrance to prevent the iron ion from entering the plane of the porphyrin ring, whereas the R–state model (oxy form in Figure 4.7) permits the iron ion into the porphyrin plane. Geometric data for hemoglobin, myoglobin and a picket-fence model compound, shown in Figure 4.17, is given in Tables 4.3 and 4.4.

In summary, researchers have found a number of methods for avoiding μ-oxo dimer formation and preserving a five-coordinate Fe(II) in iron-containing model compounds, through:

1. Modifying the imidazole to preserve the T–state (shown in Figure 4.18A).[38]
2. Modifying the porphyrin to sterically prevent addition of a large sixth ligand. The most well known version of these is the "picket-fence" porphyrin illustrated in Figure 4.17 (see also Figure 1 of reference 39), although many others have been synthesized.[8,18]
3. Attaching the five-coordinate system to a rigid support (i.e., silica gel), thereby reducing its mobility and ability to add a sixth ligand.

These strategies have been successful and have led to the X-ray crystallographic structure of the oxygenated picket-fence porphyrin described in Section 4.7.

Japanese researchers interested in a practical application for these models have synthesized one model compound, shown in Figure 4.19, that may lead to the synthesis of "artificial blood." To achieve this end, Tsuchida and co-workers synthesized the iron carrier $\alpha, \alpha, \alpha, \alpha$,-meso-*tetra*[(*o*-hydroxydimethyleicosanamido)phenyl] porphyrin. After incorporating iron into the heme, the trimethyl amine and the hydrophobic imidazole were inserted, yielding the so-called lipid-heme. Subsequently, the lipid-heme was encapsulated into phospholipid liposomes and the product tested successfully for dioxygen carrier ability in vivo.[40] The authors also concentrated the lipid-heme/liposome mixture and stored it as a freeze-dried powder. Ultimately, the researchers hope to provide the medical community with a highly concentrated, reproducible, readily transportable and easily storable heme suspension that is usable as an oxygen-carrying blood substitute in emergency situations.

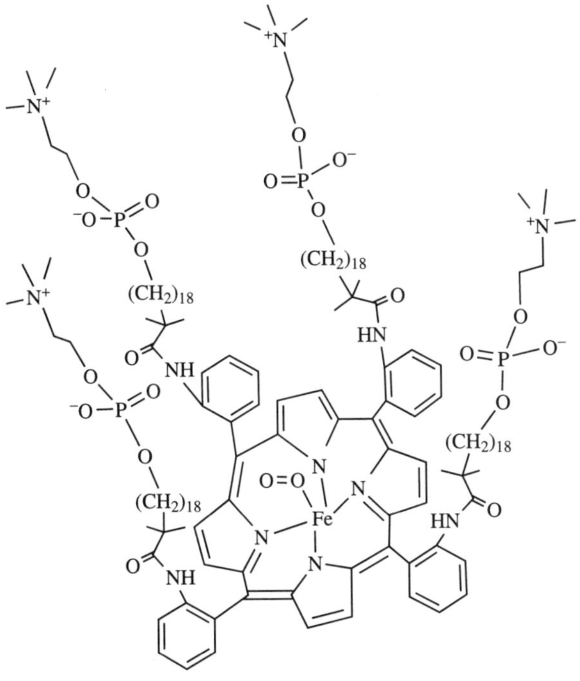

Figure 4.19 Heme incorporated into liposomes for "artificial blood." (Adapted with permission from Figure 14 of Suslick, K. S.; Reinert, T. J. *J. Chem. Ed.*, 1985, **62**(11), 974–983. Copyright 1985, Division of Chemical Education, Inc.)

4.9 BINDING OF CO TO MYOGLOBIN, HEMOGLOBIN, AND MODEL COMPOUNDS

Selective binding of dioxygen rather than carbon monoxide in wild-type biological systems is complicated by the fact that naturally occurring metalloproteins including Mb and Hb produce CO during their degradation processes.[11] Therefore, hemes must be able to carry out their oxygen transport and storage functions in the presence of significant concentrations of CO. In addition, CO binds to myoglobin and hemoglobin with affinities of, respectively, 25 and 200 times those of O_2. Collman proposes that the discrimination occurs through steric constraints imposed by amino acid residues on the distal side of the porphyrin and by selective hydrogen bonding favoring O_2 over CO coordination. Dioxygen is capable of a bent geometry when bound to the heme facilitating the hydrogen bond to the distal histidine, whereas CO's preferred linear binding mode not only prevents hydrogen bond formation but also results in steric clashes with neighboring amino acids as shown in Figure 4.20.

Not only is carbon monoxide produced during catabolism of hemes, but it is also necessary for maintenance of neuronal function and cell signaling in the vasculature and therefore is present normally in the vicinity of Mb and Hb. For synthetic systems of picket-fence, pocket, or capped porphyrins, CO affinity exceeds that for O_2 sometimes by many orders of magnitude. The in vitro preference of CO over O_2 binding to Mb and Hb are in direct conflict with the data indicating that under physiological conditions, the dioxygen affinity of Mb and Hb are orders of magnitude greater than that for CO.[20] Some of the explanations put forward to illuminate the observed affinities follow.

For small-molecule, metal–carbon monoxide complexes, the carbon monoxide ligand is almost always in a linear conformation and perpendicular to the metal. If one assumed bonding of CO to Hb or Mb in its normal linear, perpendicular mode, steric conflicts as illustrated in Figure 4.20 would occur and thus one might predict

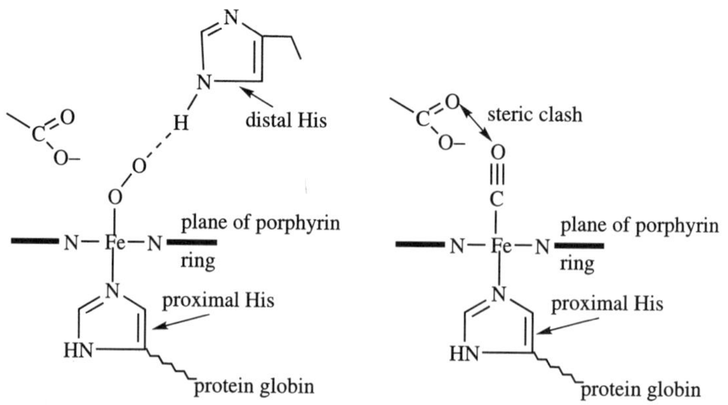

Figure 4.20 Coordination of O_2 and CO to heme iron in Mb and Hb.

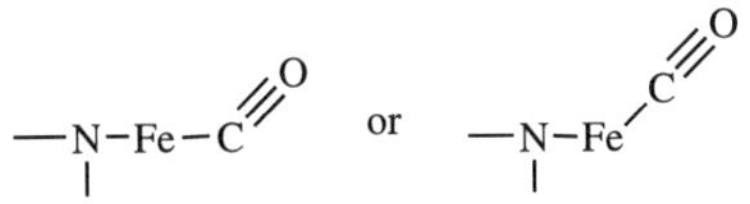

—N–Fe–C≡O or —N–Fe–C≡O

Figure 4.21 Possible Fe–CO bent conformation binding modes.

that CO binding would be less favored by physiological systems. Earlier X-ray crystallography of MbCO and $Hb(CO)_4$ had been less than clear about the carbon monoxide ligand because resolution in protein crystallographic structures was not sufficient to unequivocally distinguish between linear and bent structures. Several possible interpretations of CO binding in a bent conformation may have included, for instance, either of the binding modes illustrated in Figure 4.21.[20]

The crystal structure of MbCO at 1.5-Å resolution suggested an Fe–C–O bond angle of 120° or 140°, although a linear conformation could not be ruled out.[41] Vibrational spectroscopy confirmed the existence of two major configurations for Fe–C–O and even indicated a third minor possibility.[42] Model compounds yielded similar results. For a pocket FeCO porphyrin having $1,2\text{-}Me_2Im$ as an axial base, the Fe–C–O bond angle was found by EXAFS to be 127°.[43] The bent bonding mode for CO has been challenged in recent research, and the angle was found to be closer to 10° rather than over 100°. Some researchers believe that earlier single crystals may have been disordered or that the materials tested were impure, containing more than one type of Fe–C–O bond. More recently, Phillips and Olson[44] have discussed myoglobin's discrimination between O_2, CO, and NO on the basis of electrostatic interactions. Rather than steric interactions, the authors show that it is the polarity of the distal histidine, his64 (E7), in Mb that allows discrimination among the diatomic gases. A strong hydrogen bond exists between the his64 and bound O_2 (polarized as the $Fe^{\delta+}$–O–$O^{\delta-}$ moiety), whereas bound CO and NO have weaker interactions with the same protein chain residue. In deoxy-Mb and deoxy-Hb, an incoming ligand must displace a solvent water molecule hydrogen-bonded to the distal his64, causing a net inhibition of CO binding when compared to O_2's ability to create a replacement H-bond to that lost by water's displacement from the site pocket. Thus H-bond interactions and the kinetics of gas association and dissociation currently are thought to be much more important to Mb and Hb's diatomic gas affinity and gas molecule discrimination than bonding geometry.

In attempting to understand how the attachment and release of carbon monoxide, and ultimately dioxygen, happens on a molecular scale, Rodgers and Spiro have studied the nanosecond dynamics of the R to T transition in hemoglobin.[45] Using pulse-probe Raman spectroscopy, with probe excitation at 230 nm, these workers were able to model the R–T interconversion of the hemoglobin molecule as it moved from the R-state (HbCO) to the T-state (Hb). Under static conditions, laser excitation at 230 nm provided resonance enhancement of vibrational Raman bands of tyrosine (tyrα42) and tryptophan (trpβ37) side-chain residues that form specific hydrogen bonds across the $\alpha_1\beta_2$ interface in the T-state. These H-bonds are broken in the R-state. When the researchers collected data on Hb and HbCO under transient conditions, they found a different set of behaviors at much shorter times that reached a maximum intensity at ~50 ns. They believe these differences reflect tertiary structure changes induced by loss (deligation) of the CO ligand and before the R–T interconversion. It is known that CO molecules deligate in less than a picosecond, may recombine with the iron center in geminate fashion before leaving the binding pocket on a ~50-ns time scale, or leave the binding pocket on the same time scale. Therefore the workers believe that the 50-ns differences they detect accompany CO's leaving the binding pocket rather than CO deligation. In concert with CO leaving the binding pocket, H-bonding between residues on the A helix and E helix (the E helix being intimately connected with the heme cofactor as it contains the distal histidine, his58; see Figures 4.4 and 4.5) is detected. Also weakening of other H-bonds that would allow a shifting of the E helix toward the heme group is found. In X-ray crystallographic and molecular dynamics studies, it was found that the F helix (proximal to the heme cofactor and containing the proximal histidine ligand his87) moves as well. Taking all this evidence into account, Rodgers and Spiro propose a model for the R–T reaction coordinate that starts with CO deligation and involves movement of and strain in the E and F helices (on a subpicosecond time scale) along with movement of the Fe ion toward the proximal his residue. The "scissoring" motion of the E and F helices, illustrated in Figure 3 of reference 45, relieves the strain between the helices and allows the CO to leave the binding pocket at the same time.

Nanosecond time-resolved crystallography of MbCO has been discussed in Section 3.7.2.3 of Chapter 3.[46] After firing a 10-ns burst of laser light to break the CO–Fe bond, these researchers produced a diffraction image of the crystal through application of a 150-ps X-ray pulse. They are able to show release of the CO molecule, displacement of the Fe ion toward the proximal histidine, and recombination of the dissociated CO by about 100 μs. Essentially their results compare well with other spectroscopic studies of HbCO, MbCO and their models.

4.10 CONCLUSIONS

Myoglobin and hemoglobin have been studied exhaustively since chemists, biochemists, and biologists realized their common and abiding interest in these biochemical systems. In this chapter, a brief review of the major knowledge categories has been presented. No quantitative information has been presented on kinetic or thermodynamic aspects of O_2 binding, although much is known. Reference 8, for instance, gives an excellent review of quantitative kinetic and thermodynamic information on myoglobin, hemoglobin, and their models.

While much has been learned about myoglobin and hemoglobin, many controversies and uncertainties remain. It is certain that myoglobin stores dioxygen in muscle tissue, whereas hemoglobin carries dioxygen through the bloodstream to all parts of the body. We know the primary, secondary, tertiary, and quaternary structures of the protein matrix with a great deal of certainty, and we have much information about the cooperative nature of dioxygen binding to hemoglobin. Most known instrumental and analytical techniques have been used in the study of these important dioxygen carriers. X-ray crystallography, infrared, and high-field NMR have yielded much structural information but have not been able to unequivocally resolve a major remaining uncertainty: Why does O_2 preferentially bind to hemes in spite of the known greater affinity of Mb and Hb for CO over O_2? As Stu Borman commented in *C&E News* at the end of 1999: "For several decades, researchers have been trying to understand how carbon monoxide and molecular oxygen bind to the heme proteins myoglobin and hemoglobin. With a new millennium about to dawn, they're still at it."[47] Consensus will probably arise from a blending of several explanations: (a) hydrogen bonding and electrostatic considerations that favor O_2 coordination and (b) steric considerations that disfavor CO coordination. Certainly the questions and answers will make for interesting future reading.

REFERENCES

1. Hay, R. W. *Bio-Inorganic Chemistry*, Ellis Horwood Limited, Halsted Press, New York, 1984.
2. Nobbs, C. L.; Watson, H. C.; Kendrew, J. C. *Nature*, 1966, **209**, 339–341.
3. Fermi, G. *J. Mol. Biol.*, 1975, **97**(2), 237–256.
4. Fermi, G.; Perutz, M. F.; Shaanan, B.; Fourme, R. *J. Mol. Biol.*, 1984, **175**, 159–174. (PDB: 2HHB, 3HHB, 4HHB)
5. (a) Collman, J. P.; Gagne, R. R.; Halbert, T. R.; Marchon, J. C.; Reed, C. A.; Halbert, T. R.; Lang, G.; Robinson, W. T. *J. Am. Chem. Soc.* 1973, **95**, 7868–7870. (b) Collman, J. P.; Gagne, R. R.; Reed, C. A.; Halbert, T. R.; Lang, G.; Robinson, W. T. *J. Am. Chem. Soc.* 1975, **97**, 1427–1439. (c) Collman, J. P. J. *Acc. Chem. Res.*, 1977, **10**, 265–272.
6. Phillips, S. E. *J. Mol. Biol.*, 1980, **142**, 531–554. (PDB: 1MBO)
7. Cowan, J. A. *Inorganic Biochemistry, An Introduction*, 2nd ed., Wiley-VCH, New York, 1997.
8. Momenteau, M.; Reed, C. A. *Chem. Rev.*, 1994, **94**, 659–698.
9. Lippard, S. J.; Berg, J. M. *Principles of Bioinorganic Chemistry*, University Science Books, Mill Valley, CA, 1994.
10. Perutz, M. *Mechanisms of Cooperativity and Allosteric Regulation in Proteins*, Cambridge Press, New York, 1990.
11. Collman, J. P. *Inorg. Chem.*, 1997, **36**, 5145–5154.
12. Wilkins, P. C.; Wilkins, R. G. *Coord. Chem. Rev.*, 1987, **79**, 195–214.
13. Antanaitis, B. C.; Aisen, P. *Adv. Inorg. Biochem.*, 1983, **5**, 111–136.
14. Sjöjberg, B.-M.; Gräslund, S. A. *Adv. Inorg. Biochem.*, 1983, **5**, 87–110.
15. Toftlund, H., et al., *J. Chem. Soc. Chem. Commun.*, 1986, 191–192.
16. LeGall, J.; Prickril, B. C.; Moura, I.; Xavier, A. V.; Moura, J. J.; Huynh, B. H. *Biochemistry*, 1988, **27**, 1636–1642.
17. Theil, E. C. *Adv. Inorg. Biochem.*, 1983, **5**, 1–38.
18. Suslick, K. S.; Reinert, T. J. *J. Chem. Ed.*, 1985, **62**(11), 974–983.
19. Stryer, L. *Biochemistry*, 2nd ed., W. H. Freeman, New York, 1988, pp. 67–83.

20. Jameson, G. B.; Ibers, J. A. In Bertini, I.; Gray, H.; Lippard, S. J.; Valentine, J. S. *Bioinorganic Chemistry*, University Science Books, Mill Valley, CA, 1994.
21. Yonetani, T.; Yamamoto, H.; Itzuka, T. *J. Biol. Chem.*, 1974, **249**, 2168–2174.
22. Phillips, S. E. V. *J. Mol. Biol.*, 1980, **142**, 531–554. (PDB: 1MBO)
23. Phillips, S. E. V.; Shoenborn, B. P. *Nature*, 1981, **292**, 81–82.
24. Shaanan, B. *Nature*, 1982, **296**, 683–684.
25. (a) Wüthrich, K. *NMR of Proteins and Nucleic Acids*, Wiley-Interscience, New York, 1986. (b) Schaeffer, C.; Craescu, C. T.; Mispelter, J.; Garel, M. C.; Rosa, J.; Lhoste, J. M. *Eur. J. Biochem.*, 1988, **173**, 317–324.
26. Wuenschell, G. E.; Tetreau, C.; Lavalette, D.; Reed, C. A. *J. Am. Chem. Soc.*, 1992, **112**, 3346–3354.
27. Collman, J. P.; Gagne, R. R.; Reed, C. A.; Robinson, W. T.; Rodley, C. A. *Proc. Natl Acad. Sci.*, 1974, **71**, 1326–1329.
28. Takano, T. *J. Mol. Biol.*, 1977, **110**, 569–584. (PDB: 3MBN, 5MBN)
29. Perutz, M. F., et. al., *Acc. Chem. Res.*, 1987, **20**, 309–321.
30. Jameson, G. B., et al., *J. Am. Chem. Soc.*, 1980, **102**, 3224–3237.
31. Shaanan, B. *J. Mol. Biol.*, 1983, **171**, 31–59. (PDB: 1HHO)
32. Jameson, G. B.; Rodley, G. A.; Robinson, W. T.; Gagne, R. R.; Reed, C. A.; Collman, J. P. *Inorg. Chem.*, 1978, **17**, 850–857.
33. Jameson, G. B.; Molinaro, F. S.; Ibers, J. A.; Collman, J. P.; Brauman, J. I.; Rose, E.; Suslick, K. S. *J. Am. Chem. Soc.*, 1978, **100**, 6769–6770.
34. Hoffman, B. M.; Petering, D. H. *Proc. Natl. Acad. Sci. USA*, 1970, **67**, 637–643.
35. Basolo, F.; Hoffman, B. M.; Ibers, J. A. *Acc. Chem. Res.*, 1975, **8**, 384–392 and references therein.
36. Smith, T. D.; Pilbrow, J. R. *Coord. Chem. Rev.*, 1981, **39**, 295–383.
37. Almog, J.; Baldwin, J. E.; Dyer, R. D.; Peters, M. *J. Am. Chem. Soc.*, 1975, **97**, 226–228.
38. Collman, J. P.; Reed, C. A.; *J. Am. Chem. Soc.*, 1973, **95**, 2048–2049.
39. Collman, J. P.; Brauman, J. I.; Iverson, B. L.; Sessler J. L.; Morris, R. M.; Gibson, Q. H. *J. Am. Chem. Soc.*, 1983, **105**, 3038, 3052.
40. Tsuchida, E. (ed.). *Artificial Red Cells*, John Wiley & Sons, Chichester, 1994.
41. Kuriyan, J.; Wilz, S.; Karplus, M.; Petsko, G. A. *J. Mol. Biol.*, 1986, **192**, 133. (PDB: 1MBC)
42. (a) Braunstein, D., et al., *Proc. Natl. Acad. Sci. USA*, 1988, **85**, 8497–8501. (b) Ormos, P., et al. *Proc. Natl. Acad. Sci. USA*, 1988, **85**, 8492–8496. (c) Moore, J. N.; Hansen, P. A.; Hochstrasser, R. M. *Proc. Natl. Acad. Sci. USA*, 1988, **85**, 5062–5066.
43. Powers, L., et al. *Biochemistry*, 1984, **23**, 5519–5523.
44. Phillips, G. N.; Olson, J. S. *J. Inorg. Biol. Chem.*, 1997, **2**, 544–552.
45. Rodgers, K. R.; Spiro, T. G. *Science*, 1994, **265**, 1697–1699.
46. Srajer, V.; Teng, T.-y.; Ursby, T.; Pradervand, D.; Ren, Z.; Adachi, S.-i.; Schildkamp, W.; Bourgeois, D.; Wulff, M.; Moffat, K. *Science*, 1996, **274**, 1726–1729.
47. Borman, S. *Chem. Eng. News*, 1999, **Dec. 6**, 31–36.

5

COPPER ENZYMES

5.1 INTRODUCTION, OCCURRENCE, STRUCTURE, FUNCTION

A diverse variety of copper-containing metalloenzymes occur in plants and animals. They are utilized for electron transfer (azurin, plastocyanin, laccase), for oxygenation reactions (tyrosinase, ascorbate oxidase), and for oxygen transport (hemocyanin). Most biologically active copper centers are found in proteins outside cells or in vesicles. The enzyme superoxide dismutase is found in the cytosol as well as within cells.[1] Copper ions normally are ligated by nitrogen ligands or by a combination of nitrogen and sulfur donors. Copper-containing metalloproteins and enzymes may contain Cu(I) d^{10}, Cu(II) d^9, and Cu(III) d^8 ions. In aqueous solution, copper ions disproportionate according to reactions 5.1–5.3:

$$\mathrm{Cu^{+}_{(aq)}} + e^- \rightarrow \mathrm{Cu^0} \qquad E^0 = +0.52\,\mathrm{V} \tag{5.1}$$

$$\mathrm{Cu^{2+}_{(aq)}} + e^- \rightarrow \mathrm{Cu^{+}_{(aq)}} \qquad E^0 = +0.153\,\mathrm{V} \tag{5.2}$$

$$2\mathrm{Cu^{+}_{(aq)}} \Leftrightarrow \mathrm{Cu^{0}_{(s)}} + \mathrm{Cu^{2+}_{(aq)}} \qquad E^0 = +0.37\,\mathrm{V} \tag{5.3}$$

Using equation 5.4 in combination with equations 5.1–5.3, one finds[2]

$$\Delta G^0 = -nFE \quad \text{and} \quad \Delta G^0 = -RT\ln K \tag{5.4}$$

$$K = \frac{[\mathrm{Cu^{2+}}][\mathrm{Cu}]}{[\mathrm{Cu^{+}}]^2} = 10^6 \tag{5.5}$$

Numerical values for the equations above may change dramatically as copper ions contained in enzymes are surrounded not by water but by a variety of biological ligands. These ligands alter the copper ions' electromotive force, making the system more or less easily oxidized or reduced. The soft Cu(I) ion prefers sulfur ligands or unsaturated ligands such as *o*-phenanthroline and 2,2′-bipyridine that have large polarizable electron clouds. The relatively hard Cu(II) ion prefers the harder nitrogen ligand. Usually, copper enzymes involved in redox reactions feature both types of ligands, so that the metal centers may readily exist in either oxidation state. Outside of biological species, copper(I) d^{10} ions are often linear and two-coordinate; however, three-coordinate trigonal planar or pyramidal as well as four-coordinate tetrahedral geometries are also found. Copper(II) d^9 ions prefer tetragonally distorted octahedral complexes that exhibit strong Jahn–Teller effects. Cu(III) d^8 ions prefer square-planar geometry. As will be seen below, copper metal centers in enzymes usually exist in distorted and changeable geometries.

The Franck–Condon principle states that there must be no movement of nuclei during an electronic transition; therefore, the geometry of the species before and after electron transfer must be unchanged. Consequently, the active site geometry of a redox metalloenzyme must approach that of the appropriate transition state for the electronic transfer. Every known copper enzyme has multiple possible copper oxidation states at its active site, and these are necessary for the enzyme's function.

Copper enzymes have been classified into three types having distinctive geometries and ligand environments surrounding the metal center. The three types have distinct and identifiable properties detectable through analytical and instrumental methods. Type I copper enzymes, also called blue copper proteins, sometimes have a distorted tetrahedral center as depicted in Figure 5.1, especially when the enzyme contains copper(II) ions. In azurin, however, copper(II) adopts a trigonal bipyramidal geometry, whereas azurin and plastocyanin copper(I) metal centers adopt trigonal geometry. For Type II and Type III copper centers in Figure 5.1, dotted lines indicate ligands that may or may not form bonds to the metal center, depending on the metalloprotein under discussion and the copper oxidation state. See Figure 5.1 and Table 5.3.

Type I copper enzymes are called blue proteins because of their intense absorbance ($\varepsilon \approx 3000\,M^{-1}\,cm^{-1}$) in the electronic absorption spectrum around

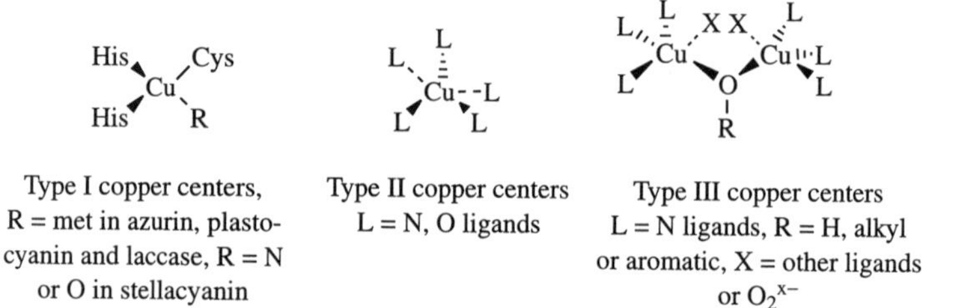

Figure 5.1 Type I, II, and III copper center geometries. Dotted lines indicate possible ligands. (Adapted from Figure 5.4 of Cowan, J. A. *Inorganic Biochemistry, An Introduction*, 2nd ed., Wiley-VCH, New York, 1997. Copyright 1997, Wiley-VCH.)

600 to 620 nm. This absorbance is assigned to a cys (S) $\rightarrow$ Cu^{2+} LMCT (ligand $\rightarrow$ metal charge transfer) band. Type I centers serve as electron-transfer sites in mobile electron-carrier proteins such as azurin and plastocyanin.[3] Their electrode potentials are tunable over a large range as shown in Table 5.2, and they also exhibit a distinctive EPR signal with small hyperfine coupling constants ($<90 \times 10^4\,cm^{-1}$). See Section 3.4.2 for a description of hyperfine coupling constants. The hyperfine coupling constants arise from interactions between Cu(II)'s unpaired *d* electron and the spin on the copper nucleus ($I = 3/2$). Four hyperfine coupling resonances should be detected for this combination (multiplicity $= 2I + 1$). Biologically, Type I copper ions prefer a combination of nitrogen (histidine) and sulfur (cysteine) ligands. The structures of azurin and plastocyanin are more fully described later in this chapter.

Type II copper enzymes generally have more positive reduction potentials, weaker electronic absorption signals, and larger EPR hyperfine coupling constants. They adopt trigonal, square-planar, five-coordinate, or tetragonally distorted octahedral geometries. Usually, type II copper enzymes are involved in catalytic oxidations of substrate molecules and may be found in combination with both Type I and Type III copper centers. Laccase and ascorbate oxidase are typical examples. Information on these enzymes is found in Tables 5.1, 5.2, and 5.3. Superoxide dismutase, discussed in more detail below, contains a lone Type II copper center in each of two subunits of its quaternary structure.

Type III copper enzymes contain binuclear Cu(II) d^9 sites that exhibit strong antiferromagnetic coupling, making them EPR silent. In coordinating dioxygen the two available Type III copper(II) ions can provide one electron each, reducing O_2 to the peroxo ligand and avoiding formation of the unwanted superoxide ion. Frequently, Type III centers are found in conjunction with Type I and II copper centers. In ascorbate oxidase the Type III center is closely associated with a Type II center, so that the three copper ions act virtually as a unit. Type III centers constitute the O_2-reducing site in many oxidases. Usually, Type III copper(I) centers in deoxyhemocyanin are coordinated to three trigonally arranged histidine ligands. When dioxygen is attached to a Type III center (in hemocyanin for instance), the copper center is oxidized to Cu(II) while the dioxygen adopts a side-on η^2:η^2-coordinated peroxo conformation. Simultaneously, each copper ion becomes five-coordinate in distorted trigonal bipyramidal or square pyramidal geometry. More detail concerning hemocyanin is found in the section discussing specific metalloproteins.

Several copper enzymes will be discussed in detail in subsequent sections of this chapter. Information about major classes of copper enzymes, most of which will not be discussed, is collected in Table 5.1 as adapted from Chapter 14 of reference 49. Table 1 of reference 4 describes additional copper proteins such as the blue copper electron transfer proteins stellacyanin, amicyanin, auracyanin, rusticyanin, and so on. Nitrite reductase contains both "normal" and "blue" copper enzymes and facilitates the important biological reaction $NO_2^- \rightarrow NO$. Solomon's *Chemical Reviews* article[4] contains extensive information on ligand field theory in relation to ground-state electronic properties of copper complexes and the application of

Table 5.1 Some Major Classes, Sources, and Functions of Copper Proteins

Protein Family	Name	Source	Biological Function
Oxygen transport (binuclear)	Hemocyanin	Mollusks and arthropods	Dioxygen transport
Copper oxygenases (binuclear)	Tyrosinase	Fungal, mammal	Tyrosine oxidation $O_2 + 2H^+ +$ monophenol $\rightarrow$ *o*-diphenol $+ H_2O$ $O_2 + 2$ *o*-diphenol $\rightarrow$ 2 *o*-quinone $+ 2H_2O$
Copper oxygenases	Dopamine β-hydroxylase	Adrenal gland, brain	Dopamine $\rightarrow$ norepinephrine Dopamine + ascorbate + $O_2 \rightarrow$ noradrenaline + dehydroascorbate + H_2O
	Peptidylglycine α-amidating monooxygenase (PAM)	Pituitary, heart	Oxidative *N*-dealkylation
	Methane monooxygenase (MMO)	Methanogenic bacteria	Methane $\rightarrow$ methanol
Copper dioxygenase	Quercetinase	Fungal	Quercetin oxidative cleavage

Copper oxidases			
"Blue" oxidases (multicopper oxidases)	Laccase	Tree, fungal	Phenol and diamine oxidation
	Ascorbate oxidase	Plants	Oxidation of L-ascorbate
	Ceruloplasmin	Human, animal serum	Weak oxidase activity
"Nonblue" oxidases	Amine oxidase	Most animals	Elastin, collagen formation
	Diamine oxidases		$R'CHNR_2 + O_2 + H_2O \rightarrow R'CHO + HNR_2$
	Galactose oxidase	Molds	Galactose oxidation $RCH_2OH + O_2 \rightarrow RCHO + H_2O_2$
	Cytochrome c oxidase	Mitochondria	Terminal oxidase $2\,O_2 + 4$ ferrocytochromec c $+ 4H^+ \rightarrow 2H_2O + 4$ ferricytochrome c
"Normal" copper enzymes	Phenoxazinone synthase	Streptomyces	Phenoxazinone formation
	Superoxide dismutase	Red blood cells	Superoxide ion detoxification $2O_2^- + 2H^+ \rightarrow H_2O_2 + O_2$

Source: Excerpted from reference 49, Table 1 and from reference 4.

Table 5.2 Spectral Properties of Some Copper Enzymes

Name of Enzyme	Type/M_r^a (daltons, Da)	Function	Absorption Bands, nm (ε, mM^{-1} cm^{-1})	Reduction Potential (mV)	Details of EPR spectrum (A × 10^{-4} cm^{-1})
Azurin	I (blue)/14 × 10^3	Electron carrier protein	625 (3.5)	330	Axial: $g_\perp = 2.052$, $g_{\parallel} = 2.29$ ($A_{\parallel} = 60$)
Plastocyanin	I (blue)/10.5 × 10^3	Electron carrier protein	597 (4.9)	370	Axial: $g_\perp = 2.053$, $g_{\parallel} = 2.26$ ($A_{\parallel} = 50$) Rhombic: $g_x = 2.042$, $g_y = 2.059$, $g_z = 2.226$ ($A_{\parallel} = 63$)
Laccase	I (blue)/60–140 × 10^3	See Table 5.1	607 (9.7)	785	Rhombic: $g_x = 2.030$, $g_y = 2.055$, $g_z = 2.300$ ($A_{\parallel} = 43$)
Ascorbate oxidase	I/145 × 10^3	See Table 5.1	614 (5.2)		Rhombic: $g_x = 2.036$, $g_y = 2.058$, $g_z = 2.227$ ($A_{\parallel} = 45.8$)
Laccase	II/60–90 × 10^3	See Table 5.1	788 (0.9)	782 (*P. versicolor*), 434 (*R. vernicifera*)	Axial: $g_\perp = 2.053$, $g_{\parallel} = 2.237$ ($A_{\parallel} = 200.6$)
Ascorbate oxidase	II/145 × 10^3	See Table 5.1	760 (3.6)		Axial: $g_\perp = 2.053$, $g_{\parallel} = 2.242$ ($A_{\parallel} = 199$)
Galactose oxidase	II/68 × 10^3	See Table 5.1		410	
Laccase	III/60–90 × 10^3	See Table 5.1	330(2.7)	570 (*P. versicolor*), 390 (*R. vernicifera*)[b]	EPR silent
Ascorbate oxidase	III/145 × 10^3	See Table 5.1	330 (2.0)		EPR silent
Tyrosinase	III/46 × 10^3	See Table 5.1	345 (20.0) 590 (1.0) 520 (CD band)	370[b]	EPR silent
Hemocyanin	III/4 × 10^5 to 9 × 10^6, subunits from 75 × 10^3	See Table 5.1	350 (20.0) 570 (1.0) 480 (CD band)	>800[b]	EPR silent

[a]Molecular weight determined by gel electrophoresis experiments.
[b]E^0 data for type III sites is subject to error.
Source: References 1 and 3–5.

these to biological copper ions. Also, electronic structure and molecular orbital theory for blue copper proteins in relation to electron transfer pathways are discussed. Excited-state spectral features are related to charge-transfer electronic transitions. Highest occupied molecular orbital (HOMO) and lowest unoccupied molecular orbital (LUMO) orbitals for end-on *cis*-μ-1,2 versus side-on μ-η_2:η_2 models of oxyhemocyanin are presented. Spectral properties and their relation to geometry for trinuclear copper cluster sites, such as found in laccase, are described as well.

Table 5.2 contains data about selected copper enzymes from the references noted. It should be understood that enzymes from different sources—that is, azurin from *Alcaligenes denitrificans* versus *Pseudomonas aeruginosa*, fungal versus tree laccase, or arthropodan versus molluscan hemocyanin—will differ from each other to various degrees. Azurins have similar tertiary structures—in contrast to arthropodan and molluscan hemocyanins, whose tertiary and quaternary structures show large deviations. Most copper enzymes contain one type of copper center, but laccase, ascorbate oxidase, and ceruloplasmin contain Type I, Type II, and Type III centers. For a more complete and specific listing of copper enzyme properties, see, for instance, the review article by Solomon et al.[4]

Typical ligands, geometries of the metal center, ligand–metal distances, and reactions catalyzed by Type I, II, and III copper enzymes are shown for several copper metalloenzymes in Table 5.3. Note that metal center geometries and ligand–metal bond lengths vary considerably based on many factors. These include, among others, metal ion oxidation state or pH and biological species from which the enzyme is extracted. For example, the copper(II) state of azurin adopts a distorted trigonal bipyramidal geometry, whereas the copper(I) state exists within a distorted trigonal planar geometry. In plastocyanin, both oxidation states adopt distorted tetrahedral geometry with long copper methionine bond-length interactions. *Limulus polyphemus* arthropod hemocyanin subunits studied by Magnus et al. (reference 37) differ substantially from those of *Octopus* hemocyanin studied by Hendrickson and co-workers.[11]

5.2 DISCUSSION OF SPECIFIC ENZYMES

5.2.1 Azurin

Azurin is a small, 128-residue, Type I copper-containing metalloenzyme involved in electron transfer in photosynthetic pathways and respiratory systems of biological organisms. Its physiological redox partners are cytochrome c_{551} and nitrite reductase. As a blue copper protein, it exhibits a large extinction coefficient in the visible spectrum around 600 nm and a small $A_{||}$ value in its EPR spectrum. Characteristics of the protein have been listed in Tables 5.1, 5.2, and 5.3. To facilitate electron transfer and cycling between Cu(II) and Cu(I), the copper coordination environment exists as a compromise between those preferred by the two oxidation states. Therefore, one would expect geometries intermediate between that expected for copper(I) (trigonal) and copper(II) (trigonal bipyramidal).

Table 5.3 Geometry of, and Reactions Catalyzed by, Some Copper Enzymes from Various Sources

Enzyme/# of Cu centers, Type, Ligands (M–L Bond Length, Å)	Oxidation State/Details of Geometry	Function or Reaction Catalyzed
Azurin (2AZA)[a] *Alicaligenes denitrificans* /1, Type I, Cu^{2+} his46 (2.08), his117(2.00), cys112(2.15), met121(3.11), gly45-O(3.13) Cu^{+} his46 (2.13), his117(2.05), cys112(2.26), met121(3.23), gly45-O(3.22)[7,13]	Cu^{2+}/tbp[b] cys, 2 his in trigonal plane, met and gly backbone amide O in axial positions Cu^{+}/trigonal cys, 2 his[7]	$Az_{(ox)} + e^- \rightleftharpoons Az_{(red)}$
Plastocyanin (7PCY)[a] *Enteromorpha prolifera*/1, Type I, Cu^{2+} his37(2.04), his87(2.10), cys84(2.13), met92(2.90) Cu^{+} @ pH 7.0/ his37(2.13), his87(2.39), cys84(2.17), met92(2.87)[5,16,17]	Cu^{2+}/Td[b] / 2 his, cys, met Cu^{+}/Td[b] 2 his, cys, met. At pH 3.8, Cu–his87 bond distance = > 4 Å[5]	$Pc_{(ox)} + e^- \rightleftharpoons Pc_{(red)}$ Couples cytochrome $b_{6/f}$ complex to P700 of photosystem I (PSI)
Laccase(1A65)[a] /1 Type I; 1 Type II; 2 Type III Type II Cu_1^{2+} his457, his396, OH^- or H_2O; Type III Cu_2^{2+} his451, his111, his401, Cu_3^{2+} his453, his109, his66 μ-OH or μ-OR[7,8]	Cu^{2+}Type I/Td[b] 2 his, cys, met Type II/trigonal 2 his, OH^- or H_2O 2 Type III/ each Cu trigonal with 3 His connected by OH^- or OR^-	$2\ RH_2 + O_2 \rightarrow 2\ R + 2\ H_2O$ (*p*-diphenols + O_2 → *p*-quinones + H_2O)
Ascorbate oxidase(1AOZ)[a] Distances given for subunit A 2 Type I his445 (2.10), his512(2.05), cys507(2.13), met517(2.90) 2 Type II Cu_1 his60(2.00), his448(2.09), OH(2.02) 2 Type III Cu_2 his62(1.98), his104(2.19), his508(2.14), μ-OH (2.02) Cu_3 his106(2.16), his450(2.06), his506(2.07), μ-OH(2.06), Cu_1–Cu_2 (3.9), Cu_1-Cu_3 (4.0), Cu_2–Cu_3 (3.4)[9]	Cu^{2+} Type I/dist. Td[b] 2 his, cys, met Cu^{2+} Type II/ trigonal/; 2 his OH^- Cu^{2+} Type III[9], each Cu trigonal with 3 his connected by OH^-	L-Ascorbate + O_2 → dehydroascorbate + H_2O

Galactose oxidase (1GOF)[a][10]		Oxidation of primary alcohols to aldehydes in sugars with reduction of O_2 to H_2O
Tyrosinase (mixed function oxidase, monooxygenase)		$O_2 + 2\ H^+ +$ monophenol $\rightarrow$ *o*-diphenol $+ H_2O$ $O_2 + 2\ H^+ + 2$ *o*-diphenol $\rightarrow$ 2 *o*-quinone $+ 2\ H_2O$
Hemocyanin *Limulus* deoxyHc Cu_A^+ his173 (2.1), his177 (2.0); his204 (1.9) Cu_B^+ his324 (2.2), his328 (2.1); his364 (1.9) Cu–his N^ε coordinated, Cu–Cu (4.6)[37]	deoxy Cu^+ 2 Type III/ each Cu trigonal with 3 his oxy Cu^{2+} 2 Type III/each Cu distorted square pyramid with 3 his and side-on $\eta^2{:}\eta^2\ O_2^{2-}$	Transport of O_2 $Hc + O_2 \rightleftharpoons Hc \cdot O_2$
Limulus (1OXY)[a] oxyHc Cu_A^{2+} his173 (2.2), his177 (2.1); his204 (2.4) Cu_B^{2+} his324 (1.9), his328 (2.4); his364 (2.1) Cu–his N^ε coordinated, Cu–Cu (3.6)[37] Octopus (1JS8)[a] oxy Cu_A^{2+} his2543, his2562, his2571; Cu_B^{2+} his2671, his2675, his2702[11]		

[a] PDB (Protein Data Bank) code.
[b] tbp = trigonal bipyramid; Td = tetrahedral.

Source: References 1–3 and 5–7.

X-ray crystallographic analyses first presented in the 1980s at 2.7-Å resolution gave coordinates for residues 3–128 and the copper atom.[12] These data are found in the Protein Data Bank (PDB) as 1AZU (reference 12b). Figure 2.10 in Section 2.2.2 shows a visualization of azurin's tertiary structure from PDB data deposited as 1JOI. Later studies of azurin mutants and the enzyme at various pH values yield data given as (a) 2AZU (a H35L mutant) and 3AZU (a H35Q mutant) (reference 15a) and (b) 4AZU and 5AZU at pH values of 5.5 and 9.0, respectively (reference 15b). The protein folding pattern, similar to those seen in plastocyanin and superoxide dismutase, is a common tertiary motif and is described as an eight-stranded beta (β) barrel. One segment (residues 54–80) lies outside the eight β-strands and contains an α-helical region (residues 55–67). A disulfide bridge between cys3 and cys26 connects the A β-strand with the loop connecting β-strands B and C. An extensive network of hydrogen bonds facilitates the tertiary protein structure. (See Figures 1.1a–f on pages 6 and 7 of reference 6 and Figure 3 of reference 15a.) Baker and co-workers have shown by X-ray crystallographic analysis at 1.9-Å resolution that the copper site in azurin undergoes minimal structural change on reduction.[13] The authors studied the native protein from *Alcaligenes denitrificans* by collecting data on both the oxidized and reduced protein. Data were collected on the reduced crystal only until a blue color, indicating oxidation, began to reappear. In the structure, three strongly bound ligands, the thiolate sulfur of cys112 and the imidazole nitrogens of his46 and his117, occur in a distorted trigonal arrangement with bond distances of 2.00–2.15 Å. The ligands constitute a compromise between the soft cysteine thiolate ligand preferred by Cu(I) and the harder histidine imidazole ligands preferred by Cu(II). The thioether of met121 and the peptide carbonyl oxygen of gly45 form longer axial bonds of ~3.1 Å. In the copper(I) enzyme each bond distance increases by 0.05 to 0.1 Å, but otherwise the coordination sphere remains quite similar. A slight increase in the radius of the copper(I) coordination site would be expected compared with the copper(II). Minimal change in bond angles were also found for the oxidized and reduced forms, leading the authors to conclude that the two structures meet the criteria for fast electron transfer; that is, a low reorganization energy is required.

Both azurin and plastocyanin have been extensively studied as models for long-distance intramolecular electron transfer (et) reactions in biological systems. In these proteins the copper center, enclosed inside the protein matrix about 7 Å from the surface, acts as the ultimate electron acceptor or donor in electron transfer. Gray and co-workers have positioned inorganic redox partners such as ruthenium compounds at an azurin surface histidine residue near the copper site to facilitate study of electron transfer.[14] His35 and a surrounding patch of residues have been implicated in electron transfer because of the pH dependence of the rate of electron transfer. At pH values above 8.0, both N^{ε} and N^{δ} of his35 would be deprontonated, allowing hydrogen bonding through water molecules to protein backbone atoms oxygen (met44) and nitrogen (gly37). The resulting structural changes would effect electron transfer kinetics as has been found experimentally. The surface structure of the proteins exhibits a negatively charged region and a hydrophobic patch around the copper ligand his117, which is positioned at the protein surface.[15] (See Figure 10 of reference 15a.) These surface structural details facilitate close approach of another protein such as cytochrome c_{551} or nitrite reductase involved in electron transfer to, from, or within azurin. The kinetics of electron transfer in this system is complex and remains under study; however, the high speed of electron transfer is underscored by the large values for k_{12} and k_{21} ($6 \times 10^6\ \mathrm{M^{-1}\ s^{-1}}$) in equation 5.6.

$$\mathrm{Az(II)} + \mathrm{Cyt(II)} \xleftrightarrow{k_{12}, -k_{21}} \mathrm{Az(I)} + \mathrm{Cyt(III)} \tag{5.6}$$

Curiously, solution structures of azurin studied by NMR are not listed in the protein data bank as of 2001, although many NMR structures of plastocyanin are available as will be discussed below.

5.2.2 Plastocyanin

The Type I copper enzyme plastocyanin is an essential electron carrier between photosystems II and I in all higher plants and some algae. More specifically, it is characterized as a mobile electron carrier from membrane-bound cytochrome $b_{6/f}$ complex to the primary donor P700 of photosystem I. It is similar to azurin in its tertiary structure but is a smaller protein (97–105 amino acid residues, depending on the plastocyanin's source). Plastocyanins have slightly higher reduction potentials when compared to azurin metalloenzymes. Plastocyanin, unlike azurin, does not contain a disulfide bond. The copper coordination sphere is distorted tetrahedral with ligands his37 and his87 and with sulfur atoms from cys84 and met92. Differing oxidation states (Cu^{2+} versus Cu^{+}) and pH variation cause coordination sphere changes as noted below. The Cu site lies about 6 Å below the protein surface and is not exposed to the surroundings. It is surrounded by eight β-strands folded up in a barrel with a hydrophobic core. X-ray crystallographic structures indicate that plastocyanin possesses two potential binding sites for physiological redox partners. These are termed the north site (a hydrophobic patch 6 Å from the Cu center) and east site (the acidic patch about 15 Å from the copper site).[16] The hydrophobic site forms a pocket in which the copper ion lies. The acidic patch consists of a surface concentration of negatively charged residues believed to be important in binding other electron transfer proteins, cofactors, or substrates having positively charged characteristics.[19] The overall dimensions of the folded protein are approximately 40 Å × 32 Å × 28 Å.[17] The bond distances given in Table 5.4 and bond angles in Table 5.5 are taken from references 17 and 18.

A solution structure of French Bean plastocyanin has been reported by Wright and co-workers,[19] using nuclear magnetic resonance techniques described in Section 3.5 of Chapter 3. The structure, determined from a plastocyanin molecule in solution rather than in a solid-state crystal, agrees well with that of reduced poplar plastocyanin X-ray crystallographic structure reported above. Conformations of protein side chains constituting the hydrophobic core of the French bean plastocyanin are well-defined by the NMR technique. Surface side chains show

Table 5.4 Bond Distances (Å) in Plastocyanins

Plastocyanin	Cu(II) *E. prolifera*	Cu(II) Poplar	Cu(I) Poplar pH = 7.0	Cu(I) Poplar pH = 3.8
Cu–S(cys84)	2.12	2.13	2.17	2.13
Cu–S(met92)	2.92	2.90	2.87	2.51
Cu–N(his37)	1.89	2.04	2.13	2.12
Cu–N(his87)	2.17	2.10	2.39	>4

Source: References 17 and 18.

more disorder, both in the NMR and X-ray diffraction structures. The solution structure indicates that a distinctive acidic surface region on plastocyanin is also disordered. It is believed that this region is important in binding other electron transfer proteins or cofactors. Of a total of 60 hydrogen bonds formed in French bean plastocyanin, 56 are found in the X-ray structure while 50 hydrogen bonds are identified in the NMR solution structure.

In reference 16 the authors have studied the NMR solution structure of reduced parsley plastocyanin. Comparisons to higher plant plastocyanins indicate that parsley plastocyanin contains fewer acidic amino acid residues in its surface acidic patch and has deleted residues at positions 57 and 58. Overall, parsley plastocyanin is found to have more similarity to algal plastocyanins than those of higher plants. In agreement with the study of poplar plastocyanin described in reference 19, the authors found that his37 and his87 are ligated to copper through the histidine δ nitrogen with the copper and the imidazole rings being coplanar. The overall coordination geometry about the copper ion was found to be approximately tetrahedral with a Cu–S^{γ} bond to cys84 and a Cu–S^{δ} bond to met92 as found in other plastocyanins. Parsley plastocyanin was found to have 54 interstrand hydrogen bonds (NH–CO) and six hydrogen bonds in turns. The molecule is described as a β-sandwich, the two faces of which are β-sheets made up by a total of eight strands. Pairs of strands are connected by loops or tight turns, and the β-sheets are separated by a hydrophobic core. There is a single helical region composed of residues 52–56, and the region resembles a 3_{10} helix ($n = 3.0$, $p = 600\,\text{pm}$) rather

Table 5.5 Bond Angles (degrees) in Green Alga *Enteromorpha prolifera*[17] and Poplar Plastocyanin[18]

Plastocyanin	Poplar	*E. prolifera*
S(cys84)–Cu–S(met92)	108	108
N(his87)–Cu–S(met92)	103	102
N(his87)–Cu–S(cys84)	123	120
N(his37)–Cu–S(met92)	85	90
N(his37)–Cu–S(cys84)	132	125
N(his37)–Cu–N(his87)	97	104

than an α-helix ($n = 3.6$, $p = 540\,\text{pm}$). The regions of the surface acidic patch in parsley plastocyanin were found to be disordered. The disorder is associated with relatively high flexibility of the region, potentially important for plastocyanin's ability to achieve a correct orientation for electron transfer with physiological partner cytochrome *f* or other redox partners.

The NMR solution structure of the blue-green Alga *Anagaena variabilis* has been studied by Led, Ulstrup, and co-workers.[20] This plastocyanin differs from the others discussed in that electron transfer is dominated by the adjacent site (the hydrophobic pocket 6 Å from the copper center) rather than the remote site (the site of negatively charged amino acid residues). The authors believe the difference is caused in part by a much smaller number of negatively charged amino acid residues at the remote site in *Anagaena variabilis* plastocyanin because overall structural differences between *A. variabilis* plastocyanin and other examples cited are minimal.

5.2.3 Superoxide Dismutase

Superoxide dismutase enzymes are functional dimers of molecular weight (M_r) of approximately 32 kDa. The enzymes contain one copper ion and one zinc ion per subunit. Superoxide dismutase (SOD) metalloenzymes function to disproportionate the biologically harmful superoxide ion–radical according to the following reaction:

$$2O_2^- + 2H^+ \rightarrow H_2O_2 + O_2 \tag{5.7}$$

One product of this reaction, H_2O_2, is also a potentially harmful substance. Hydrogen peroxide is removed by the heme iron metalloenzyme catalase according to the following equation:

$$2H_2O_2 \rightarrow 2H_2O + O_2 \tag{5.8}$$

In prokaryotes, superoxide dismutases have been found to contain redox-active manganese (MnSOD) or iron (FeSOD) metal centers. The focus here will be on copper–zinc superoxide dismutase (CuZnSOD), found in eukaryotic and some bacterial species. This enzyme, predominantly located in mammalian liver, blood cells, and brain tissue, contains two identical subunits, each of which contains one copper and one zinc atom. It is thought that the primary role of CuZnSOD is to remove superoxide ion–radical O_2^- from the cytosol. In this capacity, superoxide dismutase acts as an antioxidant inhibiting aging and carcinogenesis. Excess superoxide concentrations can inactivate enzymes containing iron–sulfur clusters and can lead to the formation of highly oxidizing species such as hydroxide radical.[21] Mutations of SOD have been associated with amyotrophic lateral sclerosis (ALS)—Lou Gehrig's disease.[3] Section 7.4.5.1 in Chapter 7 contains a discussion of ALS.

Interestingly, the $Cu^{1+/2+}$–Zn^{2+} pair of superoxide dismutase are bridged by a histidinate anion (a deprotonated histidine side chain). To date, superoxide dismutase is the only known enzyme to exhibit such a bridging motif. Either copper or zinc may be removed from the metalloprotein, forming the apoprotein. Substitution of other metals at the active site, forming M_1M_2SOD, has allowed chemists to probe metal ion coordination details, structural arrangements of amino acid residues near the active site, and the mechanistic chemistry of the enzyme. The O_2^- disproportionation reaction described for bovine erythrocyte CuZnSOD is relatively independent of pH with reaction rates typically diffusion-controlled ($k = 3.9 \times 10^9\ M^{-1}\ s^{-1}$). Overall, kinetics are dominated by diffusion of the superoxide radical into the active-site channel, guided by electrostatic charge distribution around the active site.[22]

The X-ray crystallographic structure of a Cu(II)–Zn superoxide dismutase was first presented in 1982 for bovine (ox) erythrocyte CuZnSOD.[23] Structural data are collected in the Protein Data Bank (PDB) under the code 2SOD. In this structure, two identical subunits of the protein are held together by hydrophobic interactions. Each subunit is organized as a flat cylindrical barrel of β-pleated sheet made up of eight antiparallel chains connected by irregular external loops. This common motif in protein structures is known as the β-barrel or the Greek key β-barrel fold. As discussed earlier in this chapter, this motif occurs in azurins and plastocyanins. The metal binding region binds the copper(II) and zinc(II) ions in close proximity. An imidazolate histidine side-chain bridges between the two metal centers, holding them 6.3 Å apart. Hydrogen bonding, especially between arg141 and the backbone carbonyl of cys55, in addition to a disulfide bond between cys55 and cys144, helps hold together the tertiary structure of the protein. The copper(II) ion is coordinated to the N^{ε} ligand atoms of his46 and his118, the N^{δ} ligand atom of his44, and bridges to the Zn(II) ion using the N^{ε} atom of his61. His44 and his46 are in trans position to each other in the distorted square-planar CuN_4 coordination sphere, and the tripeptide his44–val45–his46 completely blocks access to the copper ion from one side of the CuN_4 plane. The other side is accessible to solvent with a conical channel approximately 4 Å wide at the copper ion and opening to solution at the protein surface.[5] Additionally, the copper(II) ion coordinates a water molecule as its fifth ligand, yielding a distorted square-pyramidal geometry. The zinc ion's geometry is distorted tetrahedral with ligand atoms N^{δ} of his78 and his69, the N^{δ} ligand atom of the bridged his61 side chain, and the carboxyl oxygen (O^{δ}) of asp81, attached in the manner shown in Figure 5.2.

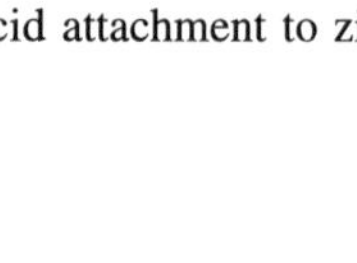

```
             O
             ||
H2N-CH-C-OH
        |
       CH2
        |
       C=O
        |
       O-Zn
```

Figure 5.2 Mode of aspartic acid attachment to zinc ion in superoxide dismutase.

The copper(II) ion of CuZnSOD can bind a variety of ligands such as CN^-, N_3^-, or halide ions in addition to the superoxide (O_2^-) substrate. Positive side chains of arg141, lys120, and lys134 are located 5, 12, and 13 Å away from the Cu(II) ion, and these are believed to play an important role in guiding anionic substrates to the active site and in hydrogen bonding with the substrates once they are in position.[24] The ε-amino group of the lys134 residue is at the far end of a channel lined with positively charged amino acid residues, presumably to attract superoxide ion and direct it down the channel to the positively charged $Cu^{1+/2+}/Zn^{2+}$ site.

The role of the zinc(II) ion in CuZnSOD is more ambiguous, and the ion may be multifunctional. One role would be stabilization of the enzyme's tertiary structure, and another possible role is to confer thermal stability to the protein. CuZn–SODs are remarkably stable to heat, with their spectral properties and activity being unchanged after 7 minutes of incubation in 10 mM pH 7.2 phosphate buffer at 75°C. Zinc's role in a possible bridge-breaking mechanism catalyzing the removal of superoxide ion has been proposed by the scheme shown as Figure 11.23 of reference 5. In this scheme the copper–histidine–zinc imidazolate bridge is broken and reformed during each catalytic cycle. Coordination of a histidine ring nitrogen to zinc(II) lowers the pK_a of the other nitrogen atom from 14 to approximately 7 to 10. At these pH values, the bridging histidine's second nitrogen atom will prefer binding to a proton rather than the Cu(I), causing the bridge to break. During reoxidation of Cu(I) to Cu(II) the bridge reforms. Concurrently, the proton formerly attached to the histidine ring nitrogen and another proton combine with a superoxide ion to produce hydrogen peroxide. Thus the reaction shown in equation 5.7 is completed. It is known, however, that removal of zinc from the enzyme does not destroy its superoxide dismutase activity, making this mechanism somewhat controversial. More discussion follows below, and a more detailed scheme illustrating superoxide's mechanism of activity is shown in Figure 5.4.

The spectral properties of CuZnSOD and many derivatives have been studied and are reported in reference 21. In these derivatives, zinc ion may be removed, thereby forming (Cu_2E_2SOD); or they may be substituted by other metals, thereby forming (Cu_2Cu_2SOD), (Cu_2Co_2SOD), and (Cu_2Cd_2SOD). Copper ion may also be substituted in derivatives such as (Zn_2Zn_2SOD), (Co_2Zn_2SOD), or (Ag_2Zn_2SOD). All derivatives having the zinc ion substituted in the native enzyme retain a similar visible absorption spectrum indicating that the copper ion environment has not changed. The visible absorption at 680 nm is indicative of a *d–d* transition of the d^9 Cu(II) ions coordinated to nitrogen ligands, and the fairly high extinction coefficient indicates a distorted geometry agreeing with X-ray crystallographic results. When Cu(II) is substituted by Ag(I) and copper is introduced to the native zinc site (Ag_2Cu_2SOD), the visible region absorbance red shifts, indicating a tetrahedral or pentagonal geometry for the copper ion. Zn_2Zn_2SOD should have no absorption bands in the visible region because Zn(II) is a d^{10} ion having no *d–d* transitions between 400 and 800 nm.

Electron paramagnetic resonance (EPR) spectra are also discussed by Valentine et al. in reference 21. The splitting of the $g_{||}$ resonance is due to hyperfine coupling between the unpaired electron on Cu(II) and the nuclear spin of the copper nucleus

($I = 3/2$). The $A_{||}$ value of 130 G ($A \times 10^{-4}\ \mathrm{cm}^{-1}$) is intermediate between larger $A_{||}$ values observed for a tetragonal Cu(II)N_4 center and the lower one observed for type I blue copper proteins (see Tables 5.2 and 5.3 and Figure 3.14). The lowering of the $A_{||}$ value indicates distortion away from a tetragonal geometry again as indicated by X-ray crystallography. The large linewidth observed in the $g_{\perp}$ region indicates that the copper is in a rhombic—that is, distorted environment. EPR spectra of a Cu_2E_2SOD shows a narrower linewidth for $g_{\perp}$ and a larger $A_{||}$ indicating that zinc's removal does change the copper ion's environment into a less distorted tetragonal arrangement.

More recent studies of superoxide dismutase systems have provided more detail about the enzyme's active site and its mechanism of dismutation. Valentine and co-workers have summarized the X-ray crystallographic studies conducted up to 1999.[25] One of these (PDB code 1XSO) determines the X-ray crystallographic structure at 1.5-Å resolution of CuZnSOD from the frog species *Xenopus laevis*.[26] The coordination sphere described is similar to that of Tainer in reference 23. In addition, these authors describe the bridging ligand his61 as defining a plane that approximately contains both metal ions—these ions being about 6 Å apart. The zinc coordination geometry is described as distorted tetrahedral and the copper ion is described as trigonal bipyramidal with the N^{δ} ligand atom of his44 and the N^{ε} ligand atom of his46 plus a water molecule as equatorial ligands. The N^{ε} ligand atom his118 and the N^{ε} atom of his61 bridging to Zn(II) would occupy the axial ligand sites in this description. Important water molecules in the active-site channel form a hydrogen-bonded network starting from the copper-ligating water molecule. This network extends in both directions in both subunits A and B of the enzyme. Water molecules hydrogen-bond with the peptide N atom of his61 as well as to carbonyl O atoms of electrostatically important resides. Thus important tertiary structural contributions are regulated not only by intraprotein interactions but also by solvent–protein interactions.

In reference 25 a reaction cycle is proposed involving inner-sphere electron transfer from superoxide to Cu(II) in the first part of the catalytic cycle and outer-sphere electron transfer in the second part of the cycle. Three new structures of yeast CuZnSOD are presented that establish proposed structures within the cycle: (1) wild type under 15 atm of oxygen pressure (PDB code 1B4L), (2) wild type in the presence of azide ion (PDB code 1YAZ, see Figure 5.3 and Color Plate 7), and (3) a his48cys mutant (PDB code 1B4T). These X-ray structures are compared to the wild-type yeast Cu(I)ZnSOD (PDB code 2JCW), the reduced copper species, known to have a broken Cu–Zn imidazolate bridge (Cu–N^{ε} distance (his63) = 3.16 Å), and a trigonal planar copper(I) coordination site (PDB 1YSO and 1JCV).[27] Note that the numbering system for amino acid residues binding to both copper and zinc in yeast SOD has changed from reference 23a (bovine erythrocyte SOD) in that each residue has a position number increased by two digits. Bond distances are collected in Table 5.6 and are further discussed below.

The authors find that the CuZnSOD protein backbone (the so-called "SOD rack") remains essentially unchanged in all structures investigated here and therefore assume that backbone changes do not play a role in the catalytic cycle.

Table 5.6 Bond Distances for CuZnSOD

	Bond Length, Å Wild-Type Yeast Cu(I) Ligands: 3 his (2JCW)[a] Cu Bridge Broken	Bond Length, Å Yeast 15 atm O_2 Ligands: 3 his (1B4L)[a] Cu Bridge Broken	Bond Length, Å Yeast-N_3^- Bound Ligands: 4 his, 1 N_3^- (1YAZ)[a] Cu Bridge Intact	Bond Length, Å Yeast, H48C Mutant Ligands: 3 his, 1 Cl^-, 1 H_2O (1B4T)[a] Cu Bridge Intact
Cu–his63 N^{ε}	3.16 (no bond)	2.73 (no bond)	2.04	2.09
Cu–his48 N^{ε}	2.06	2.24	2.76	No bond to cys48 replacing his48
Cu–his46 N^{ε}		2.06	2.05	1.99
Cu–his120 N^{ε}		2.10	2.11	2.12
Cu–N_3^-			2.15	
Cu–Cl^-				2.13
Cu–OH_2				Not given

[a]Protein Data Bank (PDB) code.

Source: References 25 and 27.

When the wild-type Cu(I)ZnSOD enzyme colorless crystal was kept under 15 atm of oxygen, several significant changes were noted. The crystal color became blue, seen as evidence for Cu(II) character. The copper ion remained in a trigonal planar arrangement with his46, his48, and his120 ligands at distances of 2.06, 2.24, and 2.00 Å, respectively. No electron density was noted between the copper ion and the N^{ε} of his63, now 2.73 Å from the copper ion (Cu–N^{ε} distance (his63) = 3.16 Å for the wild type). In the 15-atm oxygen crystal, the copper ion moved toward his63 and the copper–his48 distance became longer (2.24 Å versus 2.06 Å in the wild type). The authors believe that the 15-atm crystal exhibits a mixture of two states: (1) the imidizolate bridge broken and (2) the bridge-intact conformations.

To attempt to answer the question of whether the superoxide substrate first enters the Cu(II) or Cu(I) coordination sphere before its reduction or oxidation, the authors studied the azide ion bound CuZnSOD (PDB code 1YAZ). This molecule is visualized in Figure 5.3 using Wavefunction, Inc. Spartan '02 for Windows™. The protein chain is shown in brown ribbon form with the β-barrel portion of the molecule shown to the bottom right of the figure. One short α-helix appears above the copper ion. The copper ion (in blue/green space-fill form) is coordinated to his120 (in ball-and-spoke form) to the right, and his46 underneath. The copper ligand his48 is hidden behind the copper ion. Azide ion appears as three, blue

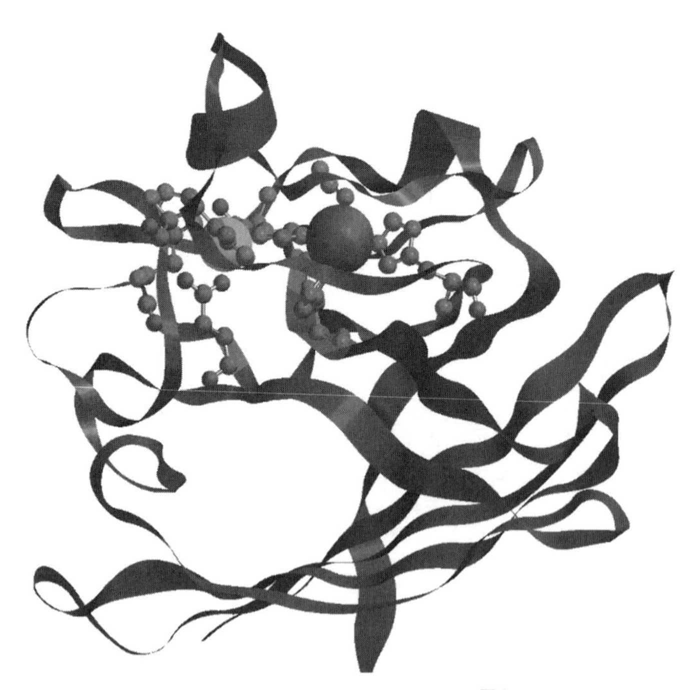

Figure 5.3 Wavefunction, Inc. Spartan '02 for Windows™ visualization of superoxide dismutase with azide ion from PDB data deposited as 1 YAZ from reference 25. See text for visualization details. Printed with permission of Wavefunction, Inc., Irvine, CA. (See color plate.)

ball-and-spoke form atoms coordinated to the copper ion. The his63 ligand coordinates to copper and bridges to the smaller green zinc ion seen to the left in space-fill form. The zinc ion's remaining three ligands are also shown in ball-and-spoke form—asp83 below zinc, his 71 in front and his80 at top left.

Analysis of this crystal found azide-bound copper ion in square-pyramidal geometry with the his63 bridge intact and his46, his63, his120, and azide acting as equatorial ligands. Bond distances were Cu–his46, Cu–his63, Cu–his120, and Cu–azide at 2.05, 2.04, 2.11, and 2.15 Å, respectively. The copper ion has moved 1.28 Å from its position in the Cu(I) wild-type structure, reforming the copper–imidazolate bridge. His48 meanwhile has become loosely associated with the copper ion in its axial position with Cu–N^{ε} distance (his48) = 2.76 Å. The azide ion is further bound in the active site through interaction with arg143 guanidinium nitrogen and an invariant ordered water molecule. Citing NMR evidence gathered by other researchers who found that azide ion does not directly bind to Cu(I) in the reduced enzyme,[28] the authors indicate that azide could potentially be in the same position in both Cu(II) and Cu(I) SOD, but in the Cu(I) moiety the azide interacts weakly with Cu(I) as the copper ion moves away from azide toward the trigonal plane. His120 appears to retain its position in all three moieties discussed, while the copper ion moves away from and toward formation of the his63 imidazolate bridge. It is also noted that the his63 ring tilts dramatically on going from the oxidized to reduced species with the his63 N^{ε} atom moving 0.66 Å during this process. To study the effect of changes undergone by his48 during the oxidation and reduction, the authors investigated an H48C CuZnSOD mutant (PDB code 1B4T). In the yeast wild-type structure, his48 is a tightly associated equatorial ligand (Cu–N^{ε} distance (his48) = 2.06 Å), whereas in the azide bound species it becomes a weakly associated axial ligand (Cu–N^{ε} distance (his48) = 2.76 Å). The authors found that the cys48 ligand was ignored by the copper in the H48C mutant, leaving the copper ion in square-pyramidal geometry with his46, his63, his120, and chloride in equatorial positions at distances of 1.99, 2.09, 2.12, and 2.13 Å, respectively, and a water molecule as the axial ligand.

The authors conclude that superoxide ion probably binds in a similar fashion to the azide and that conserved water ligands in the enzyme structure both hydrogen-bond with and help guide the substrates toward the copper ion. If this is the case, then superoxide binds directly to Cu(II) (inner-sphere electron transfer) in the following reaction:

$$O_2^- + \text{Cu(II)ZnSOD} \rightarrow \text{Cu(I)ZnSOD} + O_2 \qquad (5.9)$$

However, while remaining in the binding site, superoxide accepts its electron when the oxygen is more than 3 Å from the copper ion (outer-sphere electron transfer) in the following reaction:

$$O_2^- + \text{Cu(I)ZnSOD} + 2H^+ \rightarrow \text{Cu(II)ZnSOD} + H_2O_2 \qquad (5.10)$$

The authors summarize their proposed mechanism in the following manner as illustrated in Figure 5.4 (adapted from reference 25). Starting at the top in Figure 5.4 and moving clockwise, superoxide is electrostatically guided into the active site channel and then associates with arg143, displacing a water molecule. Superoxide directly binds to Cu(II) and delivers its electron to the copper ion via an inner-sphere electron transfer. The Cu–his63 bridge breaks. Dioxygen diffuses out of the active site channel and is replaced by a hydrogen-bonding water molecule. In the

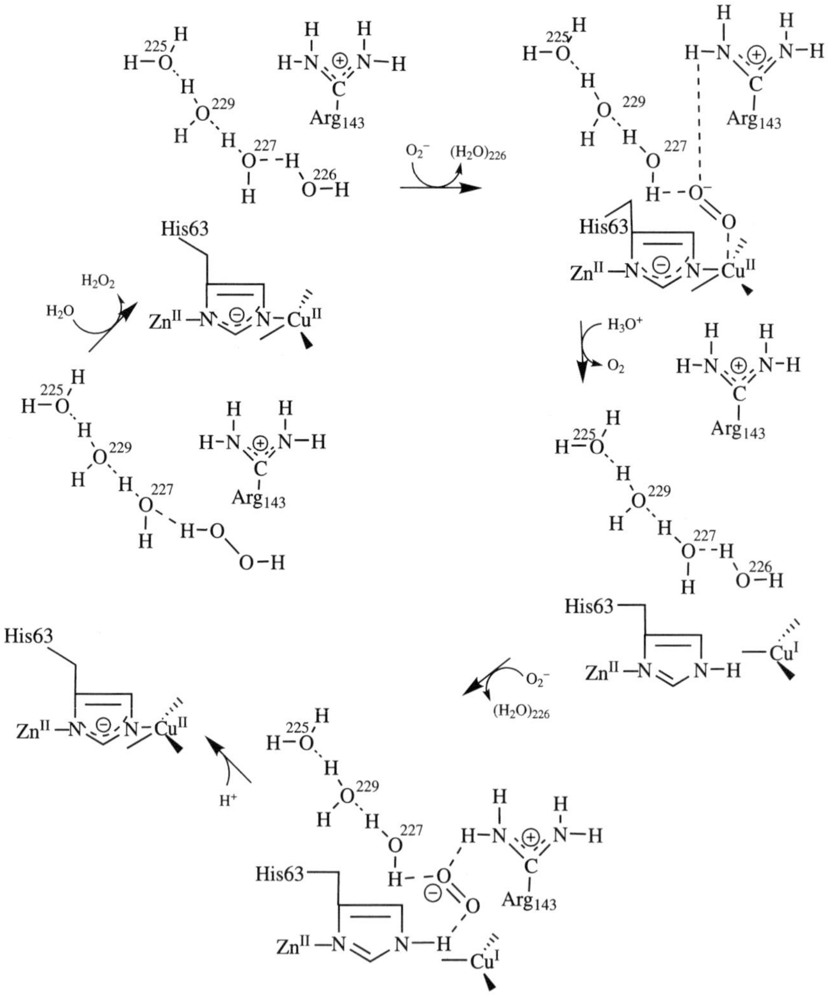

Figure 5.4 Detailed mechanism for superoxide dismutase activity. (Adapted with permission from Figure 8 of Hart, P. J.; Balbirnie, M. M.; Ogihara, N. L.; Versissian, A. M.; Weiss, M. S.; Valentine, J. S.; Eisenberg, D. *Biochemistry*, 1999, **38**, 2167–2178. Copyright 1999, American Chemical Society.)

second half of the reaction cycle, another superoxide enters the active site channel, displaces a water molecule, and hydrogen-bonds to arg143 and another water molecule. As the electron is accepted from Cu(I) via an outer-sphere electron transfer, superoxide simultaneously takes a proton from a H-bonded water molecule and from the imidazole bound proton of the tilted bridging his63. The Cu(II) ion then reforms the (deprotonated) his63 bridge. Electrically neutral hydrogen peroxide diffuses out of the active site channel, again being replaced by a water molecule. Significant protein backbone changes do not take place and do not appear to play a role in the catalytic cycle.

Support for this mechanism has been provided by EXAFS studies indicating three-coordinate Cu(I) having three histidine ligands.[29] X-ray crystallographic support for the three-coordinate copper site has also been found by Hough and Hasnain in a crystal containing both five-coordinate and three-coordinate copper sites.[30] These are collected under the PDB code 1CBJ. These authors solved crystal structures for bovine CuZnSOD in two different space groups. In the $P2_12_12_1$ form, subunit A appears to be a reduced three-coordinate copper site, whereas subunit B appears to be an oxidized five-coordinate copper site. The active sites for subunits A and B are described in Figure 5.5. From the X-ray structure, it is evident that his61 rotates away from the copper ion in the reduced form (subunit A).

Bertini and co-workers have studied the solution structure of human CuZn-SOD.[31] In reference 31b (PDB code 1BA9), substitution of two hydrophobic residues, phe50 and gly51, with two glu residues disrupted the quaternary structure of the protein, producing a soluble monomeric form. Additionally, glu133 has been changed to gln, and the mutated form is given the title Q133M2SOD. This form has only 10% of the SOD activity of the native enzyme and contains copper(I) ion, the reduced form of the enzyme. Although the eight-stranded β-barrel fold is retained in this mutant, monomerization causes large disorder in the subunit–subunit interface and produces significant changes in the conformation of the electrostatic loop forming one side of the active site channel. Conformation of this loop is fundamental in determining the optimal electrostatic potential necessary for driving superoxide ions to the copper site. The authors believe that the changes in electrostatics, along with movement of the copper ion away from the facilitating

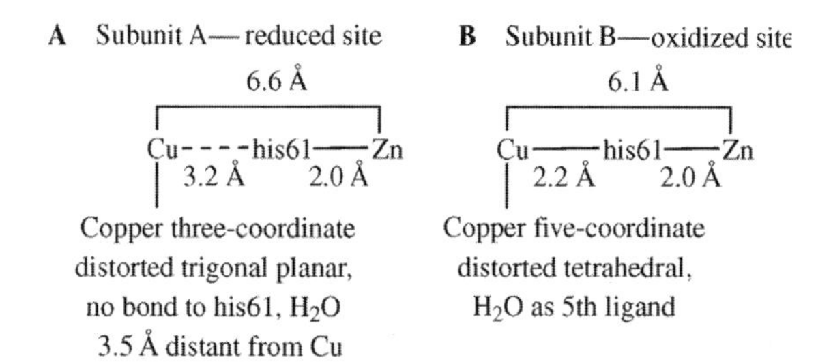

Figure 5.5 CuZn–SOD active sites in (A) reduced and (B) oxidized forms as described in reference 30.

arg143 in the reduced monomeric enzyme, are the causes of low enzyme activity. In Q133M2SOD, the zinc ion lies completely buried within the protein whereas the copper ion lies at the bottom of the active site channel, a crevice 13 Å deep and 22 Å across. The crevice narrows to 3 Å in proximity to the copper ion. At the active site in the reduced monomer, the copper(I) ion is tricoordinated to three histidines (his 46, 48, 120) whereas the zinc(II) ion is tetracoordinated to three histidines (his 63, 71, 80) and asp83. The imidazolate bridge between the copper and zinc is clearly broken in this structure. His63 is protonated at the N^{ε} position and is bound to zinc ion at the histidine's N^{δ} atom. These authors postulate that protonated his63 transfers this proton to superoxide ion in agreement with the second half of the mechanism described in Figure 5.4. The distance between the copper(I) ion and the N^{ε} of his63 is 3.5 Å, while it is 3.2 Å both in reduced yeast SOD in reference 27 and in reduced bovine SOD.[30] In reduced yeast SOD the Cu–Zn distance is 6.7 Å, whereas in the reduced monomeric species being discussed here the distance is 7.2 Å. The stereoview in Figure 5 of reference 31b indicates that the solution structure of reduced monomeric human CuZnSOD is very similar, but not identical, to that of the reduced yeast CuZnSOD determined by X-ray crystallography.[27]

These solution NMR and X-ray crystallographic findings have been contradicted by X-ray structures solved by Rypniewski et al.[32] The results show a reduced active site unchanged from the oxidized state and let these authors to propose a five-coordinate copper ion that exists throughout the oxidation and reduction process. In 2001 the Protein Data Bank listed 39 X-ray crystallographic and NMR solution structures for CuZnSOD, including oxidized, reduced, genetically modified, and other species with or without attached substrates or substrate mimics such as azide ion. The reader is advised to search the Protein Data Bank for additional and more up-to-date structural depositions and search the literature for further discussion of mechanism.

Geometric optimizations, energetics calculations, redox potential calculations by density functional, and electrostatic methods have been carried out on oxidized CuZnSOD[33] starting with X-ray crystallographic data from oxidized bovine erythrocyte CuZnSOD described in reference 23b. The model for the active site consisted of 65 atoms (66 for the protonated structure), 37 heavy atoms, and 28 (29) hydrogen atoms, including one water molecule weakly bonded to Cu in the apical position. Once this geometry was optimized, the active site was docked to the protein using the oxidized form of *Xenopus laevis* (PDB code 1XSO) in reference 26 or the reduced form of yeast CuZnSOD (PDB code 1YSO) in reference 27. The active site model was geometry optimized in oxidized, oxidized protonated, reduced, and reduced nonprotonated forms. The rms deviations in Cu–ligand and Zn–ligand bond distances agreed with experimental data within 0.09 Å for the oxidized structure[26] and 0.19 Å for the reduced.[27] Cu–O (water) bond distances had more discrepancy as is the case for both X-ray and EXAFS experimental data. The his61 N^{ε}–Cu bond is elongated from 2.04 Å in the oxidized form to 2.95 Å in the reduced nonprotonated form and finally to 3.39 Å in the reduced protonated form indicating breaking of the his61 N^{ε}–Cu bond. During the elongation the his61 ring tilts away from the copper ion and the position of the copper ion shifts relative to the his61. The authors explain the his61 N^{ε}–Cu bond elongation in molecular orbital terms as an antibonding interaction of the Cu d_{z^2}, $d_{x^2-y^2}$ and his61 N^{ε} p_z orbitals. No significant changes in Zn site geometry take place during the reduction. Energetic calculations indicate that the bond breaking can take place without large potential energy changes, again favoring the Cu–his61–Zn bridge broken mechanistic explanation and consistent with high reaction rates observed for superoxide dismutation.

Structural and mechanistic studies of superoxide dismutases have led to great advances in understanding the function of this complex enzyme. The Cu(I)–his–Zn bridge-broken mechanism seems well-supported at this time. Explanation of the role of zinc ion in the process seems less quantified, although it is believed to modulate the pK_a of the bridging histidine in such a way as to lead to facilitated protonation of the second incoming superoxide ion. Models that include solvent and H-bonding participation of other amino acid residues (such as arg143 or arg122) around the active site may lead to a greater understanding of zinc's role.

5.2.4 Hemocyanin

Hemocyanin (Hc) is a large multimeric protein (mol. wt. 4×10^5 to 9×10^6 Da) that binds molecular oxygen at dinuclear Type III copper sites. The metalloprotein is an extracellular dioxygen carrier in arthropods and mollusks, binding O_2 as peroxide ion at a dinculear copper site with concomitant oxidation of copper(I) to copper(II). Molluscan hemocyanins are cylindrical molecules having up to 20 subunits and high molecular weights (9×10^6 Da). In arthropod hemocyanin, six subunits containing one dinuclear copper center per subunit aggregate to form a hexamer. In the final quaternary structure, eight hexamers combine to contain a total of 48 dinuclear copper centers (molecular weights up to 3.5×10^6 Da).[34] Each subunit with M_r = approximately 7.5×10^4 Da contains one dioxygen binding site. Cu(II)'s presence in oxyhemocyanin causes species using this metalloenzyme as their dioxygen carrier to have blue blood.

Hemocyanin copper centers, in the absence of O_2, are described as (1) deoxy—colorless form having (Cu^+ Cu^+); (2) semi-met—having (Cu^{2+} Cu^+); and (3) met—blue unoxygenated form (Cu^{2+} Cu^{2+}). Deoxy-Cu(I) hemocyanin subunits of mol. wt. 5×10^4 to 7.4×10^4 bind dioxygen cooperatively through a homotropic allosteric effect. The allosteric effect of binding dioxygen to a number of subunits is expressed as a change in the quaternary structure and enhanced binding of O_2 for the remainder of the subunits. Calcium also binds to the protein, changing intersubunit interactions and O_2 affinity. This latter effect is a heterotropic allosteric effect.[34] In deoxy-Hc two Cu(I) ions, at variable internuclear distances depending on the species under study, are each trigonally coordinated by three histidines. These histidines are conserved in all known arthropod hemocyanins. An empty cavity exists between the two copper ions to accommodate the dioxygen molecule. Oxyhemocyanin has a characteristic intense optical transition at 340 nm and has a weak 580-nm transition. Both are assigned as ligand–metal charge transfer bands. The resonance Raman spectrum contains an O–O stretching vibration around

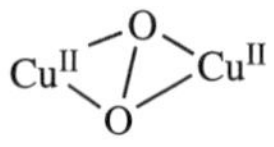

Cu^{II} (O–O) Cu^{II}

Figure 5.6 $[Cu_2\text{-}\mu\text{-}\eta^2{:}\eta^2\ O_2]$ peroxo binding mode found in hemocyanin.

740–750 cm^{-1}. This very low energy stretching vibration is now assigned to the peroxide ion O–O bond attached in a so-called side-on $\mu\text{-}\eta^2{:}\eta^2$ mode as illustrated in Figure 5.6. Other modes of peroxo ligand attachment are discussed in the model compounds section below. In this structure the two d^9 Cu(II) ions in oxy-Hc are strongly coupled magnetically, with the result that the dinuclear center is essentially diamagnetic.

Volbeda and Hol[35] analyzed the crystal structure of a single hemocyanin hexamer of $M_r = 4.6 \times 10^5$ Da from the arthropod spiny lobster, *Panulirus interruptus*, at 3.2-Å resolution in its deoxygenated form. The data are collected as PDB code 1HC1 (subunit 1) through 1HC6 (subunit 6) and 1HCY (refined using constrained 32-point group symmetry). They describe the hexamer as a trimer of "tight dimers." Contacts in the tight dimer, consisting of hydrogen bonding and hydrophobic interactions, are more numerous and better conserved during evolution than those of the trimer. The polypeptide fold of *P. interruptus* Hc has three domains. Domain 1 with 170 amino acid residues contains seven α-helical regions and one β-strand. Domain 2, also mainly helical and globular, contains the copper ions near its center, with each copper ion being coordinated by three histidyl residues along the α-helices surrounding the copper ions. Cu(A)–Cu(B) distances vary between 3.3 and 3.9 Å, depending on the subunit studied. Cu(A) coordinates the N^ε of his194 and his198 at contact distances near 2 Å (the authors modeled these at 1.95 Å based on EXAFS results)[36] and the N^ε of his224 at 2.4 to 3.1 Å, whereas Cu(B) coordinates N^ε of his344 at 2 Å or less, the N^ε of his348 at 2.4 to 3.1 Å, and the N^ε of his384 at 2.0–2.3 Å. The N^ε atoms of his194, his198, his344, and his384 along with Cu(A) and Cu(B) are approximately contained in a plane. His224 and his348 are arranged perpendicular to this plane and on opposite sides of it. The coordination core of two coppers and six histidines are surrounded by a large number of hydrophobic residues. Domain 2 is the most important domain for inter-subunit contacts stabilizing the hexamer and for transmitting conformational changes that occur upon dioxygen binding. Domain 3 has a more irregular shape, containing a seven-stranded β-barrel and several long loops that reach out into domain 1.

The seven-stranded β-barrel structure of hemocyanins is quite similar to that found in CuZnSOD discussed previously in Section 5.2.3. Several internal cavities are found in the Hc hexamer. The authors speculate that one cavity separating two tight dimer contact areas (surrounded by mainly hydrophilic residues such as asp152 and glu267) may bind a Ca^{2+} ion, known to affect dioxygen binding cooperativity. Another cavity is located in a region where the subunit's three domains come together. This cavity, which is exposed to the outside of the hexamer, may provide dioxygen's entry path. The crystal structure described by Volbeda and

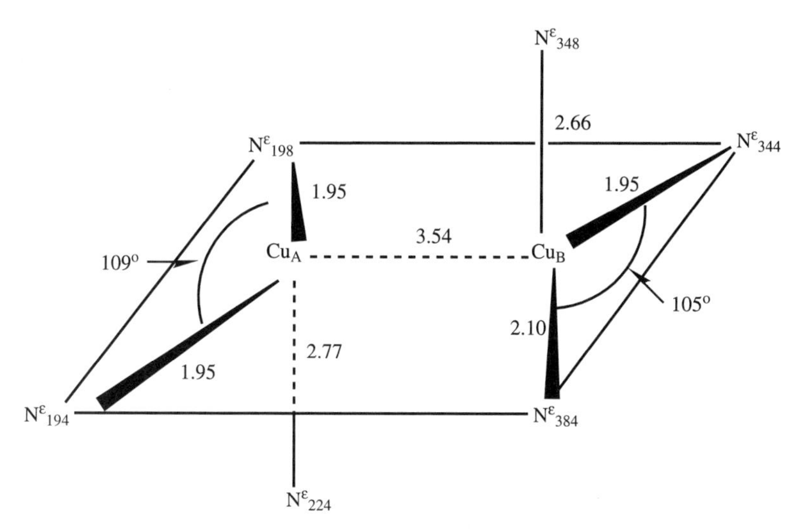

Figure 5.7 Schematic diagram of copper coordination in *Panulirus* Hc. (Adapted with permission from Figure 16 of Volbeda, A.; Hol, W. G. J. *J. Mol. Biol.*, 1989, **209**, 249–279. Copyright 1989, Academic Press.)

Hol did not identify a bridging ligand between the two copper ions of each subunit, although discussion by many authors had postulated this as an explanation for the strong antiferromagnetic coupling of the two copper ions and the distorted nature of the identified copper coordination site. One might have expected a more planar arrangement of all histidine ligands around the copper ion instead of the distorted geometry observed as shown in Figure 5.7, adapted from Figure 16 in reference 35a. As will be seen below in Section 5.3, reversible dioxygen binding by antiferromagnetically coupled Cu(II) ions does not require the presence of a bridging ligand besides the peroxo ligand arising from O_2 binding and copper(I) oxidation. Geometry of both the copper(I) and copper(II) sites are found to be variable and distorted in these models.

The X-ray structure of an oxygenated hemocyanin from the arthropod horseshoe crab, *Limulus polyphemus*, was reported by Magnus and co-workers in 1994 at 2.4-Å resolution (PDB code 1OXY).[37] Figure 5.8 visualizes this molecule using Wavefunction, Inc. Spartan ′02 for Windows™ software. The two blue/green copper ions with red dioxygen between them are visualized in space-fill form. Three histiding ligands coordinating to the top copper ion are—counterclockwise from the right—his364, his324 and his328. The second (lower) copper ion is coordinated—counterclockwise from the left—to his173, his177, and his204. Domain 1 is shown in blue ribbon form at the left; domain 2, containing the coordinated metals and dioxygen, is in red ribbon form at the center and top of the figure; and domain 3, in yellow, is shown at the right-hand side. Note the strand from domain 3 reaching through domain 2 into domain 1.

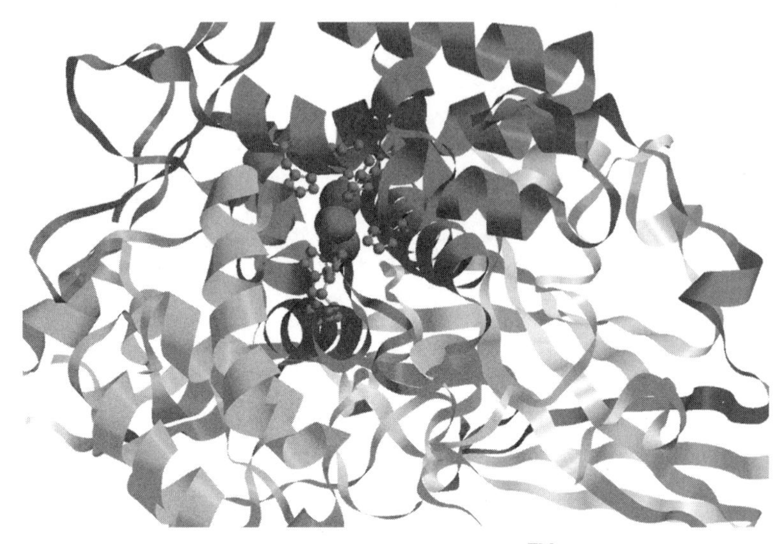

Figure 5.8 Wavefunction, Inc. Spartan '02 for Windows™ visualization of oxygenated hemocyanin from PDB data deposited as 1 OXY from reference 37a. See text for visualization details. Printed with permission of Wavefunction, Inc., Irvine, CA. (See color plate.)

The hemocyanin occurs in octahexameric form containing eight immunologically distinct subunit types. The homohexamer studied here, subunit II hemocyanin (*Limulus* II), comprises 628 residues and is not cooperative in dioxygen binding. In reference 37, the structure of oxygenated *L. polyphemus* Hc is compared to that of the deoxygenated subunit II structure carried out by the same group earlier (PDB code 1NOL).[38] Deoxygenated *Limulus* II contains two coppers coordinated by conserved histidines in an approximate trigonal plane at the active site with a long Cu–Cu distance of 4.6 Å. This distance is compared to that for the deoxy *Panulirus* Hc at approximately 3.5 Å (reference 35a). Oxygenated *Limulus* II contains copper ions at a distance of 3.6 Å and the same coordinating histidines only slightly displaced from their positions in the deoxy form. The oxygen atoms lie in a plane orthogonal to the Cu–Cu axis in η^2:η^2 geometry. The two copper ions in the oxy form lie in an approximate square plane formed by two oxygens and four N^{ε} atoms of histidines 173, 177, 324, and 364, while the N^{ε} atoms of his204 and his328 coordinate above and below the plane to complete the axial positions of square-pyramidal geometry.

To explain cooperative dioxygen binding in hemocyanin, the authors postulate conformational states of high (R-state) or low (T-state) oxygen affinity similar to states described for hemoglobin in Chapter 4. Under physiological conditions, deoxygenated Hc is found in the low-affinity T-state, although it is possible to find oxygenated Hc in this state as well. In the T-state the phenyl ring of phe49 and the imidazole ring of his328 are stacked with parallel rings 3.5 Å apart. Also a chloride ion is bound between ser47 (belonging to domain 1) and arg333 (belonging to domain 2) in the interface between domains 1 and 2. It had been found previously that the oxygen affinity of *Limulus* II Hc is greatly reduced by chloride ion binding. In the high-affinity deoxygenated or oxygenated R-state, phe49 moves away from his328 by almost 3 Å and the chloride ion is lost. Thus ser47 can move away from arg333, separating domains 1 and 2 through the operation of an 8° rotation. The oxygen-binding site is affected by removal of a constraint on a copper ligand (his328), the chloride-binding site is removed by separation of ser47 and arg333, and the domain 1–domain 1 interface between two subunits of the hexamer triggers a change in hexameric quaternary structure.

The X-ray crystallographic structure of mollusk *Octopus* hemocyanin has been studied at 2.3-Å resolution by Hendrickson and co-workers in reference 11. Molluscan hemocyanins are hollow cylindrical molecules with $M_r = 9 \times 10^6$ Da. Subunits associate as decamers, di-decamers, or larger assemblies. In the absence of magnesium or other cations, assemblies may dissociate into subunits and further into "functional units" capable of reversibly binding dioxygen. The internal functional units from *Octopus* are labeled "a" through "h," and it is the g functional unit (Odg) that is studied in reference 11. Primary sequence similarities between arthropodan and molluscan hemocyanins are minimal and limited to residues surrounding the Cu_B-binding site. The arthropodan hemocyanin subunit has an M_r of approximately 7.5×10^4 Da, whereas the *Octopus* subunit is approximately 3.5×10^5 Da and the Odg functional unit is 5.0×10^4 Da. The *Limulus* subunit studied by Magnus et al. in reference 37 is a three-domain protein (amino-terminal α-helical, copper-binding, and C-terminal β-sandwich), whereas Odg has two domains (N-terminal copper-binding and β-sandwich).

The Odg copper-binding domain consists of nine α-helices, six 3_{10} helices, six short β-strands, and a branched carbohydrate. The three longest α-helices—α4, α8, and α10—contain four of the six copper ligands, with the copper atoms lying near the start of α10 and toward an antiparallel β-sandwich domain (see Figure 4 of reference 11) consisting of residues 2793 to 2892. The blue color of Odg crystals, the aerobic experimental conditions, and indications of oxygen atoms between the copper ions indicate the oxygenated state for the crystal structure. The *Octopus* oxygen-binding site is similar to that seen for *Limulus* oxy-hemocyanin described in reference 37. The Cu_A site is ligated by N^{ε} atoms of his2543, his2562, and his2571, while the Cu_B site is ligated by N^{ε} atoms of his2671, his2675, and his2702. The Cu_A–Cu_B distance is approximately 3.5 Å. When viewed down the copper–copper axis, the histidine ligands adopt trigonal geometry partially staggered with respect to each other. The peroxo group placed between the copper atoms refines to an orthogonal bridge, with the oxygen atoms completing a distorted, flattened tetrahedron about each copper. The μ-η_2:η_2 mode of peroxo binding (see Figure 5.6) is that proposed for all other hemocyanin structures.

Figure 7 of reference 11 compares similarities between *Limulus* and *Octopus* hemocyanins by illustrating the overlap of 30 amino acids about the Cu_B region. The authors find that all residues within 5.0 Å of the Cu–Cu midpoint are conserved between arthropodan and molluscan hemocyanins and occupy analogous positions except for one (see the following paragraph). The three histidines about Cu_B overlap almost exactly in orientation and are less than 1 Å apart. The histidines about Cu_A have contacts less than 1 Å apart but are oriented at quite different angles about the copper ion. Another conserved structure noted for all hemocyanins whose structures are known is the sequence motif F(phe)xxxH(his). These authors, following the experimental determination of others, extend the hemocyanin and tyrosinase motif to read F(phe)xxxH(his)xxxxxxxE(asp)/D(glu). For Odg the residues are glu2579 on $\alpha 4$ of Cu_A and asp2706 on $\alpha 10$ of the Cu_B site.

In reference 37 the authors describe the *Limulus* hemocyanin low affinity T-state in which the phenyl ring of phe49 and the imidazole ring of his328 are stacked with parallel rings 3.5 Å apart. These residues move apart during the rotation of domain 1 away from domain 2 during the change to the high affinity R-state. In Odg a leucine residue, leu2830, occupies an analogous position to phe49. Leu2830, part of a β-sandwich domain, positions itself in the interdomain region of Odg known to be sensitive to active site events. Because phe49 is conserved in all arthropodans and leu2830 is conserved in all molluscans, the authors of reference 11 believe that the leucine residue fulfills a "sensor" function similar to that described for the *Limulus* phe49.

5.3 MODEL COMPOUNDS

5.3.1 Introduction

Synthesis of copper-containing model complexes to mimic the properties of copper metalloproteins poses a different problem than that described for myoglobin and hemoglobin model systems. Copper enzymes and dioxygen carriers utilize amino acid side chains for their ligand systems almost exclusively; no system comparable to iron heme systems exist for copper in the biological environment. Initially, chemists had difficulty handling the kinetically labile Cu(I) and Cu(II) compounds and in finding spectroscopic techniques useful for studying the d^{10} diamagnetic copper(I) ion. Successful stabilization of appropriate Cu(I) complexes and their O_2 adducts came about through the use of macrocyclic ligand systems beginning in the 1980s. New and better biomimetic ligand systems for Types I, II, and III copper-containing metalloenzymes continue to be discovered and studied by bioinorganic researchers. A small sampling of these model systems will be presented here.

The basic premises of biomimetic inorganic chemistry applied to copper complexes led Karlin to outline the following concepts in 1985.[50a] Model complexes should be found that mimic the spectral and structural properties of the natural system (and ideally its reactivity). The model should elucidate details of the natural system's characteristics and correlate well with the metalloprotein's function. The experimentalist's approach should include design of the proper ligand system with appropriate ligand donors—that is, O and/or N and/or S, mono- or multifunctional ligation, and incipient bridging of metal centers. Systematic investigation of ligand donor types, coordination geometry, variation in M–M distances, and variation in chelate ring size for polydentate ligand systems should be carried out. In all cases, spectroscopic investigation by X-ray crystallography,

NMR, EPR, EXAFS, infrared, resonance Raman, and ultraviolet-visible spectroscopy should follow. Kinetic and thermodynamic information about the model complexes in comparison to that known for natural systems should be gathered. These concepts were updated in 1999 by Karlin, writing in reference 49. Model studies should provide reasonable bases for hypotheses about a biological structure and its reaction intermediates. Researchers should determine the model's competence in carrying out reactions that mimic metalloprotein chemistry. Using these methods and criteria, researchers may hope to exploit Cu–oxygen systems as practical dioxygen carriers or oxidation catalysts for laboratory and industrial purposes.

5.3.2 Type I Copper Enzyme Models

Examples of successful model ligand systems synthesized to mimic Type I copper redox enzymes such as plastocyanin or azurin have been presented by Malachowski et al.[39] Ligand systems as shown in Figure 5.9 were synthesized to force copper metal centers into a distorted tetrahedral geometry common to the blue copper proteins. These proved to be successful in producing copper(I) complexes from the reactions of hindered N,S ligand systems with copper(II) salts. The ligand systems were designed to prevent disulfide formation and exploited properties of the biphenyl moiety in which the dihedral angle between phenyl rings (normally 42°) could be increased by bulky substituents. It was thought that resultant copper complexes would be forced out of planar and toward more tetrahedral geometry as found in the native enzymes. In order to assess whether a linking biphenyl ring would be necessary for copper complexation, the ligand N,S-mpy, 1-methyl-4-(2-pyridylmethylsulfanyl)benzene (Figure 5.9A), was reacted with $[Cu(H_2O)_6][ClO_4]_2$, yielding the copper(II) complex $[Cu(N,S\text{-mpy})_2][ClO_4]_2$. To assess whether sulfur ligands were required for complexation, the ligand N_4-mpy, 2,2′-bis(2-pyridylmethylamino)biphenyl (Figure 5.9B), was reacted with $[Cu(H_2O)_6][ClO_4]_2$, yielding the copper(II) complex $[Cu(N_4\text{-mpy})_2][ClO_4]_2$. In comparison, the ligands N_2S_2-mpy, 2,2′-bis(pyridylmethylsulfanyl)biphenyl (Figure 5.9C), and N_2S_2-mim, (2,2′-bis(4-methylimidazol-5-yl)methylsulfanyl)biphenyl (Figure 5.9D), when reacted with copper(II) salts, yielded the copper(I) complexes $[Cu(N_2S_2\text{-mpy})]X$ and $[Cu(N_2S_2\text{-mim})]X$, respectively, where $X = ClO_4^-$ or BF_4^-. The X-ray crystallographic structure of $[Cu(N_2S_2\text{-mpy})][ClO_4] \cdot MeCN$ indicated complexation by two pyridyl nitrogens and two thioether sulfurs in a distorted tetrahedral geometry with bond angles ranging from 87.5 to 136.3°. The Cu–N bond lengths are typical to those found for other four-coordinate copper(I)–pyridine systems. A 83° dihedral angle between the biphenyl aromatic rings indicates the twist possible in this ligand system facilitating the tetrahedral geometry about the metal. Oxidation of the $[Cu(N_2S_2\text{-mpy})][ClO_4]$ via cyclic voltammetry yields a potential of +770 mv versus SCE, reproducing the very high reduction potentials found for the blue copper proteins (see Table 5.2). The potential is much lower for the $[Cu(N_4\text{-mpy})_2][ClO_4]_2$ complex (+210 mV) and intermediate for the $[Cu(NS\text{-mpy})_2][ClO_4]_2$ (+530 mV). The voltammetric experiment indicated chemical

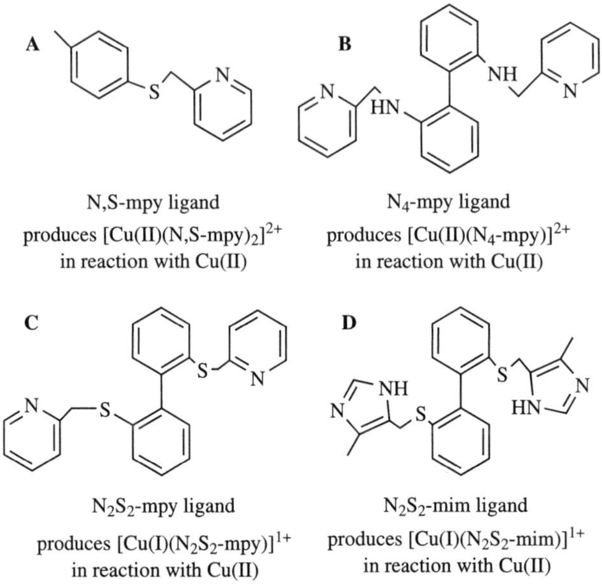

A

N,S-mpy ligand
produces $[Cu(II)(N,S\text{-}mpy)_2]^{2+}$
in reaction with Cu(II)

B

N_4-mpy ligand
produces $[Cu(II)(N_4\text{-}mpy)]^{2+}$
in reaction with Cu(II)

C

N_2S_2-mpy ligand
produces $[Cu(I)(N_2S_2\text{-}mpy)]^{1+}$
in reaction with Cu(II)

D

N_2S_2-mim ligand
produces $[Cu(I)(N_2S_2\text{-}mim)]^{1+}$
in reaction with Cu(II)

Figure 5.9 Ligands (A) N,S-mpy, (B) N_4-mpy, (C) N_2S_2-mpy, and (D) N_2S_2-mim designed to mimic azurin and plastocyanin active sites. (Adapted with permission of The Royal Society of Chemistry from Malachowski, M. R.; Adams, M.; Elia, N.; Rheingold, A. L.; Kelly, R. S. *J. Chem. Soc., Dalton Trans.*, 1999, 2177–2182.

reversibility for the $[Cu(N_2S_2\text{-}mpy)][ClO_4]$ complex also, illustrating the stability achieved with the N_2S_2-mpy ligand system. The authors conclude that the biphenyl backbone is necessary for the observed spontaneous reduction (only Cu(II) products observed for the NS-mpy ligand) and that sulfur ligands are needed for production of the reduced species (only Cu(II) products observed for the N_4-mpy ligand). Observation of the high potential for the $[Cu(N_2S_2\text{-}mpy)][ClO_4]$ system in comparison to the others is an additional indication that the geometry of the complexes is shifting from distorted planar to that of a distorted tetrahedron.

5.3.3 Type II Copper Enzyme Models

Models for SOD-like activity require flexible ligands of medium-strength donor power because both Cu(II) (preferring four- or five-coordinate geometry) and Cu(I) (preferring four-, three-, or even two-coordinate geometry) must be accommodated. Jan Reedijk and co-workers have described a mixed pyrazole–imidazole ligand donor set having an intramolecularly imidazolato-bridged asymmetric dicopper(II) site.[40] The imidazolato-bridged model compound $[Cu_2(bpzbiap)Cl_3]$, where Hbpzbiap = 1,5-bis(1-pyrazolyl)-3-[bis-(2-imidazolyl)methyl] azapentane, in solution at low concentration, catalyzes the dismutation of superoxide at biological pH. The authors present X-ray crystallographic data for two complexes, one of which contains an imidazolato bridging ligand. These are illustrated in Figure 5.10. At pH

Figure 5.10 Models for superoxide dismutase as illustrated in reference 40. (A) $[Cu_2(bpz\text{-}biap)Cl_3]$ **(1)** and (B) $[Cu_2(Hbzbiap)Cl_4]$ **(2)**.

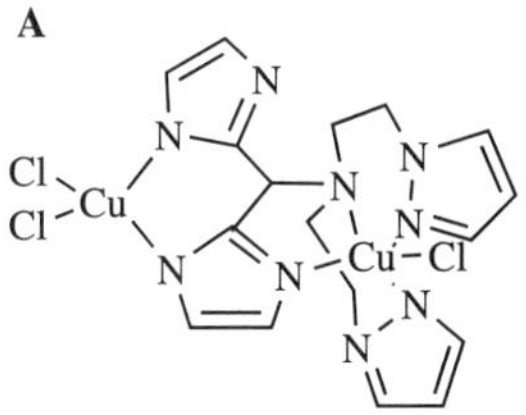

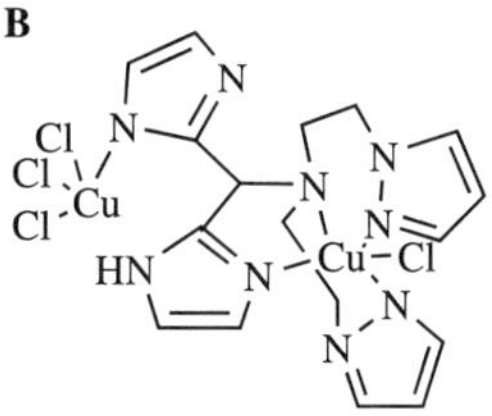

values (between 5.8 and 8.2) that include the physiological region, the imidazolato bridged species shows both EPR and UV-vis indications of copper(II)–copper(II) coupling through the imidazolato bridge. The copper(II)–copper(II) bond distances found to range from 5.566(1) to 6.104(1) Å are similar to those found for bovine erythrocyte SOD in reference 23a. Complex **(1)** in Figure 5.10 exhibits significant catalytic activity toward superoxide anion of the same order of magnitude as the best SOD analogs described in the literature. In complex **(1)**, one copper ion's coordination geometry is intermediate between square-planar and tetrahedral while the other copper ion is found to be in a distorted square-pyramidal geometry. For complex **(2)**, one copper ion is tetrahedral while the other is also in a distorted square-pyramidal geometry.

5.3.4 Type III Copper Enzyme Models

The model compound situation for Type III copper enzymes is complex with many observed and characterized O_2 binding modes. Some of these are shown in Figures 5.12 to 5.14 and Figures 5.17 to 5.20. Hemocyanin models must exhibit the bonding pattern shown in Figure 5.11 (see also Figure 5.12D) as found in the native enzymes. Other model compound bonding modes mimic tyrosinase-type oxidase reactions, as illustrated in Figures 5.14 and 5.21. These important models may lead to the design of new oxygenation catalysts. A sampling of the Type III copper enzyme model research completed over the past 10–15 years is presented in Sections 5.3.4.1 through 5.3.4.4.

5.3.4.1 Karlin Group Tridentate Model Compounds. Kenneth Karlin and co-workers have summarized their research concerning formation, stabilities, and structures of copper–dioxygen complexes.[41] The model compounds for Type III copper enzymes are relevant to the dioxygen carrier hemocyanin (Hc) and enzymes that "activate" O_2 facilitating incorporation of oxygen into biological substrates. These include tyrosinase (tyr) and other monooxygenases as well as "blue" multi-copper oxidases—that is, laccase and ascorbate oxidases (see Tables 5.1 and 5.2). In 1997, at the time of reference 41a's publication, direct spectroscopic evidence for

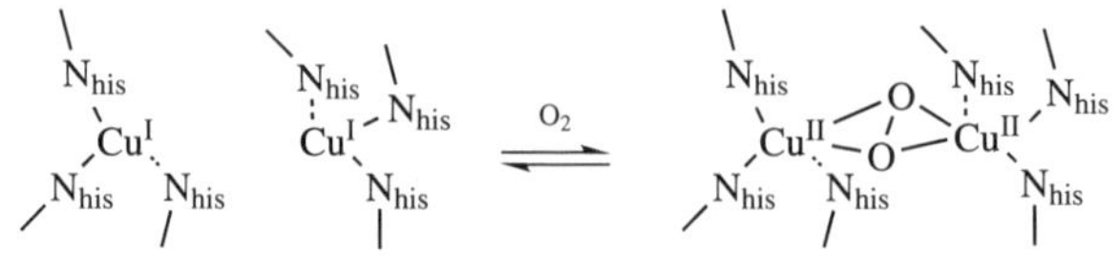

Figure 5.11 Side-on peroxo binding mode in hemocyanin. (Adapted with permission from Karlin, K. D.; Kaderli, S.; Zuberbühler, A. D. *Acc. Chem. Res.*, 1997, **30**, 139–147. Copyright 1997, American Chemical Society.)

dioxygen adducts of hemocyanin, tyrosinase, and laccase had been gathered. Hc exhibits Cu(I):O_2 = 2:1 stoichiometry with binding in a side-on μ-η^2:η^2-bridging peroxo (O_2^{2-}) as indicated in Figure 5.11, while Tyr models take on the end-on, μ-η^1:η^1 peroxo binding mode shown in Figures 5.12C, and 5.14. This structure has been determined, as discussed above, by X-ray crystallographic analyses of oxyHc. The electronic absorption spectrum of oxyHc has bands at 345 nm (ε = 20,000 M^{-1} cm^{-1}) and 570 nm (ε = 1000 M^{-1} cm^{-1}) with peroxide $\nu_{O\text{–}O}$ stretch at 740 cm^{-1}. Models for oxyHc would be expected to conform to the data found for the biological species. However, possibilities for dioxygen binding to copper and other metal centers include, among others, those illustrated in the Figure 5.12. Many of these O_2 binding modes have been found in model compounds as will be illustrated.

As early as 1985, Karlin's group had synthesized dicopper(I) model compounds that would add dioxygen and carbon monoxide reversibly at low temperatures in dichloromethane.[50a] These are shown in Figures 5.13 and 5.14. The macrocyclic ligand system, including aliphatic and 2-pyridyl (py) nitrogen ligands and a phenoxo bridge, was found to coordinate O_2 in an end-on η^1:η^1 peroxo manner. Resonance Raman spectroscopic evidence included that for a peroxo O–O and Cu-peroxo bonds ($\nu_{O\text{–}O}$ = 803 cm^{-1}, $\nu_{Cu\text{–}O_2^{2-}}$ = 488 cm^{-1}). The electronic spectrum has only two ligand $\rightarrow$ metal charge transfer transitions (505 nm, ε = 6300 M^{-1} cm^{-1}, 610(sh) nm, ε = 2400 M^{-1} cm^{-1}) assigned to $\pi_{\sigma^*} \rightarrow d_{x^2-y^2}$ and $\pi_{v^*} \rightarrow d_{x^2-y^2}$, respectively. These absorbances are consistent with unsymmetrical end-on peroxo coordination with the peroxo oxygen atoms inequivalent as indicated in Figure 5.12C (inequivalent binding mode). End-on η^1:η^1 peroxo ligation of O_2 in model compounds can be contrasted with that observed for side-on η^2:η^2 peroxo coordination in the Kitajima models to be described below.[45]

The oxygenation rate for the model compound $[Cu_2^I(XYL\text{–}O^-)]^+$ in Figure 5.13 is too fast to be measured even at 173 K, leading the authors to conclude that the bridging phenoxo group preorganizes the dicopper(I) complex for oxygenation. Oxygenation at −80°C led to a deep purple compound. The compound exhibited resonance Raman stretching frequencies typical of peroxo coordination $\nu_{O\text{–}O}$ = 803 cm^{-1} and $\nu_{Cu\text{–}O_2^{2-}}$ = 488 cm^{-1}. Note the inequivalent peroxo binding mode as illustrated in Figure 5.12C.

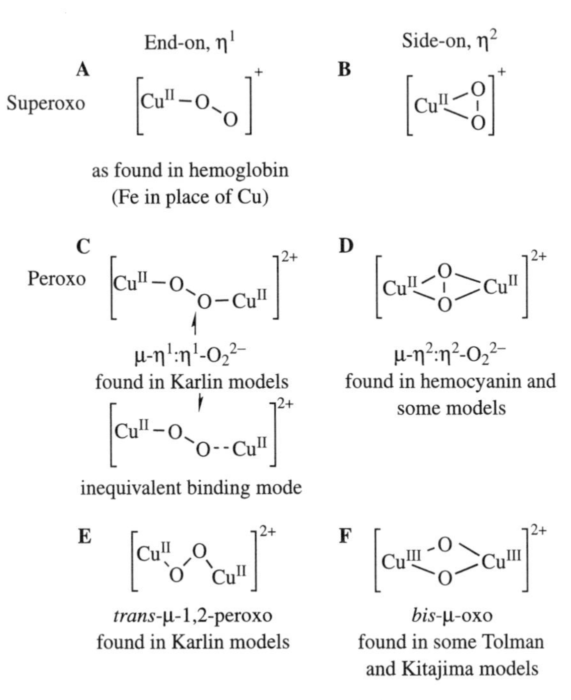

Figure 5.12 Some possible modes for metal–O_2 binding. (A) superoxo end-on, η^1; (B) superoxo side-on, η^2; (C) peroxo end-on μ-η^1:η^1-O_2^{2-}; (D) peroxo side-on μ-η^2:η^2-O_2^{2-}; (E) *trans*-μ-1,2-peroxo end-on; (F) *bis*-μ-oxo side-on.

Oxygenation rates were first examined for the system $[Cu^I_2(R–XYL–H)]^{2+}$. In the model described here, two bis[2-(2-pyridyl)ethyl]amine (PY2) units are linked by a xylyl spacer group (R = H). Although initially proposed as a crude hemocyanin model, this system now is studied as a model for tyrosinase and is an example of hydrocarbon oxygenation taking place under mild conditions—that is, <1 atm O_2

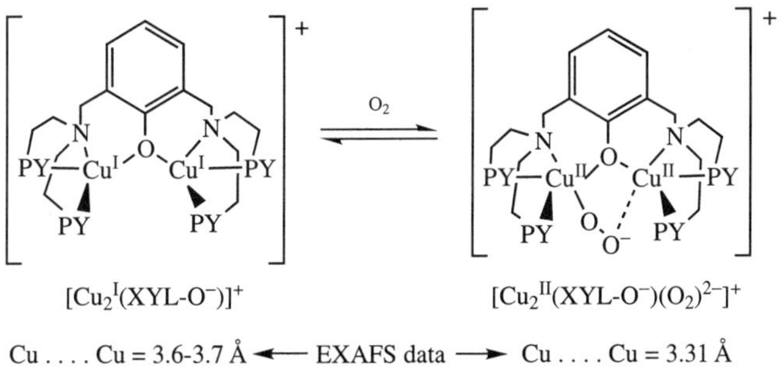

$[Cu_2^I(XYL\text{-}O^-)]^+$ $\quad$ O_2 $\quad$ $[Cu_2^{II}(XYL\text{-}O^-)(O_2)^{2-}]^+$

Cu . . . . Cu = 3.6-3.7 Å ← EXAFS data → Cu . . . . Cu = 3.31 Å

Figure 5.13 Model compound showing reversible O_2 binding. (Adapted with permission from Karlin, K. D.; Kaderli, S.; Zuberbühler, A. D. *Acc. Chem. Res.*, 1997, **30**, 139–147. Copyright 1997, American Chemical Society.)

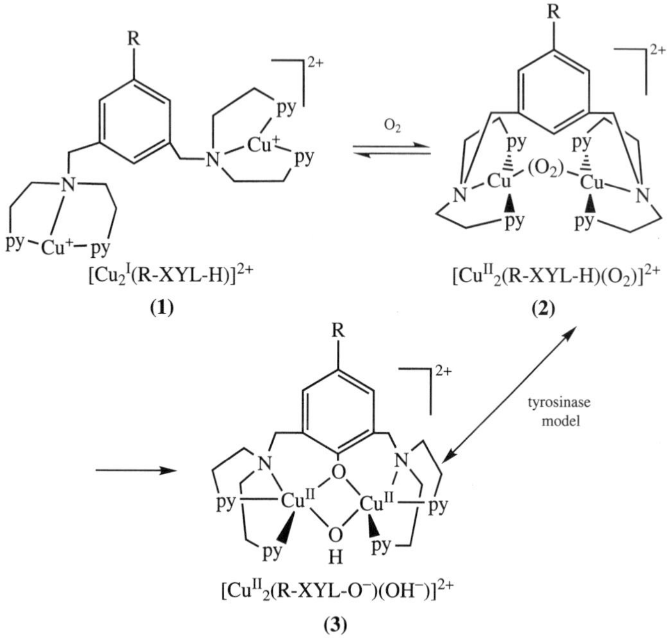

Figure 5.14 Model compound for kinetic study of reversible oxygenation and tyrosinase model. (Adapted with permission from Scheme 1 of Karlin, K. D.; Kaderli, S.; Zuberbühler, A. D. *Acc. Chem. Res.*, 1997, **30**, 139–147. Copyright 1997, American Chemical Society.)

and at room temperature or below. The detailed proposed mechanism involves formation of the $Cu_2–O_2$ end-on peroxo moiety followed by electrophilic attack on the arene substrate, held in close proximity as shown in Figure 5.14.

A thorough kinetic and thermodynamic analysis of this model system (small positive or negative enthalpies of formation are canceled by more negative entropies of formation) led Karlin's group to conclude that the stability of dioxygen binding is driven by favorable enthalpies, but unfavorable reaction entropies preclude observation of $Cu_2–O_2$ at room temperatures.[41a]

Another tridentate ligand system, reported by the Karlin group in 1998 has the structure shown below in Figure 5.15.[42] This ligand reacts with copper(I) salts in acetonitrile to produce $[(MePY2)Cu^{I}(CH_3CN)]^+$, and then it reacts with O_2 at $-80°C$ in acetonitrile to produce both the side-on peroxo copper(II) species $[\{(MePY2)Cu\}_2(\mu\text{-}\eta^2{:}\eta^2\text{-}O_2)]^{2+}$ (Figure 5.12D), along with a small amount (1–10%) of a bis-μ-oxo-dicopper(III) species $[\{(MePY2)Cu\}_2(O)_2]^{2+}$ similar to that shown as **(3)** in Figure 5.14 or as bonding mode B in Figure 5.18.[42] In later studies the amount of the bis-μ-oxo-dicopper(III) species was found to be around 20%.[43] This combination of products was found to effect clean H-atom abstraction reactions with exogenously added hydrocarbon substrates, a tyrosinase-type behavior not observed with a similar ligand system N_nPY2 where $n = 3–5$ (see Figure 5.16).[43]

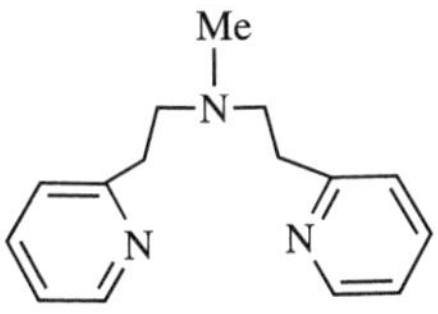

Figure 5.15 Ligand methylbis[2-(2-pyridyl)ethyl]amine (MePY2).

Another ligand system, bis[2-(2-pyridyl)ethyl]amine (PY2) having a $(CH_2)_n$ linker, was also prepared and studied.[41a] If $n = 4$, as shown in Figure 5.16, the ligand is called N4; that is, N4 = N_4PY2. Upon oxygenation, peroxo complexes would have the formula $[Cu_2(N4)(O_2)]^{2+}$.

It was presumed that the more flexible $(CH_2)_n$ linker and the absence of an obvious position for electrophilic attack such as is found in system $[Cu^I_2$(R–XYL–H)$]^{2+}$, as seen in Figure 5.14, would lead to more rapid oxygenation reactions and a more stable complex than that formed from $[Cu^I_2$(R–XYL–H)$]^{2+}$. Instead the researchers found a more complex system in kinetically controlled oxygenation studies. Two different peroxo complexes form via a postulated open-chain superoxo species as shown in the scheme shown in Figure 5.17.[41a]

Depending on the length of the alkyl linker in the ligand system N_nPY2, oxygenation of the Cu(I) complexes leads to butterfly (nonplanar) or planar $Cu_2(N_nPY2)O_2$ cores. Frontier molecular orbital calculations for both butterfly and planar side-on peroxo cores indicate that hydrogen atom abstraction will not be a favorable reaction for these species. The investigators conclude that H-atom abstraction reactions are either due to the Cu(III)$(\mu\text{-O})_2$ complex present in $[(MePY2)Cu^I(CH_3CN)]^+$ oxygenation reactions or due to increased accessibility of the Cu_2O_2 core for $[\{(MePY2)Cu\}_2(\mu\text{-}\eta^2\text{:}\eta^2\text{-}O_2)]^{2+}$ (MePY2 shown in Figure 5.15) over $[\{(N_nPY2)Cu\}_2(\mu\text{-}\eta^2\text{:}\eta^2\text{-}O_2)]^{2+}$ (N_nPY2 shown in Figure 5.16) cores.[43] Tolman's group describes the Cu(III)$(\mu\text{-O})_2$ complex as the *bis*-μ-oxo dimer as is discussed in the next section and as shown in Figure 5.12F.

5.3.4.2 Tolman Group Tetradentate Model Compounds. Research into copper model compounds in William B. Tolman's laboratory identified the binding mode called the *bis*-μ-oxo isomer (Figure 5.12F). Instead of the pyridine-based ligands used by Karlin's group, Tolman's group used amine-based ligands such as

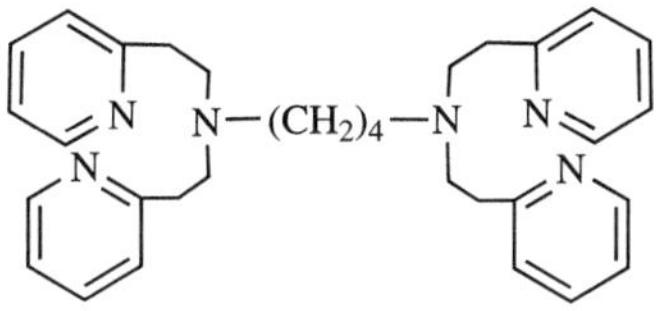

Figure 5.16 Ligand bis[2-(2-pyridyl)ethyl]amine with $(CH_2)_n$ linker (N_4PY2 or N4).

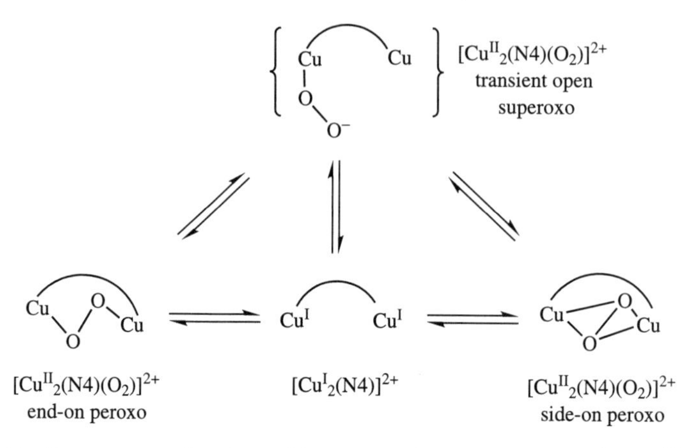

Figure 5.17 Superoxo transient formed in oxygenation reactions of $[Cu^{I}_{2}(N4)]^{2+}$. (Adapted from Scheme 2 of Karlin, K. D.; Kaderli, S.; Zuberbühler, A. D. *Acc. Chem. Res.*, 1997, **30**, 139–147. Copyright 1997, American Chemical Society.)

L^{iPr_3} = 1,4,7-triisopropyl-1,4,7-triazacyclononane or L^{Bn_3} = 1,4,7-tribenzyl-1,4,7-triazacyclononane. In Figure 5.18, R is either the isopropyl or the benzyl group. When R is benzyl, the *bis*-μ-oxo dimer is favored; however, when R is isopropyl the authors believe that both the μ-oxo dimer and the μ-η^2:η^2 peroxo dimer coexist in the crystalline state. In solution, structural preference for the μ-oxo (Figure 5.12F) or the μ-η^2:η^2 peroxo dimer (Figure 5.12D), in model complexes with the isopropyl group as part of the ligand system, was found to be dependent on the solvent in which oxygenation is carried out.[44] In reference 43, researchers have studied both the Karlin and Tolman ligand systems using instrumental methods and density functional theory. Conclusions state that $N_{eq}CuN_{eq}$ bite angles of 89° or less in the Cu_2O_2 cores favor the *bis*-μ-oxo binding mode (bonding mode B in Figure 5.18, 512F) whereas $N_{eq}CuN_{eq}$ bite angles of 93° or more (bite angles of 94–99° are found in oxyHc) favor the side-on η^2:η^2 peroxo modes (bonding mode A of Figure 5.18, 5.12D).

5.3.4.3 Kitajima Group Tetradentate Model Compounds. A successful model compound for hemocyanin's side-on peroxo bridged copper ions was synthesized by Kitajima and co-workers and first reported in 1988.[45] In this work, a copper(I) complex $Cu(PPh_3)[HB(3,5\text{-}Me_2pz)_3]$ was oxidized by PhIO to yield a μ-oxo dinuclear copper(II) complex $[Cu(PPh_3)[HB(3,5\text{-}Me_2pz)_3]]_2O$. Treatment of this complex with H_2O_2 yielded a μ-peroxo complex. Subsequently, the same workers used a more hindered hydrotris(pyrazolyl)borate ligand ($[HB(3,5\text{-}i Pr_2pz)_3]$ = hydrotris-(3,5-diisopropylpyrazolyl)borate anion) to form a similar complex having enhanced stability and crystallinity. The reaction is illustrated in Figure 5.19.

Another gradient-corrected density functional study of reversible dioxygen binding and reversible O–O bond cleavage has been carried out for $Cu_2(\mu$-η^2:η^2

Figure 5.18 *bis* μ-oxo (bonding mode B) or side-on η^2:η^2 peroxo (bonding mode A) dimers in model complexes. (Adapted with permission from Halfen, J. A.; Mahapatra, S.; Wilkinson, E. C.; Kaderli, S.; Young, V. G., Jr.; Aue, L., Jr.; Zuberbühler, A. D.; Tolman, W. B. *Science*, 1996, **271**, 1397–1400. Copyright 1996, American Association for the Advancement of Science.)

O_2), mode A, and $Cu_2(\mu\text{-}O)_2$, mode B, core structures of $\{[LCu]_2O_2\}^{2+}$, where L = 1,4,7-triazacyclonanane **(1)** (see bonding mode A structure of Figure 5.18) or L = hydrotris(pyrazolyl)borate **(2)** (see bonding mode A structure of Figure 5.19).[46] From the literature it was known that **(1)** would bind dioxygen reversibly and undergo core isomerization from mode A to mode B, whereas **(2)** would bind O_2 irreversibly and exhibit only mode A binding. The reference 46 author also found that while the model system $\{[(NH_3)_3Cu]_2O_2\}^{2+}$ **(3)** could be a qualitative model for bonding in hemocyanin model compounds, this simplified system could not explain the differences in chemistry between **(1)** and **(2)**. The author calculated the reaction internal energy (ΔE) for **(1)** and **(2)** according to the following reaction:

$$2Cu^{I}L + O_2 \rightarrow Cu_2O_2L_2 \tag{5.11}$$

Results indicated a ΔE of −60 kJ/mol for **(1)**, in contrast to ΔE of −184 kJ/mol for **(2)**. Comparison to $\Delta E = -50$ to -80 kJ/mol for other model compounds

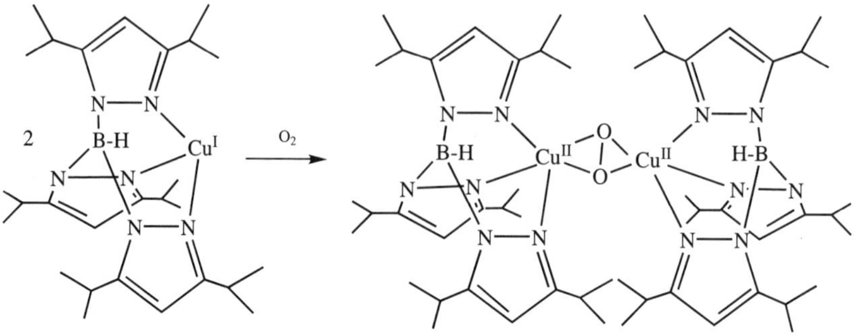

μ-η^2:η^2-peroxo
$[Cu^{II}_2L(O_2)]^{2+}$
Bonding mode A
(2)

Figure 5.19 Kitajima $Cu_2(\mu$-η^2:η^2 $O_2)$ mode A hemocyanin model compound. (Adapted with permission from Kitajima, N.; Moro-oka, Y. *Chem. Rev.*, 1994, **94**, 737–757. Copyright 1994, American Chemical Society.)

exhibiting reversible dioxygen binding (such as that described in Figures 5.13 and 5.14 above and TMPA ligands described below) led the authors to conclude that a ΔE of -60 kJ/mol for **(1)** was reasonable for reversible O_2 binding, whereas ΔE of -184 kJ/mol for **(2)** was reasonable for irreversible O_2 binding. In addition, calculated core isomerization energies were calculated and found to be $+1$, $+12$, and $+49$ kJ/mol for **(1)**, **(2)**, and **(3)** respectively. The transformations are shown schematically in the following equation:

$$Cu_2(\mu\text{-}\eta^2\text{:}\eta^2 O_2),\ \text{mode A} \Leftrightarrow Cu_2(\mu\text{-}O)_2,\ \text{mode B} \qquad (5.12)$$

Calculated barrier heights for formation of $Cu_2(\mu$-η^2:η^2 $O_2)$, mode A, were found to be $+33$ and $+37$ kJ/mol for **(1)** and **(2)**, respectively. Both the negligible isomerization energy ($+1$ kJ/mol) and the low barrier of **(1)** ($+33$ kJ/mol) are in line with observed fast interconversion of the core isomers. However, the $+12$-kJ/mol isomerization energy of **(2)** does not explain why $Cu_2(\mu$-$O)_2$, mode B, was not observed. The author postulates that, as described by Kitajima, decomposition to the μ-oxo dinuclear copper(II) complex $[Cu(PPh_3)[HB(3,5\text{-}Me_2pz)_3]]_2O$ prevents observation of mode B. And the author concludes that orbital interactions of **(1)** and **(2)** are similar, leading to similar structural, magnetic, and spectroscopic properties as shown in Table 5.7. Description of the transition state of the core isomerization of **(1)** and **(2)** show a small energy barrier for both. Core isomerization for **(2)** is not observed because isomerization leads to decomposition of the complex. The theoretical system **(3)**, while successful for interpretation of binding and spectra, is inappropriate for prediction of reactivity and especially core isomerizations.

Table 5.7 Spectroscopic and Physical Properties of Cu_2–O_2 Adducts

Core Structure	Compound/Ligands (Solvent)	λ_{max} (nm)	ε (M^{-1} cm^{-1})	Resonance Raman Spectrum (cm^{-1}) (^{18}O Data)	Cu–Cu Distance (Å)
$Cu^{II}(\mu\text{-}\eta^2{:}\eta^2\text{-}O_2)Cu^{II}$	OxyHc 3his per Cu (H_2O)	340 580	20,000 1,000	748 (708)	3.6(2) (X-ray) reference 37, 38
$Cu^{II}(\mu\text{-}\eta^2{:}\eta^2\text{-}O_2)Cu^{II}$	L^{iPr_3} (CH_2Cl_2)	366 510	22,500 1,300	722 (680)	
$Cu^{III}(\mu\text{-}O)_2Cu^{III}$	L^{iPr_3} (THF)	324 448	11,000 13,000	600 (580)	
$Cu^{III}(\mu\text{-}O)_2Cu^{III}$	L^{Bn_3} (CH_2Cl_2)	318 430	12,000 14,000	602–608 (583)	2.794(2) (X-ray)

Source: Reference 44.

In summary, a number of dioxygen binding modes have been found for tridentate ligand systems binding to copper(I) metal centers. Ligand geometry and solvent are two important considerations in determining whether the *bis*-μ-oxo (Figure 5.12F or Figure 5.18, bonding mode B) or the μ-η^2:η^2 peroxo dimer (Figure 5.12D or Figure 5.18 and 5.19, bonding mode A) geometry predominates (as illustrated in equation 5.13). The so-called core interconversions have important implications for oxygenation catalysts and are the subject of continuing interest in these research groups.

$$[M_2(\mu\text{-}\eta^2\text{:}\eta^2\text{-}O_2]^{n+} \Leftrightarrow [M_2(\mu\text{-}O)_2]^{n+} \qquad (5.13)$$

In reference 44 the authors found that the $[Cu_2(\mu\text{-}O)_2]^{2+}$ core with either the L^{iPr_3} or L^{Bn_3} ligand system will activate C–H bonds for cleavage and oxidation, again having important implications for oxygenation catalysis. These authors provide a useful summary for the two core structures, which are replicated in Table 5.7.

5.3.4.4 Karlin Group Tetradentate Model Compounds. While the tridentate ligand systems discussed above were being described, tetradentate ligands were also being studied. In 1993, Karlin's group synthesized the ligand TMPA (tris(2-pyridylmethyl)amine) and studied its complexation with Cu(I) followed by its oxygenation. Pseudoreversible O_2 binding going through a superoxocopper(II) transient intermediate to the thermodynamic product $[\{(TMPA)Cu\}_2(O_2)]^{2+}$, a *trans*-μ-1,2-$Cu_2O_2^{2+}$ species (shown in Figure 5.12E), was found for the TMPA and other quinolyl-substituted analogs according to Figure 5.20. The kinetics and thermodynamics of these systems were studied and the complex's X-ray crystallographic structure was determined at −90°C.[47] It was found that the copper atoms were pentacoordinate with distorted trigonal bipyramidal geometry and had the peroxo oxygen atoms at axial positions. The Cu–Cu distance and the O–O

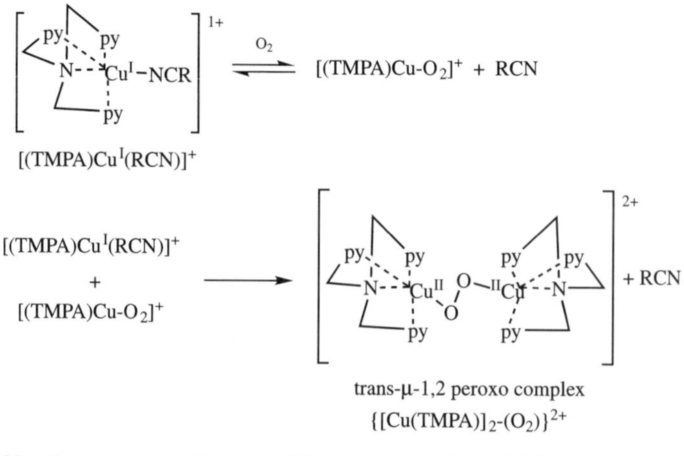

$$[(TMPA)Cu^I(RCN)]^+ \underset{}{\overset{O_2}{\rightleftharpoons}} [(TMPA)Cu\text{-}O_2]^+ + RCN$$

$$[(TMPA)Cu^I(RCN)]^+ + [(TMPA)Cu\text{-}O_2]^+ \longrightarrow \{[Cu(TMPA)]_2\text{-}(O_2)\}^{2+} + RCN$$

trans-μ-1,2 peroxo complex
$\{[Cu(TMPA)]_2\text{-}(O_2)\}^{2+}$

Figure 5.20 Square-pyramidal copper(II) peroxo complex exhibiting reversible oxygenation/deoxygenation behavior. (Adapted from references 47 and 41a.)

distance were found to be 4.359 Å and 1.432 Å, respectively. The complex was EPR silent, exhibited three visible absorbances assigned to $O_2^{2-} \rightarrow$ Cu charge transfer bands (440 nm ($\varepsilon = 2000\ M^{-1}\ cm^{-1}$), 525 nm ($\varepsilon = 11,500\ M^{-1}\ cm^{-1}$), and 590 nm sh ($\varepsilon = 7600\ M^{-1}\ cm^{-1}$)), and exhibited resonance Raman assigned to O–O at 832 cm^{-1} and Cu–O at 561 cm^{-1}. The complex indicated as $\{[Cu(TMPA)]_2(O_2)\}^{2+}$ was intensely purple in color. The complex was found to decolorize under vacuum and then, if rechilled to −80°C, to reform the purple peroxo complex.

5.3.5 Summary

William B. Tolman has summarized the knowledge gained by study of copper model complexes and applied the results to propose mechanisms for C–H bond scission and O atom transfer by bis(μ-oxo), $[Cu_2(\mu\text{-}O)_2]$, cores.[48] These results are pertinent for both biological studies and industrial processes in which catalytic mono- and dioxygenation reactions are important. Writing in "Bioinorganic Catalysis", edited by Jan Reedijk and Elisabeth Bouwman, Kenneth Karlin related the biological activities of hemocyanin and tyrosinase.[49] It is suggested that tyrosinase's active site is highly accessible to exogenous ligands compared to that found in hemocyanins. In a simplistic view, Hc's active site consists of a tyrosinase-type protein with additional protein sequences making up a domain that shields the active site and prevents the binding of large ligand substrates. The relationship of the two catalytic cycles can be seen in Figure 5.21 as adapted from reference 49.

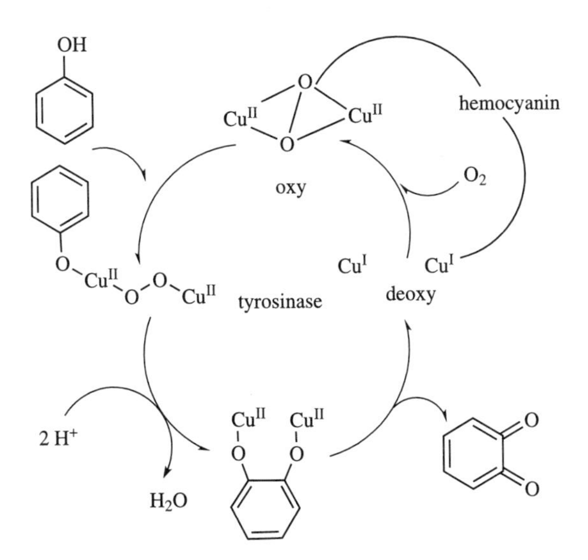

Figure 5.21 Catalytic cycles for hemocyanin and tyrosinase. (Adapted with permission from Karlin, K. D. In Reedijk, J.; Bouwman, E. eds. *Bioinorganic Catalysis*, 2nd ed., Marcel Dekker, New York, 1999, pp. 469–534. Copyright 1999, Marcel Dekker.)

5.4 CONCLUSIONS

This discussion of copper-containing enzymes has focused on structure and function information for Type I blue copper proteins azurin and plastocyanin, Type III hemocyanin, and Type II superoxide dismutase's structure and mechanism of activity. Information on spectral properties for some metalloproteins and their model compounds has been included in Tables 5.2, 5.3, and 5.7. One model system for Type I copper proteins[39] and one for Type II centers[40] have been discussed. Many others can be found in the literature. A more complete discussion, including mechanistic detail, about hemocyanin and tyrosinase model systems has been included. Models for the blue copper oxidases laccase and ascorbate oxidases have not been discussed. Students are referred to the references listed in the reference section for discussion of some other model systems. Many more are to be found in literature searches.[50]

REFERENCES

1. Frausto da Silva, J. R. R.; Williams, R. J. P. *The Biological Chemistry of the Elements: The Inorganic Chemistry of Life*, Clarendon Press, New York, 1991, pp. 40–41, 389–399.
2. Hay, R. W. *Bio-Inorganic Chemistry*, Ellis Horwood Limited, Halsted Press, New York, 1984.
3. Cowan, J. A. *Inorganic Biochemistry, An Introduction*, 2nd ed., Wiley-VCH, New York, 1997.
4. Solomon, E. I.; Baldwin, M. J.; Lowery, M. D. *Chem. Rev.*, 1992, **92**, 521–542.
5. Lippard, S. J.; Berg, J. M. *Principles of Bioinorganic Chemistry*, University Science Books, Mill Valley, CA, 1994.
6. Harrison, P. M., ed., *Metalloproteins Part 1: Metal Proteins with Redox Roles*, Verlag Chemie GmbH, Weinheim, 1985.
7. Fenton, D. E. *Biocoordination Chemistry*, Oxford University Press, New York, 1995.
8. Ducros, V.; Brzozowski, A. M.; Wilson, K. S.; Brown, S. H.; Ostergaard, P.; Schneider, P.; Yaver, D. S.; Pedersen, A. H.; Davies, G. J. *Nat. Struct. Biol.*, 1998, **5**, 310.
9. (a) Messerschmidt, A.; Rossi, A.; Ladenstein, R.; Huber, R.; Bolognesi, M.; Gatti, G.; Marchesini, A.; Petruzzelli, R.; Finazzi-Agro, A. *J. Mol. Biol.*, 1989, **206**, 513. (b) Messerschmidt, A.; Ladenstein, R.; Huber, R.; Bolognesi, M.; Avigliano, R.; Petruzzelli, R.; Rossi, A.; Finazzi-Agro, A. *J. Mol. Biol.*, 1992, **224**, 179–205 (PDB: 1AOZ).
10. Ito, N.; Phillips, S. E.; Stevens, C.; Ogel, Z. B.; McPherson, M. J.; Keen, J. N.; Yadav, K. D.; Knowles, P. F. *Nature*, 1991, **350**, 87 (PDB: 1GOF).
11. Cuff, M. E.; Miller, K. I.; vanHolde, K. E.; Hendrickson, W. A. *J. Mol. Biol.*, 1998, **278**, 855–870 (PDB: 1JS8).
12. (a) Adman, E. T.; Canters, G. W.; Hill, H. A. O.; Kitchen, N. A. *FEBS Lett.*, 1982, **143**, 287. (b) Adman, E. T.; Jensen, L. H. *Israel J. Chem.*, 1981, **21**, 8 (PDB: 1AZU). (c) Adman, E. T.; Watenpaugh, K. D.; Jensen, L. H. *Proc. Natl. Acad. Sci.*, 1975, **72**, 4854.
13. (a) Shepard, W. E. B.; Anderson, B. F.; Lewandoski, D. A.; Norris, G. E.; Baker, E. N. *J. Am. Chem. Soc.*, 1990, **112**, 7817–7819. (b) Baker, E. N. *J. Mol. Biol.*, 1988, **203**, 1071–1095 (PDB: 2AZA).

14. (a) Gray, H. B. *Chem. Soc. Rev.*, 1986, **15**, 17–30. (b) Faham, S.; Mizoguchi, T. J.; Adman, E. T.; Gray, H. B.; Richards, J. H.; Rees, D. C. *J. Biol. Inorg. Chem.*, 1997, **2**, 464–469.
15. (a) Nar, H.; Messerschmidt, A.; Huber, R.; van de Kamp, M.; Canters, G. W. *J. Mol. Biol.*, 1991, **221**, 427–447 (PDB: 2AZU, 3AZU). (b) Nar, H.; Messerschmidt, A.; Huber, R.; van de Kamp, M.; Canters, G. W. *J. Mol. Biol.*, 1991, **221**, 765–772 (PDB: 4AZU, 5AZU).
16. Bagby, S.; Driscoll, P. C.; Harvey, T. S.; Hill, A. O. *Biochemistry*, 1994, **33**, 6611–6622 (PDB: 1PLA, 1PLB).
17. Collyer, C. A.; Guss, J. M.; Sugimura, Y.; Yoshizaki, F.; Freeman, H. C. *J. Mol. Biol.*, 1990, **211**, 617–632 (PDB: 7PCY).
18. (a) Guss, J. M.; Freeman, H. C. *J. Mol. Biol.*, 1983, **169**, 521–563. (b) Guss, J. M.; Harrowell, P. R.; Murata, M.; Horris, V. A.; Freeman, H. C. *J. Mol. Biol.*, 1986, **192**, 361–387.
19. Moore, J. M.; Lepre, C. A.; Gippert, G. P.; Chazin, W. J.; Case, D. A.; Wright, P. E. *J. Mol. Biol.*, 1991, **221**, 533–555 (PDB: 9PCY).
20. Badsberg, U.; Jorgensen, A. M. M.; Gesmar, H.; Led J. J.; Hammerstad, J. M.; Jespersen, L.; Ulstrup, J. *Biochemistry*, 1996, **35**, 7021–7031 (PDB: 1NIN).
21. Valentine, J. S.; Wertz, D. L.; Lyons, T. J.; Liou, L. L.; Goto, J. J.; Gralla, E. B. *Curr. Opin. Chem. Biol.*, 1998, **2**, 253–262.
22. (a) Getzoff, E. D.; Tainer, J. A.; Stempien, M. M.; Bell, G. I.; Hallewell, R. A. *Proteins*, 1989, **5**, 322–336. (b) Bordo, D.; Djinovic, K.; Bolognesi, M. *J. Mol. Biol.*, 1994, **238**, 366–386.
23. (a) Tainer, J. A.; Getzoff, E. D.; Beem, K. M.; Richardson, J. S.; Richardson, D. C. *J. Mol. Biol.*, 1982, **160**, 181–217 (PDB: 2SOD). (b) McRee, D. E.; Redford, S. M.; Getzoff, E. D.; Lepock, J. R.; Hallewell, R. A.; Tainer, J. A. *J. Biol. Chem.*, 1990, **265**, 14234 (PDB: 3SOD).
24. Valentine, J. S.; DeFreitas, D. M. *J. Chem. Ed.*, 1985, **62**(11), 990–997.
25. Hart, P. J.; Balbirnie, M. M.; Ogihara, N. L.; Versissian, A. M.; Weiss, M. S.; Valentine, J. S.; Eisenberg, D. *Biochemistry*, 1999, **38**, 2167–2178 (PDB: 1B4L, 1B4T, 1YAZ, 2JCW).
26. Carugo, K. D.; Battistoni, A.; Carri, M. T.; Polticelli, F.; Desideri, A.; Rotilio, G.; Coda, A.; Wilson, K. S.; Bolognesi, M. *Acta Crystallogr.*, 1996, **D52**, 176–188 (PDB: 1XSO).
27. Ogihara, N. L.; Parge, H. E.; Hart, P. J.; Weiss, M. S.; Goto, J. J.; Crane, B. R.; Tsang, J.; Slater, K.; Roe, J. A.; Valentine, J. S.; Eisenberg, D.; Tainer, J. A. *Biochemistry*, 1996, **35**, 2316–2321 (PDB: 1YSO, 1JCV).
28. Leone, M.; Cupane, A.; Militello, V.; Stroppolo, M. E.; Desideri, A. *Biochemistry*, 1998, **37**, 4459–4464.
29. Blackburn, N. J.; Hasnain, S. S.; Binsted, N.; Diakun, G. P.; Garner, C. D.; Knowles, P. F. *Biochem. J.*, 1984, **219**, 985–990.
30. Hough, M. A.; Hasnain, S. S. *J. Mol. Biol.*, 1999, **287**, 579–592 (PDB: 1CBJ).
31. (a) Banci, L.; Bertini, I.; Del Conte, R.; Piccioli, M.; Viezzoli, M. S. *Biochemistry*, 1998, **37**, 11780–11791. (b) Banci, L.; Benedetto, M.; Bertini, I.; Del Conte, R.; Fadin, R.; Magnani, S.; Viezzoli, M. S. *J. Biol. Inorg. Chem.*, 1999, **4**, 795 (PDB 1BA9).
32. Rypniewski, W. R.; Mangani, S.; Bruni, B.; Orioli, P. L.; Casati, M.; Wilson, K. S. *J. Mol. Biol.*, 1995, **251**, 282–296.

33. Konecny, R.; Li, J.; Fisher, C. L.; Dillet, V.; Bashford, D.; Noodleman, L. *Inorg. Chem.*, 1999, **38**, 940–950.

34. Holm, R. H.; Kennepohl, P.; Solomon, E. I. *Chem. Rev.*, 1996, **96**, 2239–2314.

35. (a) Volbeda, A.; Hol, W. G. J. *J. Mol. Biol.*, 1989, **209**, 249–279 (PDB: 1HC1–1HC6, 1HCY). (b) Gaykema, W. P. J.; Hol, W. G. J.; Vereijken, J. M.; Soeter, N. M.; Bak, H. J.; Beintema, J. J. *Nature*, 1984, **309**, 23–29 (PDB 1HC1–1HC6, 1HCY).

36. Woolery, G. L.; Powers, L.; Winkler, M.; Solomon, E. I.; Spiro, T. G. *J. Am. Chem. Soc.*, 1984, **106**, 86–92.

37. (a) Magnus, K. A.; Hazes, B.; Ton-That, H. Bonaventura, C.; Bonaventura, J. *Proteins*, 1994, **19**, 302–309 (PDB: 1OXY). (b) Magnus, K. A.; Ton-That, H.; Carpenter, J. E. *Chem Rev.*, 1994, **94**, 727–735.

38. Hazes, B.; Magnus, K. A.; Bonaventura, C.; Bonaventura, J.; Dauter, Z.; Kalk, K. H.; Hol, W. G. J. *Prot. Sci.*, 1993, **2**, 597–619 (PDB: 1NOL).

39. Malachowski, M. R.; Adams, M.; Elia, N.; Rheingold, A. L.; Kelly, R. S. *J. Chem. Soc., Dalton Trans.*, 1999, 2177–2182.

40. Tabbi, G.; Driessen, W. L.; Reedijk, J.; Bonomo, R. P.; Veldman, N.; Spek, A. L. *Inorg. Chem.*, 1997, **36**, 1168–1175.

41. (a) Karlin, K. D.; Kaderli, S.; Zuberbühler, A. D. *Acc Chem. Res.*, 1997, **30**, 139–147. (b) Jung, B.; Karlin, K. D.; Zuberbühler, A. D. *J. Am. Chem. Soc.*, 1996, **118**, 3763–3764.

42. Obias, H. V.; Lin, Y.; Murthy, N. N.; Pidcock, E.; Solomon, E. I.; Ralle, M.; Blackburn, N. J.; Neuhold, Y.-M.; Zuberbühler, A. D.; Karlin, K. D. *J. Am. Chem. Soc.*, 1998, **120**, 12960–12961.

43. Pidcock, E.; DeBeer, S.; Obias, H. V.; Hedman, B.; Hodgson, K. O.; Karlin, K. D.; Solomon, E. I. *J. Am. Chem. Soc.*, 1999, **121**, 1870–1878.

44. Halfen, J. A.; Mahapatra, S.; Wilkinson, E. C.; Kaderli, S.; Young, V. G., Jr.; Aue, L. Jr.; Zuberbühler, A. D.; Tolman, W. B. *Science*, 1996, **271**, 1397–1400.

45. (a) Kitajima, N.; Moro-oka, Y. *Chem. Rev*, 1994, **94**, 737–757. (b) Kitajima, N.; Fujisawa, K.; Fujimoto, C.; Moro-oka, Y.; Hashimoto, S.; Kitagawa, T.; Toriumi, K.; Tatsumi, K.; Nakamura, A. *J. Am. Chem. Soc.*, 1992, **114**, 1277–1291. (c) Kitajima, N.; Koda, T.; Moro-oka, Y. *Chem. Lett.*, 1988, 347. (d) Kitajima, N.; Koda, T.; Hashimoto, S.; Kitagawa, T.; Moro-oka, Y. *J. Chem. Soc., Chem. Commun.*, 1988, 151. (e) Kitajima, N.; Koda, T.; Hashimoto, S.; Kitagawa, T.; Moro-oka, Y. *J. Am. Chem. Soc.*, 1991, **113**, 5664. (f) Kitajima, N.; Fujisawa, K.; Moro-oka, Y. *J. Am. Chem. Soc.*, 1989, **111**, 8975–8976.

46. Bèrces. A. *Inorg. Chem.*, 1997, **36**, 4831–4837.

47. Tyeklar, Z.; Jacobson, R. R.; Wei, N.; Murthy, N. N.; Zubieta, J.; Karlin, K. D. *J. Am. Chem. Soc.*, 1993, **115**, 2677–2689.

48. Tolman, W. B. *Acc. Chem. Res.*, 1997, **30**(6), 227–237.

49. Karlin, K. D. In Reedijk, J.; Bouwman, E. eds. *Bioinorganic Catalysis*, 2nd ed., Marcel Dekker, New York, 1999, pp. 469–534.

50. (a) Karlin, K. D.; Gultneh, Y. *J. Chem. Ed.*, 1985, **62**(11), 983–990. (b) Karlin, K. D.; Kaderli, S.; Zuberbühler, A. D. *Acc. Chem. Res.*, 1997, **30**, 139–147. (c) Hüber, M.; Bubacco, L.; Beltramini, M.; Salvato, B.; Elias, H.; Peisach, J.; Larsen, E.; Harnung, S. E.; Haase, W. *Inorg. Chem.*, 1996, **35**, 7482–7492. (d) Lubben, M.; Hage, R.; Meetsma, A.; Bÿma, K.; Feringa, B. L. *Inorg. Chem.*, 1995, **34**, 2217–2224.

6

THE ENZYME NITROGENASE

6.1 INTRODUCTION

The complex enzyme nitrogenase is the only natural substance now known that reduces the "inert" molecule dinitrogen (N_2) to ammonia under mild conditions (290 K and 0.8 atm). This important biological process, providing reduced nitrogen for incorporation in nucleic and amino acids, cannot be carried out by most organisms even though an abundant supply of the starting material exists in earth's atmosphere. The N_2 to NH_3 reduction and protonation takes place at a high energy cost because up to 40% of a bacterium's ATP production may be used for this purpose when it is actively fixing nitrogen.[21] A diagram indicating important nitrogenase features (to be discussed in detail below) is shown in Figure 6.1, as adapted from Figure 5.21 of reference 9. Reference 22 (PDB code: 1N2C) contains visualizations of nitrogenase's topography as assembled from X-ray structural data.

Industrially, ammonia has been produced from dinitrogen and dihydrogen by the Haber–Bosch process, which operates at very high temperatures and pressures, and utilizes a promoted iron catalyst. Millions of tons of ammonia are generated annually for incorporation into agricultural fertilizers and other important commercial products. The overall reaction is exergonic, as indicated in equation 6.1:

$$N_2 + 3H_2 \rightarrow 2NH_3 \qquad \Delta G^0 = -33.2\text{kJ/mol} \tag{6.1}$$

However, the first step of the process, forming diazene (N_2H_2), has a large positive free energy change, as shown below in equation 6.4.

As nitrogenase's structure and function have become known at the molecular level through the work of many groups over the past 20 years, chemists have

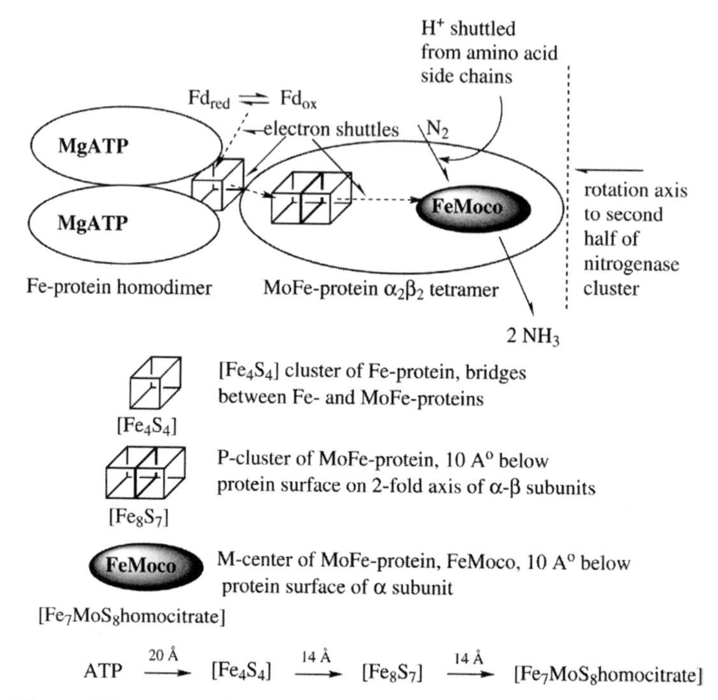

Figure 6.1 Cartoon illustrating the structure of the enzyme nitrogenase.

continually attempted to mimic its behavior in the laboratory. Often the goal has been the design of more efficient catalytic systems for industrial ammonia production. However, preparation of synthetic chemical systems exhibiting efficient catalytic dinitrogen reduction to hydrazine or ammonia under mild conditions of temperature and pressure have continued to elude chemists up to the present time. Some of the accomplishments in modeling nitrogenase's structure and function are discussed in Section 6.6.

The catalytic reduction of dinitrogen to ammonia in biological systems is thermodynamically favorable overall and takes place according to equation 6.2:

$$N_2 + 8H^+ + 8e^- + 16MgATP \rightarrow 2NH_3 + H_2 + 16MgADP + 16P_i \qquad (6.2)$$

Reaction 6.2 is coupled to the oxidation of the electron transfer protein ferredoxin (Fd), with an estimated $\Delta G^0 = -\ 65.6$ kJ/mol, as shown in equation 6.3[1]:

$$N_2 + 8H^+ + 8Fd_{red} \rightarrow 2NH_3 + H_2 + 8Fd_{ox} \qquad \Delta G^0 = -65.6\,\mathrm{kJ/mol} \qquad (6.3)$$

The first step in dinitrogen reduction is highly endothermic because the dinitrogen triple bond is stable both kinetically and thermodynamically. For example, if the first product in N_2 reduction was the production of diazene (N_2H_2), the process would be endergonic by +220 kJ/mol, as shown in equation 6.4[2]:

$$N_2 + H_2 \rightarrow N_2H_2 \qquad \Delta G^0 = +220\,\text{kJ/mol} \tag{6.4}$$

It has been estimated[3] that approximately 175 million metric tons of dinitrogen are fixed annually by nitrogenase (at 290 K and 0.8 atm pressure) in contrast to the 50 million metric tons produced by the iron-catalyzed Haber–Bosch process (at 600–800 K and 500 atm pressure). Another author reports that each route—Haber–Bosch and nitrogenase—annually converts about 108 tons of N_2 to NH_3.[4] Although the mechanism of the Haber–Bosch process has not been completely clarified, evidence exists that the rate-determining step involves the activation of dinitrogen on the catalytic iron surface leading to the breaking of the strong triple bond of dinitrogen. This event is followed by reaction with molecular hydrogen that is also activated on the catalytic surface. In nitrogenase, protons are believed to be shuttled to the dinitrogen substrate at the active site through amino acid side-chain interactions while electrons are accumulated by oxidation and reduction of the iron ions in the enzyme's three different iron–sulfur clusters. The rate-determining step is thought to be dissociation of the two protein complexes making up the enzyme. Figure 6.2 presents a schematic diagram illustrating biological, industrial, and laboratory methods for NH_3 production as adapted from reference 32.

Many nitrogenase enzyme assemblages capable of reducing nitrogen to ammonia occur in organisms called diazotrophs and contain the second-row transition metal molybdenum, even though second-row elements are used rarely in biological enzyme systems. Although only molybdenum-containing enzyme will be discussed in any detail here, other nitrogenases are known that substitute either vanadium or another iron for the molybdenum atoms in the enzyme. All molybdenum-containing nitrogenases are composed of two separable components, the Fe–protein containing four iron atoms and having a molecular weight (M_r) of approximately 68 kDa and the MoFe–protein containing 30 iron atoms and two molybdenum atoms and having a molecular weight (M_r) of approximately 240 kDa. (See Figure 6.1.) The iron and molybdenum atoms are contained in metal–sulfur clusters that are more fully described in Sections 6.3 to 6.5. As expected, the amino acid residues that are most conserved in nitrogenase enzymes from different species are those serving as metal–sulfur cluster ligands or those implicated in subunit interactions and transport of protons and electrons to the active site.

Nitrogenase's smaller component, Fe–protein, is a γ_2 homodimer of two identical subunits. Fe–protein transfers electrons in an ATP-dependent manner through its iron–sulfur cluster (two ATPs per one [4Fe–4S] cluster) to MoFe–protein, an $\alpha_2\beta_2$ tetramer. Within the larger and more complex nitrogenase subunit—MoFe–protein—electrons are transferred first through the so-called P-cluster, an unusual iron–sulfur cluster containing eight irons and seven sulfurs as inorganic sulfide ions. Subsequently, electrons are transferred to the substrate dinitrogen that is bound, reduced, and protonated at a third metal–sulfur cluster of

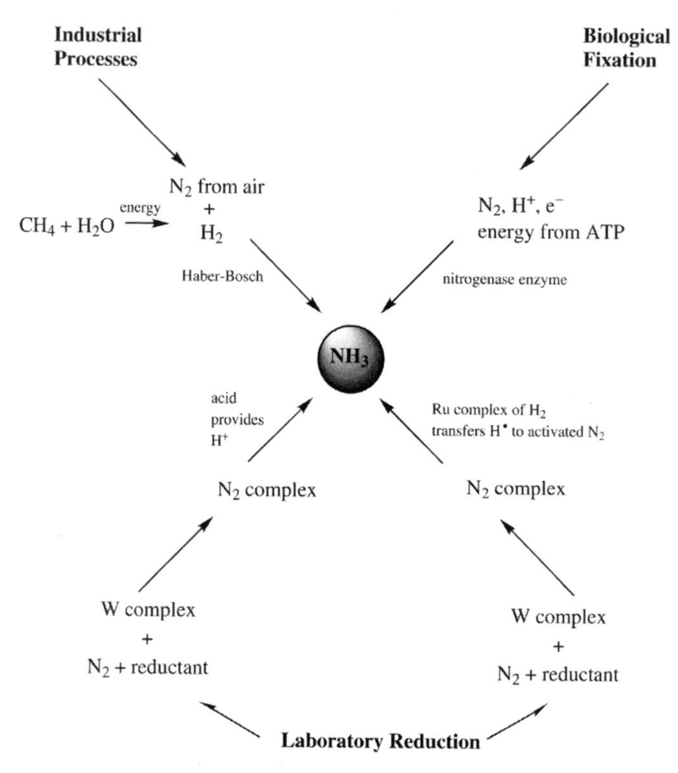

Figure 6.2 Industrial, biological, and laboratory methods for NH_3 production. (Adapted with permission from Leigh, G. J. *Science*, 1998, **279**, 506–507. Copyright 1998, American Association for the Advancement of Science.)

stoichiometry 7Fe:1Mo:8S:1 homocitrate ion. This cluster is called the M center or the iron–molybdenum cofactor (FeMoco). It is believed that nitrogenase's FeMoco binds substrate N_2, that electrons reach N_2 in a cascade through the other iron–sulfur clusters, and that protons reach the active site from solution through amino acid side-chain hydrogen bonding shuttles.[5] See Figure 6.1 for an illustration of nitrogenase's subunits and cofactors.

In summary, reduction of dinitrogen by nitrogenase requires cooperativity among nitrogenase's subunits and involves three basic types of electron transfer steps:

1. Reduction of Fe–protein by electron carriers such as ferredoxin or flavodoxin or by inorganic dithionite ion.
2. Transfer of single electrons from Fe–protein to MoFe–protein in an MgATP-dependent process. Two molecules of MgATP per one electron transferred are required, reflecting the dimeric nature of the Fe–protein. (See Figure 6.1.) When MgATP binds to the Fe–protein, its reduction potential is lowered, facilitating electron transfer from Fe–protein to MoFe–protein.

3. Actual electron transfer to the dinitrogen substrate at the MoFe–protein, with electrons first passing through the MoFe–protein's P-cluster. During this process, dinitrogen is most probably bound to the iron–molybdenum cofactor (FeMoco) of the MoFe–protein.[6]

All nitrogenase substrates are reduced by at least two electrons, so steps 1 and 2 above must be repeated until enough electrons are accumulated in the MoFe–protein where substrate reduction takes place. In the reaction reducing dinitrogen to two moles of ammonia, one mole of dihydrogen is also produced, requiring a total accumulation of eight electrons at the catalytic site. Electron transfer involves a series of sequential associations and dissociations of the Fe–protein/MoFe–protein complex with dissociation believed by most researchers to be the rate-determining step. These processes are controlled and timed by MgATP binding and hydrolysis. Two MgATP molecules are hydrolyzed for each association–dissociation event, and one electron is transferred concomitantly from Fe–protein to MoFe–protein.[7]

The mechanism and sequence of events that control delivery of protons and electrons to the FeMo cofactor during substrate reduction is not well understood in its particulars.[8] It is believed that conformational change in MoFe–protein is necessary for electron transfer from the P-cluster to the M center (FeMoco) and that ATP hydrolysis and P_i release occurring on the Fe–protein drive the process. Hypothetically, P-clusters provide a reservoir of reducing equivalents that are transferred to substrate bound at FeMoco. Electrons are transferred one at a time from Fe–protein but the P-cluster and M center have "electron buffering" capacity, allowing successive two-electron transfers to, and protonations of, bound substrates.[8] Neither component protein will reduce any substrate in the absence of its catalytic partner. Also, apoprotein (with any or all metal–sulfur clusters removed) will not reduce dinitrogen.

In addition to dinitrogen, nitrogenase can reduce a variety of small, unsaturated molecules; some of these are illustrated in reactions 6.5–6.7[9]:

$$C_2H_2 \rightarrow C_2H_4 \tag{6.5}$$

$$N_2O \rightarrow N_2 \tag{6.6}$$

$$HCN \rightarrow CH_3NH_2 \tag{6.7}$$

6.2 DETAILED MECHANISTIC STUDIES

Seefeldt and Dean[7] describe the nitrogenase enzyme's so-called Fe cycle in the following steps:

1. Two moles of MgATP bind to the reduced Fe–protein containing the [4Fe–4S] cluster in the 1^+ oxidation state as $[Fe_4S_4]^+$. During this step, Fe–protein polypeptide conformation changes take place, altering the electronic properties of the [4Fe–4S] cluster.

2. Fe–protein interacts with MoFe–protein. Correct docking of Fe–protein to MoFe–protein is associated with conformational changes taking place during step 1. Steps 1 and 2 are prerequisites for all following nitrogenase reactions and for substrate reduction.
3. Intercomponent electron transfer takes place from reduced Fe–protein to MoFe–protein as Fe–protein's [4Fe–4S] cluster moves to the 2^+ oxidation state, $[Fe_4S_4]^{2+}$.
4. Two moles of MgATP are hydrolyzed. Hydrolysis of MgATP to MgADP is involved in timing intercomponent electron transfer.
5. Fe–protein dissociates from MoFe–protein (the rate-determining step).
6. Two moles of MgADP and two P_i are released.
7. Fe–protein is reduced to complete the cycle.

The so-called midpoint potential, E_m, of protein-bound [Fe–S] clusters controls both the kinetics and thermodynamics of their reactions. E_m may depend on the protein chain's polarity in the vicinity of the metal–sulfur cluster and also upon the bulk solvent accessibility at the site. It is known that nucleotide binding to nitrogenase's Fe–protein, for instance, results in a lowering of the redox potential of its [4Fe–4S] cluster by over 100 mV. This is thought to be essential for electron transfer to MoFe–protein for substrate reduction.[11b]

Researchers Jang, Seefeldt, and Peters[10] found that changes in polarity near the $[Fe_4S_4]$ cluster of Fe–protein, in its nucleotide-free form, caused changes in its midpoint potential, E_m. In this case the replacement of phe135 by tryptophan, a F135W mutation, changed E_m from -310 mV for the native protein to -230 mV for the F135W mutant. The change took place in spite of the fact that X-ray crystallographic structure of the mutant shows very small differences when compared to that of the native Fe–protein. Several small structural changes in the $[Fe_4S_4]$ cluster's locale are believed responsible. These include trp's indole amide nitrogen atom residing $\sim$4.4 Å from the inorganic sulfur of the $[Fe_4S_4]$ cluster, inclusion of a water molecule between the trp indole amide and the inorganic S atoms, and movement of the trp peptide amide to within $\sim$3.6 Å of the $[Fe_4S_4]$ cluster's inorganic sulfur. The authors conclude that orientation and proximity of dipoles in the vicinity of the $[Fe_4S_4]$ cluster contribute to a mechanism for E_m modulation.

Lanzilotta and co-workers have elucidated electron transfer scenarios from Fe–protein to MoFe–protein and within MoFe–protein itself using EPR and changes in midpoint potentials (E_m) of the metal–sulfur clusters.[11] In direct EPR evidence for electron transfer from Fe–protein's $[4Fe–4S]^+$ cluster to P-clusters of MoFe–protein, the authors found that the EPR signal for Fe–protein's $[4Fe–4S]^+$ disappears as MoFe–protein's P-cluster changes from its P^{2+} to its P^+ state. The electron transfer appears to take place only in the presence of MgATP and not MgADP. The authors conclude that the first electron transferred goes into MoFe–protein's P-cluster and that the MgATP-*bound* protein conformation is necessary for the process but that MgATP hydrolysis is *not* required for the electron transfer.

In reference 11a the authors present evidence for significant shifts in midpoint potentials for nitrogenase metal centers as a result of Fe–protein binding to MoFe–protein. The midpoint potentials for the three nitrogenase metal centers, namely the [4Fe–4S] cluster of the Fe–protein, the [8Fe–7S] P-cluster, and FeMoco or M center of the MoFe–protein, were determined within a nondissociating nitrogenase complex prepared with a site-specifically altered Fe–protein (leu at position 127 deleted, L127Δ). The use of a leu127Δ deletion mutation mimics the change induced by MgATP binding to the protein complex. The midpoint potential for each metal center was determined by mediated redox titrations, with the redox state of each center being monitored by parallel and perpendicular mode EPR spectroscopy. The midpoint potential of the Fe–protein $[4Fe\text{–}4S]^{2+/1+}$ cluster couple was observed to change by −200 mV from −420 mV in the uncomplexed L127Δ Fe–protein to −620 mV in the L127Δ Fe–protein/MoFe–protein complex. Presumably the E_m change would facilitate electron transfer into MoFe–protein. Also, the midpoint potential of the two-electron oxidized couple of the P-cluster (P^{2+}) of MoFe–protein was observed to shift by −80 mV upon protein–protein complex formation. No significant change in the midpoint potential of an oxidized state of FeMoco (M^{OX}) was observed upon complex formation. The results suggest that the energy of protein–protein complex formation is coupled to an increase in the driving force for electron transfer by 80 mV without involvement of the FeMoco cluster.

The authors of reference 12 have presented spectroscopic evidence for changes in the redox state of nitrogenase's P-cluster. It is known that serβ188 is an important ligand for one iron atom in MoFe–protein's P-cluster. This amino acid residue coordinates to an iron atom through its γ O in the P^{OX} (oxidized) state of the P-cluster but not in the P^{N} (reduced) state. Further description and discussion of the P^{OX} and P^{N} forms are found in Section 6.5.2. These researchers found that a $\beta188^{cys}$ mutant (cysteine substituted for serine at position β188) would reduce substrates H^+, C_2H_2, and N_2 at only 30% of the wild-type protein's rate and that, under the same conditions, the rate of MgATP hydrolysis was 80% of the wild-type protein's rate. The authors concluded that reduction and/or protonation of substrates and MgATP hydrolysis were somewhat decoupled. Stopped-flow experiments concurrent with EPR studies during dinitrogen reduction indicated that the P^{N}-cluster was oxidized following or during a four-electron reduction of FeMoco. This experiment led to the conclusion that P-clusters undergo redox changes during nitrogenase turnover in the $\beta188^{cys}$ mutant studied, supporting the nitrogenase model in which P-clusters function as intermediate electron transfer sites for electrons that ultimately reach substrates.

Researchers studying the stepwise kinetics of nitrogenase electron transfer using stopped-flow kinetic techniques have presented other scenarios. One hypothesis presents kinetic evidence that dissociation of Fe–protein from MoFe–protein is not necessary for re-reduction of Fe–protein by flavodoxins.[13] These authors state that the possibility of ADP–ATP exchange while Fe–protein and MoFe–protein are complexed with each other cannot be excluded and that dissociation of the complex during catalysis may not be obligatory when flavodoxin is the Fe–protein reductant. This leads to the hypothesis that MgATP binds to the preformed Fe–protein/

MoFe–protein complex, causing the essential conformational changes necessary for electron transfer.

Because the six-electron reduction of N_2 to NH_3 is thermodynamically favorable at pH 7 when coupled to ferredoxin or flavodoxin reductants, other researchers believe that the MgATP may be required for some kinetic reason such as overcoming a high activation energy or ensuring irreversibility of the electron transfer steps.[6] Other possibilities are that MgATP hydrolysis provides energy to stabilize a partially reduced intermediate during sequential delivery of electrons from Fe–protein to MoFe–protein or that MgATP binding and hydrolysis open and close electron gates, ensuring that multiple electrons are accumulated within MoFe–protein prior to substrate reduction.[7]

The so-called Fe cycle of nitrogenase must be repeated in two-electron increments for each step of N_2 reduction, assuming that the first step would be formation of the thermodynamically unstable diazene (N_2H_2) species followed by its reduction to hydrazine (N_2H_4), then finally to ammonia. Intermediate structures have been proposed in which a partially reduced and protonated N_2 (as N_2H_4 perhaps) migrates to the molybdenum atom of the M center for final reduction, cleavage of the N–N bond, and protonation to yield NH_3.[14] Other experimental evidence from EPR, to be discussed in Section 6.4, also indicates that ATP hydrolysis and electron transfer are not directly coupled. Figure 6.3 indicates the possibilities for N_2 reduction in the absence (through N_2H_2 generation) or presence (through M–N_2H_2 generation) of dinitrogen–metal cluster interaction.

Chemists believe that understanding nitrogenase's mechanism for small molecule reductions, along with the designing of model compounds to mimic its activity, will be important in finding efficient catalysts for industrial reductive preparations under milder conditions than that of the Haber–Bosch process. However, 30 years of work in this field has not yielded a viable alternative to the Haber–Bosch process; and although much more is known presently about nitrogenase's structure and function, many questions remain concerning the exact mechanism of activation and

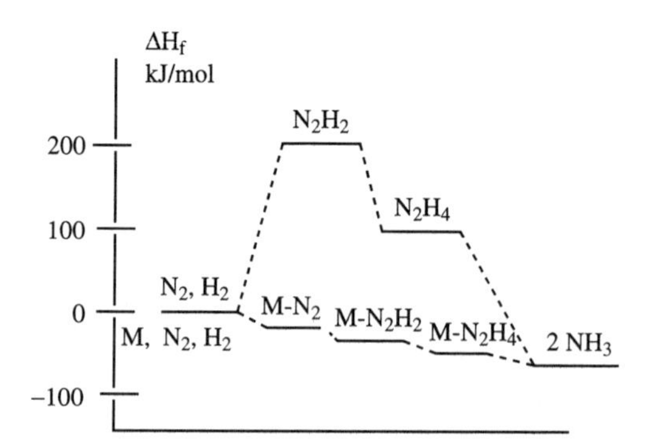

Figure 6.3 Energy changes for N_2 reduction in the absence (top) or presence (bottom) of metal–sulfur complexes. (Adapted with permission from Figure 7b of Sellman, D., Sutter, J. *Acc. Chem. Res.*, 1997, **30**, 460–469. Copyright 1997, American Chemical Society.)

reduction of dinitrogen in this complex enzyme. All researchers involved in the study of nitrogenase's mechanism of electron transfer and substrate reduction believe that further experimentation will be necessary to define the correct mechanistic steps for the process. Specific structural details of nitrogenase's metal–sulfur clusters are discussed in greater detail in the following sections.

6.3 IRON–SULFUR CLUSTERS

To successfully describe the structure and function of nitrogenase, it is important to understand the behavior of the metal–sulfur clusters that are a vital part of this complex enzyme. Metal–sulfur clusters are many, varied, and usually involved in redox processes carried out by the protein in which they constitute prosthetic centers. They may be characterized by the number of iron ions in the prosthetic center; that is, rubredoxin (Rd) contains one Fe ion, ferredoxins (Fd) contain two or four Fe ions, and aconitase contains three Fe ions.[7] In reference 18, Lippard and Berg present a more detailed description of iron–sulfur clusters; only the $[Fe_4S_4]$ cluster typical of that found in nitrogenase's Fe–protein is discussed in some detail here. The P-cluster and M center of MoFe–protein, which are more complex metal–sulfur complexes, are discussed in Sections 6.5.2. and 6.5.3.

Helmut Beinert, Richard Holm, and Eckard Münck have discussed iron–sulfur clusters as modular, multipurpose structures and have included a useful reaction scheme illustrating synthetic processes, interconversions and forms found in proteins.[15] Metal–sulfur clusters are held in place within proteins by coordination of the thiolate sulfur of cysteine residues to the iron ions. The basic structure of iron–sulfur clusters can be visualized as distorted cubes of interpenetrating Fe_4 and S_4 (as inorganic S^{2-} ions) tetrahedra.[18] (See Figure 1.10 of Chapter 1 for a representative structural drawing.) Iron–iron and sulfur–sulfur distances of approximately 2.75 Å and 3.55 Å, respectively, are found. Iron(II) and iron(III) oxidation states in tetrahedral coordination spheres usually generate high-spin d^5 and d^6 systems. Multiple ferro- and antiferromagnetic couplings of iron spin states existing between iron ions, and through the sulfur bridging ligands, lead to a rich variety of electron paramagnetic resonance (EPR) spectra for the clusters. These EPR spectra are often characteristic for given oxidation states of the iron ions within the cluster and are used to identify them. Mössbauer spectroscopy, (see Section 3.6), also may be used to determine the oxidation state of iron within the clusters. Note that lower iron oxidation states exhibit larger isomer shifts. Table 6.1 contains some typical 4Fe:4S cluster data.

The fourth state with $[Fe_4S_4]^0$ shown in Table 6.1 was recently described as the most reduced form possible for the Fe–protein's $[Fe_4S_4]$ cluster.[16] Usually, only two oxidation states for a given metal–sulfur cluster are stable. Therefore a stable $[Fe_4S_4]^0$ state in Fe–protein's iron–sulfur cluster (as appears likely from experimental evidence presented in reference 16) would be unique because the cluster would then have three stable oxidation states, $[Fe_4S_4]^{2+/1+/0}$. It appears also that the all-ferrous state is only stable in the protein-bound cluster and not for model

Table 6.1 Characteristics of 4Fe–4S Iron–Sulfur Clusters

Oxidation State	Formal Valence	EPR *g* Values (Temperature)	Mössbauer Isomer Shift, δ (mm/s)	λ_{max} (nm), (Extinction Coefficient, $\times 10^{-3}$, per Fe)
Oxidized $[Fe_4S_4]^{3+}$	3 Fe^{3+} 1 Fe^{2+}	2.04, 2.04, 2.12 (<100 K)	0.31	325 (8.1) 385 (5.0) 450 (4.6)
Intermediate $[Fe_4S_4]^{2+}$	2 Fe^{3+} 2 Fe^{2+}	None	0.42	305 (4.9) 390 (3.8)
Reduced $[Fe_4S_4]^{+}$	1 Fe^{3+} 3 Fe^{2+}	1.88, 1.92, 2.06 (<20 K)	0.57	Unfeatured
$[Fe_4S_4]^0$	4 Fe^{2+}	Complex	0.68	

Source: Adapted from reference 18.

4Fe–4S complexes. The P-cluster in MoFe–protein is a variant of a 4Fe–3S cluster having the formula Fe_8S_7. The P-cluster's structure is best described as two cuboidal Fe_4S_3 fragments bridged by a central sulfur atom. The M center of MoFe–protein, usually described as the iron–molybdenum cofactor or FeMoco, has an $MoFe_7S_9$ core with cuboidal Fe_4S_3 and $MoFe_3S_4$ clusters bridged by three μ_2-S^{2-} ions. These clusters will be discussed in Section 6.5.

Iron–sulfur clusters have been "self-assembled" by chemists since the 1980s according to the following reactions.[17] This work has continued to the present time as discussed below in Section 6.6.

$$4FeCl_3 + 12RS^- \rightarrow 4Fe(SR)_3 + 12Cl^- \tag{6.8}$$

$$4Fe(SR)_3 \xrightarrow{4OMe^-, 4HS^-} [Fe_4S_4(SR)_4]^{2-} + RSSR + 6RS + 4MeOH \tag{6.9}$$

Core extrusion studies—removal of the iron–sulfur cluster intact from the enzyme surroundings—have been carried out and the iron-cluster types in proteins identified through the process shown in equation 6.10.[18] DMSO/H_2O is the protein unfolding solvent for this process. By this method, Fe–protein and MoFe–protein metal–sulfur clusters have been removed from the holoenzyme for separate analysis by many instrumental techniques.

$$\text{holoprotein} + \text{xsRSH} \xrightarrow{80\%\text{v/v DMSO/H}_2\text{O}} [Fe_nS_n(SR)_4]^{2-} (n = 2, 4) + \text{apoprotein} \tag{6.10}$$

In the fall of 2000 an international workshop on the chemistry and biology of iron–sulfur clusters was held at Virginia Polytechnic Institute and State University, Blacksburg.[19] Many functions for iron–sulfur clusters besides electron transfer and binding and activating of substrate were discussed at this conference. Researchers have discovered that clusters may regulate gene expression, serve as sensors for iron, oxygen, superoxide and other molecules, deliver sulfur and possibly iron for the synthesis of other biomolecules, and play a role in generating radicals. It is also now realized that biologically active metal clusters do not spontaneously self-assemble as described above but are synthesized by proteins expressed by very specific genes and inserted into the proper proteins by other specific biological mechanisms. However, the synthetic work of the Holm group has been extremely useful in understanding the biological synthetic process because every cluster conversion previously found in the laboratory has been found to also operate in biological species.

Two genes coding for protein products required to synthesize iron–sulfur cluster cores for nitrogenase are called NifS and NifU, where Nif indicates that they are needed for nitrogen fixation. Another set of genes, or operon, has been identified that produces six proteins, all of which appear to be necessary for iron–sulfur cluster formation in diverse species including bacteria, yeast, plants, and humans. One protein product, called IscU where Isc stands for iron–sulfur cluster, is central to the assembly process for Fe_2S_2 and Fe_4S_4 clusters. Another protein product, IscS, appears to be a major trafficker of intracellular sulfur, providing the element to many biochemical entities including iron–sulfur clusters. Metallochaperone proteins, similar to those discussed for copper ions in Section 7.4, appear to be part of the cluster synthesis, transport, and protein insertion process. At the same conference[19] it was reported that the bacterial genes needed to synthesize iron–sulfur clusters have homologs in yeast and other eukaryotes, and in fact an essential function of mitochondria in yeast is production of iron–sulfur clusters. A hypothetical process, presented in this article, for Fe–S cluster formation in mitochondria includes (1) an iron transport protein to bring iron into the mitochondria, (2) an iron–sulfur cluster assembly apparatus that obtains sulfur atoms from cysteine (producing alanine), and (3) an iron–sulfur cluster transporter protein that carries the completed cluster out of the mitochondria and inserts it into the proper apoprotein. Researchers attending the conference agreed that inorganic chemists, biologists, biochemists, geneticists, and spectroscopists must work closely together to further elucidate the mechanisms of biological iron–sulfur cluster synthesis, transport, and utilization.

6.4 Fe–PROTEIN STRUCTURE

Fe–protein, the unique, highly specific electron donor to MoFe–protein, mediates coupling between ATP hydrolysis and electron transfer to MoFe–protein and also participates in the biosynthesis and insertion of FeMoco into MoFe–protein. Fe–protein contains one ferredoxin-like $[Fe_4S_4]^{2+/1+}$ cluster as its redox center. There is now evidence for an $[Fe_4S_4]^0$ "super-reduced" state in which four high-spin iron(II) ($S = 2$) sites are postulated. These were previously discussed in Section 6.3 and illustrated in Table 6.1.[16] The $[Fe_4S_4]$ cluster in this state bridges a dimer of identical subunits in Fe–protein. Dithionite ($S_2O_4^{2-}$)-reduced proteins can be isolated and are paramagnetic, exhibiting electron paramagnetic resonance (EPR) spectra with g factors characteristic of $[Fe_4S_4]^+$ centers.[20] (See Table 6.1.) The cluster exhibits a spin mixture of $S = 1/2, 3/2$. When ATP is bound, the $g = 1.94$ signal changes from rhombic ($x \neq y \neq z$) to axial ($x = y \neq z$).

Fe–protein belongs to a large class of proteins that couple binding and hydrolysis of nucleotide triphosphates (ATP in this case) to changes in protein conformation, an allosteric interaction. Because the phosphate groups of the ATP nucleotide are ~20 Å from the 4Fe:4S cluster, it is unlikely that ATP hydrolysis and electron transfer are directly coupled. Rather, allosteric changes at the subunit interface with participation of amino acid residues are thought to be responsible for connecting electron transfer with ATP hydrolysis. In nitrogenase the conformational changes are used to drive electron transfer from the 4Fe–4S cluster of Fe–protein to the substrate reduction site of MoFe–protein with electrons passing through the P-cluster of MoFe–protein. Intimate details and timings for assembly and disassembly of the Fe–protein/MoFe–protein complex are not well understood at this time. However, as stated above, it is believed that dissociation of the protein complexes is the rate-determining step in electron transfer.

The X-ray crystallographic structure of the Fe–protein from *Azobacter vinelandii*, called Av2, was first solved at 2.9-Å resolution.[6] The designation Av2 denotes the genus, the species, and Fe–protein homodimer as component 2 of the complete nitrogenase protein. This structure is found in the Protein Data Bank (PDB) as 1NIP[6] and 2NIP (the updated structure).[21] Figure 6.4 shows the stereodiagram of Cα trace (the trace of the alpha carbon atoms of each amino acid residue of the protein chains) for the Av2 Fe–protein looking down the dimer twofold axis.

The two protein subunits found in the holoenzyme, each containing an iron–sulfur cluster, fold into interconnected α-helices and β-pleated sheet that, at their interface, ligate the 4Fe–4S cluster through four cysteine residues, two from each

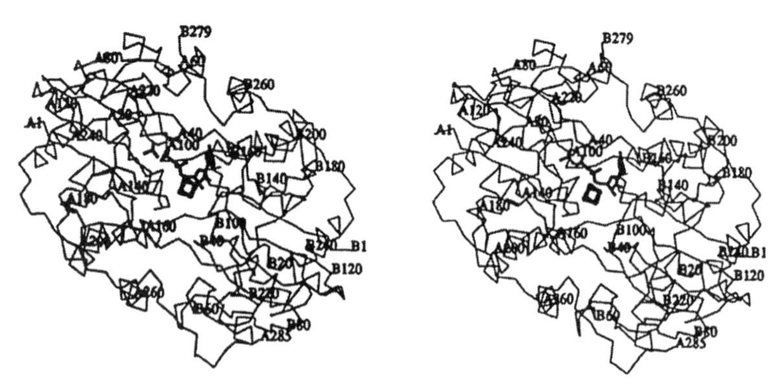

Figure 6.4 Stereodiagram of the Cα trace for Av2 Fe–protein. (Reprinted with permission from Georgiadis, M. M.; Komiya, H.; Chakrabarti, P.; Woo, D.; Kornuc, J. J.; Rees, D. C., *Science*, 1992, **257**, 1653–1659. Copyright 1992, American Association for the Advancement of Science.)

Fe–protein subunit. Specifically, these would be identified as (1) cysα97 and cysα132 and (2) cysβ97 and cysβ132. The shape of the overall fold is said to resemble an iron butterfly[2] with the iron–sulfur cluster representing the butterfly's head. This type of fold is common to many nucleotide-binding proteins. Both cysteine ligands are located near the NH_2-terminal end of α-helices, and these and other side chain interactions provide at least six potential NH–S hydrogen bonds stabilizing the iron–sulfur cluster's position. The residues arg100 and glu112 have been identified on the surface of Fe–protein as likely salt bridge formation sites with MoFe–protein these residues are also close to the 4Fe–4S cluster. Such salt bridges are known to be good stabilizers of protein quaternary structure. In reference 6, the authors modeled an ADP molecule also coordinated at the subunit interface at a distance of 20 Å from the iron–sulfur cluster, too far away to permit direct chemical coupling of electron transfer and ATP hydrolysis. The authors believe that conformational changes in Fe–protein influence its ability to receive electrons from electron donors (such as ferredoxin) and transfer them to electron acceptors (MoFe–protein), and they describe Fe–protein's function as that of a "molecular clock." Fe–protein's cofactor, the 4Fe–4S cluster, may be extracted intact for analysis as described in Section 6.3, and the extracted clusters may be added to apoenzyme to regenerate active enzyme.

In another refinement of the Fe–protein's X-ray structure,[22] PDB code 1N2C, it is shown that structural changes within Fe–protein facilitate docking with MoFe–protein and expose Fe–protein's iron–sulfur cluster to MoFe–protein for electron transfer. This structure contains an $ADP{\cdot}AlF_4^-$ molecule that models ATP/ADP interactions within Fe–protein. Adopting the nomenclature of reference 22, residues from both subunits of Fe–protein are labeled γ whereas MoFe–protein residues are labeled with α or β according to their subunit of origin. Interactions of MoFe–protein valα158 and leuβ157 with sulfurs of the Fe–protein 4Fe–4S cluster link its surface to MoFe–protein. Carbonyl oxygens of leuα123 and valα124 as well as leuβ123 and valβ124 form main-chain hydrogen bonds to both cysγ97's, which are in turn ligands of the 4Fe–4S cluster. The two Argγ100's of the α and β subunits of Fe–protein are of particular importance in side-chain interactions between it and MoFe–protein. These residues protrude into the MoFe–protein surface forming ionic and hydrogen bonds. Within Fe–protein, interactions between the $ADP{\cdot}AlF_4^-$ molecule (that lies parallel to the interface between Fe–protein subunits) and Fe–protein take place through side-chain oxygens of serγ16 and thrγ17. Negative charges on the phosphates and AlF_4^- are compensated by lysγ15 and lysγ41. The adenosine portion of the $ADP{\cdot}AlF_4^-$ is fixed through ribose O2′ and O3′ hydrogen bonds to the gluγ221 side chain, and the adenine N1 and N7 atoms are hydrogen bonded to the main-chain amide of aspγ214 and to the aspγ185 side chain, respectively. Several hydrophobic residues have stacking interactions with the adenine ring and the exocyclic amino group of adenine is hydrogen bonded to the main-chain oxygen of proγ212.

In reference 21, the Fe–protein X-ray structures of *A. vinelandii*, Av2, PDB code 2NIP, at 2.2-Å resolution is compared to that of *C. pasteurianum*, Cp2, PDB code 1CP2, at resolution 1.93 Å as well as to Fe–protein aa sequences and structures of other closely related protein species that similarly bind nucleotides. As described by these authors, nucleotide-binding proteins have common core elements that are characterized by (1) predominantly parallel β-sheet (eight-stranded in Av2) flanked by α-helices (numbering nine in Av2), (2) a phosphate binding loop (P-loop) containing GXXXXGKS/T consensus sequences (residues 9–16 in Av2), and (3) Switch I and Switch II regions (residues ~38–43 and 125–135, respectively, in Av2). The Switch I and Switch II regions undergo conformational changes upon hydrolysis of ATP to ADP and pass this information along to other regions of the molecule. The P-loop, Switch I, and Switch II regions include key aa residues involved in nucleotide binding to Fe–protein.

As mentioned above for the 1NIP structure, the formation of NH–S hydrogen bonds between main-chain amide groups and sulfur atoms of the cluster and thiol groups of cysteine ligands are important for maintaining Fe–protein structure. The NH–S bonds, however, do vary in number, and residues involved, when considering the different Fe–proteins studied, that is, 2NIP,[21] 1CP2,[21] and 1N2C.[22] Differences are especially notable between the two nucleotide-free Fe–protein structures (2NIP, 1CP2) and the 1N2C structure that contains Fe–protein and MoFe–protein as well as nucleotide. The reference 21 authors conclude that Fe–protein is a molecule of multiple functions and conformations, that a continuum of Fe–protein structures exist, and that both these facts reinforce the perception of nitrogenase as a dynamic enzyme system.

6.5 MoFe–PROTEIN STRUCTURE

6.5.1 Overview

MoFe–protein, a tetramer containing two α and two β protein subunits, is the site of dinitrogen (substrate) binding and reduction. The complete unit is called component 1, and therefore references to it would take the form Av1 for the MoFe–protein component of *A. vinelandii*, for instance. Two distinct types of redox centers are present in MoFe–protein: the P-cluster and the iron–molybdenum cofactor (FeMoco or M center). The P-cluster exists as an unusual double iron–sulfur cluster. It is believed to function as an electron mediator and accumulator, taking electrons from Fe–protein's iron–sulfur cluster and passing them along to the site of N_2 reduction at the M center. Knowledge of the P-cluster's structure has evolved since the initial X-ray crystallographic structure was described in reference 6. Details will be discussed in Section 6.5.2. The M center is unusual in that it contains three-coordinate iron ions in its cluster rather than the usual four-coordinate metal centers. Researchers believe that the coordinate unsaturation in the M center may facilitate attachment of N_2 for its reduction and protonation. Further details are discussed below in Section 6.5.3. Proposed structures of N_2–M center complexes are also described further in Section 6.6.

The P-cluster and M centers are about 14 Å apart; therefore, electron transfer between them is not easy to explain. Electron transfer is believed to take place through electron tunneling involving the protein's amino acid side chains. Each

MoFe–protein αβ dimer is believed to function independently, and each contains an independently operating M center and P-cluster.[4] (See Figure 6.1.) Because the two FeMoco's contained in the αβ dimer function separately, the possibility of binuclear coordination of substrate dinitrogen to two Mo metal centers is eliminated. This mode of dinitrogen attachment was proposed before the X-ray crystallographic structure of FeMoco was published. Binuclear N_2 coordination to molybdenum was a logical assumption for nitrogenase because an Mo–μ–N_2–Mo attachment is a well-known and researched phenomenon in small-molecule organometallic chemistry.

It is now believed that the MoFe–protein's P-cluster contains a [4Fe–3S] cuboid joined to a [4Fe–4S] cuboid, although, as discussed below, it was first reported crystallographically as two [4Fe–4S] clusters.[8] Uncertainty existed for sometime as to exact nature of bridging disulfide or sulfide ligand joining the two Fe:S clusters but it is now known that the P-cluster does NOT contain a disulfide bond. This is important because the all-ferrous structure $[4Fe–4S]^0$ proposed from Mössbauer studies then becomes more possible for the P-cluster's [4Fe–4S] cube. In 1993 Bolin et al.[1] proposed a six-coordinate S for the P-cluster's center as in Figure 1a,b of Thorneley's article.[8] This is now believed to be the correct conformation. A central six-coordinate S makes this cluster much harder to synthesize in the laboratory, and this feat has not been accomplished as of the date of this text's publication. Whatever its oxidation state or structure, the P-cluster mediates electron transfer from Fe–protein to the M center of MoFe–protein, and it must be reduced at some point to allow transfer of its electron(s).

As has been stated previously, a hypothesis put forward to explain MoFe–protein's activity[8] is that a protein conformation change in MoFe–protein is necessary for electron transfer from P-cluster to its associated FeMoco (M center) and that this is driven by ATP hydrolysis and P_i release. The latter processes are believed to occur on the Fe–protein. The fact that release of inorganic phosphate occurs after oxidation of Fe–protein supports this hypothesis. Available data do not suggest that substrates bind to P-clusters, but mutagenesis of βser188 (shown to coordinate to the P-cluster in the discussion below) to gly or βcys153 to ser decreases reduction activity in the enzyme. Therefore there is linkage between the protein environment of the P-cluster and the protein interface between the Fe– and MoFe–proteins. Supporting evidence for this hypothesis is suggested by transient absorbance changes in MoFe–protein that are linked to P-cluster oxidation. Thorneley's suggestion is that P-clusters provide a reservoir of reducing equivalents that are transferred to the substrate bound at the FeMoco. Electrons are transferred one at a time from Fe–protein, while the P-cluster and FeMoco each have "electron buffering" capacity allowing successive two electron transfers to and protonations of substrates bound to FeMoco.

The same author[8] discusses hypotheses put forward to explain the M center's contribution as the substrate binding site. Some of the known behaviors include the following:

1. Residues in vicinity of the M center modulate its spectroscopy.

2. An extensive H-bonding network exists in the vicinity of homocitrate and nearby bridging sulfides.
3. The MoFe–protein's environment is "tuning" the cofactor FeMoco to maximize N_2 reduction and minimize H_2 production.

Relative rates of N_2 and H_2 production are dependent on the metal coordinated to αhis422 (Mo, V, or Fe). V and Fe nitrogenases produce more H_2 relative to the amount produced by a Mo-containing enzyme. High electron flux through MoFe–protein favors NH_3 production, whereas at low electron flux, H_2 is much favored. Relative and absolute concentrations of Fe– and MoFe–proteins, nature and concentration of reductant (flavodoxin or dithionite), and other parameters such as temperature, pH, and ionic strength modulate dinitrogen reduction. To maximize dinitrogen reduction and ammonia production, one must have a very specific environment and a high electron flux. Thorneley's steady-state kinetic diagram indicates that states E_3 or E_4 must be reached to bind or reduce dinitrogen as shown in Figure 3 of reference 8 and as shown as Figure 6.5. In Figure 6.5 the various nitrogenase-reduced states (E_n) indicate the number of completed Fe–protein cycles during which one electron is transferred to the MoFe–protein. E_3 and E_4 are states in which N_2 binds. States E_2–E_4 may release H_2; and with acid quenching, NH_3 is released from E_5. The binding site and mode of N_2 binding are extremely important to the function of the enzyme. These will be discussed in Sections 6.5.2 and 6.5.3.

EPR, ENDOR, Mössbauer, EXAFS, and MCD spectroscopies to further elucidate intimate details of electron and proton transfer within nitrogenase are difficult

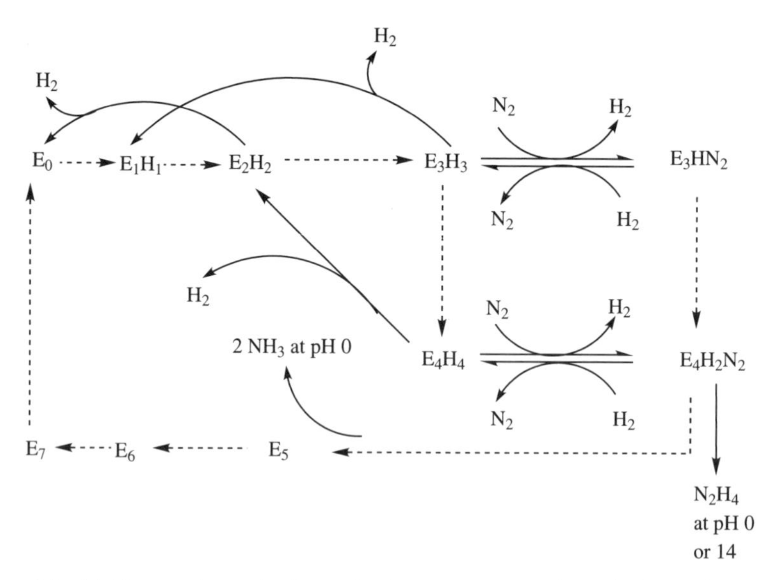

Figure 6.5 MoFe–protein cycle for reduction and protonation of N_2. (Adapted from Figure 3 of Thorneley, R. N. F.; Lowe, D. J. *J. Biol. Inorg. Chem.*, 1996, **1**, 576–580. Copyright 1996, Society of Biological Inorganic Chemistry.)

because the important intermediates are only present at low concentrations for 1–5 seconds and require rapid-freezing or chemical-quench techniques for their detection. In the future, stopped-flow FTIR may provide time-resolved spectra of intermediates, products, and inhibitors. In spite of these difficulties, structural details of the P-cluster and M center in MoFe–protein have evolved since the first X-ray crystallographic data were published in 1992–1993. Incorporation of EXAFS, EPR, Mössbauer, and other analytical data plus information gathered from model compound syntheses have led to the structural refinements outlined in the following two sections.

6.5.2 Details of the P-Cluster

The P-cluster, located at the interface of MoFe–protein's α- and β-subunits, is believed to function as the electron transfer mediator between Fe–protein and the N_2 reduction site at the M center. The P-cluster is contained within a hydrophobic environment and located approximately 10 Å below the MoFe–protein surface. Three cysteine side chains from each subunit bind to iron ions in the P-cluster. The cluster is now known to exist in P^{OX} and P^{N} forms in active enzyme, both with stoichiometry Fe_8S_7. The P^{N} form, with its octahedrally coordinated central sulfur, has the structure shown in Figure 6.6. As can be seen in Table 6.3, the P^{N} form contains all ferrous irons, corresponding to the P ($S = 0$) state, whereas the P^{OX} form corresponds to the P^{2+} ($S = 3$ or 4) form.

Dithionite-reduced P-clusters are diamagnetic and consistent with an all-ferrous state having $S = 0$. Oxidation of the reduced P-cluster results in iron–sulfur clusters with the following spin identities:

1. P ($S = 0$)
2. P^{+} ($S = 1/2, 5/2$)
3. P^{2+} ($S = 3$ or 4)
4. P^{3+} ($S = 1/2, 7/2$)

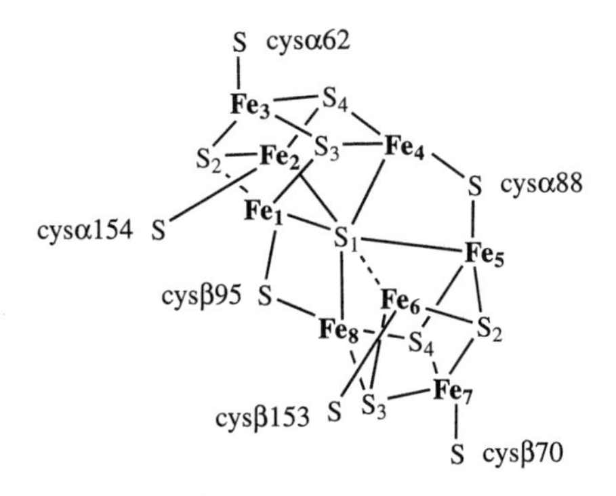

Figure 6.6 P^{N} form of nitrogenase's P-cluster.

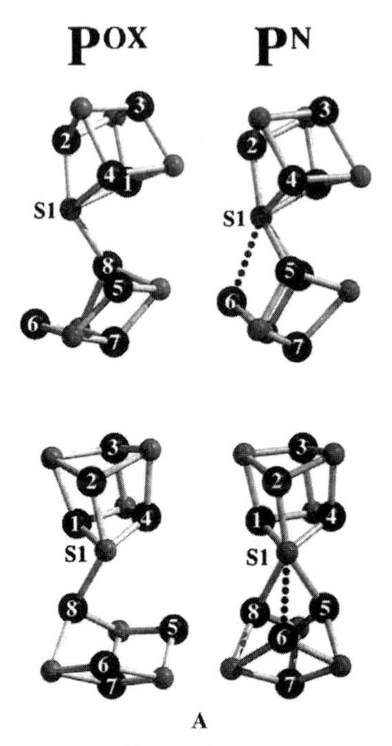

Figure 6.7 (A) Structural models of P^N and P^{OX} as reprinted from Figure 2 of reference 24. (B) Comparison of P^N and P^{OX} structures as reprinted from Figure 4 of reference 24. (Both figures reprinted with permission from Peters, J. W.; Stowell, M. H. B.; Soltis, S. M.; Finnegan, M. G.; Johnson, M. K.; Rees, D. C. *Biochemistry* 1997, **36**, 1181–1187. Copyright 1997, American Chemical Society.)

For P^{2+} detection by parallel mode EPR shows an unusual $g = 12$ excited state providing a spectroscopic signature for the presence of P-clusters.[8] The ability to undergo multielectron oxidation–reduction reactions assists the P-cluster's electron transfer capability. Many polar and charged groups surround the metal centers stabilizing anionic species generated during reduction. Some of these may also serve as a "bucket brigade" for protons, which must also arrive at the substrate reduction site. The edge–edge distance between FeMoco (M center) and the P-cluster is approximately 14 Å. Neither cluster has been chemically synthesized, although some model clusters have been prepared (see the work of Holm's group in Section 6.6).

As stated above, the P-cluster pair was originally described as containing two conjoined Fe_4S_4 cubane clusters coupled by two bridging cysteine sulfurs and linked by a disulfide bridge.[23] Further structure refinement of the X-ray crystallographic structure to 2.0-Å resolution led to the realization that the original crystals studied contained co-crystallized P^{OX} (oxidized) and P^N (reduced) forms of the cofactor.[24] With the iron atoms labeled as in Figures 2 and 4 from reference 24, one sees that cysα62 and cysα154 coordinate Fe3 ($Fe^{2.5+}$ in P^{OX}) and Fe2 ($Fe^{2.5+}$ in P^{OX}), respectively, while cys α88 provides a bridging cysteinyl ligand coordinating

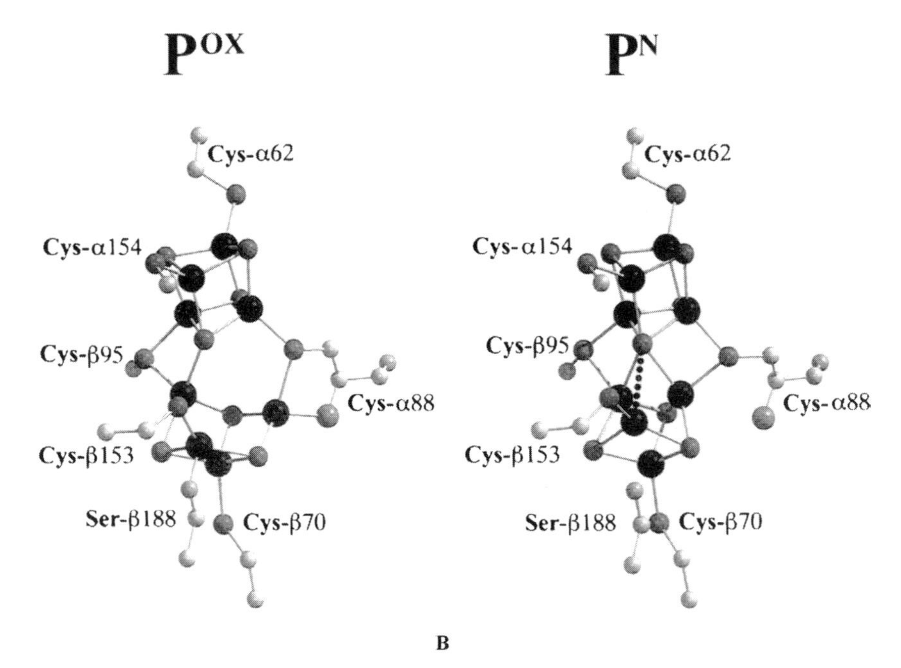

Figure 6.7 (*Continued*)

Fe4 and Fe5 (both Fe^{2+} in P^{OX}). P-cluster structural models are reprinted from reference 24 in Figure 6.7 (see also Figure 6.6).

Cysβ70 and cysβ153 coordinate Fe7 (Fe^{2+} in P^{OX}) and Fe6 (Fe^{3+} in P^{OX}), respectively, while cysβ95 bridges Fe1 and Fe8 (both Fe^{2+} in P^{OX}). Reduction of P^{OX} to P^{N} results in movement of the Fe5 and Fe6 atoms and rearrangement of the covalent ligands. In the P^{OX} state six cysteine ligands coordinate the P-cluster to the protein. Additionally, cysα88 coordinates Fe5 through a backbone amide ligand and the Oγ of serβ188 coordinates Fe6 (in addition to Fe6's ligation to cysβ153). In P^{N} the two noncysteinyl ligands are replaced by interactions with the central S atom (which then assumes distorted S_6 octahedral geometry). In both oxidized and reduced forms, all irons in the P-cluster remain four-coordinate. Metal–metal bond distances become shorter in the P-cluster reduced form (P^{N}), although the opposite trend might be expected (addition of electrons to the valence shell of metals usually causes ligands or other metals to move away unless Me–Me bonding is invoked). The bond shortening is confirmed by X-ray as well as EXAFS data. Selected bond lengths for oxidized and reduced forms of the P-cluster from X-ray data[24] are collected in Table 6.2. All Fe–Fe distances are shorter in the reduced P-cluster form; movement of Fe_4, Fe_5, and Fe_6 are especially pronounced.

EXAFS data published in 1995 indicate the same contraction in metal–metal distances as the P-cluster is reduced.[25] In the thionine-oxidized P^{2+} form, short Fe–Fe distances are found at 2.74 Å; in the resting P form, short Fe–Fe distances are around 2.67 Å. Long Fe–Fe distances (cross cluster) of approximately 3.75 Å

Table 6.2 Bond Lengths for P^N and P^{OX} in Å

Bond	P^N	P^{OX}	Difference $P^{OX}-P^N$
Fe_1–Fe_2	2.42	2.43	+0.01
Fe_2–Fe_3	2.76	2.78	+0.02
Fe_3–Fe_4	2.59	2.69	+0.10
Fe_4–Fe_5	3.03	3.77	+0.74
Fe_5–Fe_6	2.56	3.88	+1.32
Fe_6–Fe_7	2.65	2.77	+0.12
Fe_7–Fe_8	2.62	2.72	+0.10
Fe_5–S_{center}	2.43	3.81	+1.38
Fe_6–S_{center}	2.92	3.86	+0.94

Source: Reference 24.

remain similar in oxidized, resting, and reduced forms. These authors also note expansion of Fe–S distances in the reduced enzyme.

In references 8 and 24 the authors speculate that the coordination sites of the Oγ of serβ188 and the amide nitrogen of cysα88 will be protonated in their free state (P^N) and deprotonated in their bound states (P^{OX}). Thus simultaneous release of two protons to be transferred to the M center site of substrate reduction and protonation may take place. This would couple proton and electron transfer in the P-cluster, agreeing with the stoichiometry of the nitrogenase reaction. Analysis of EPR and Mössbauer data confirm the presence of the P^{OX} and P^N P-cluster forms and result in a clearer picture of this complex cofactor.[26] Mössbauer data indicate that all atoms in the native P^N state are ferrous.[27] In spite of the fact that all Fe–S clusters contain high-spin iron, the cluster is EPR-silent. This indicates that, as with other biological iron–sulfur clusters, the P-cluster exhibits strong antiferromagnetic coupling between the metal sites mediated by the bridging sulfur ligands. One model for P^{OX} and P^N oxidation states and spin states for the two-cubane halves of the P-cluster is shown in Table 6.3.

The 1.6-Å resolution X-ray crystal structure of *Klebsiella pneumoniae* MoFe–protein, called Kp1, was solved in 1999 using the coordinates of the 2.2-Å Av1

Table 6.3 Oxidation and Spin States for Nitrogenase P-Cluster Iron Atoms

Name of State	Iron Oxidation States	$[Fe_4S_4]$ Core Oxidation State	Total Spin
P^N (P or P^0)	($4\ Fe^{2+}$), ($4\ Fe^{2+}$)	$[Fe_4S_4]^0$ $S_1=0$ $[Fe_4S_4]^0$ $S_2=0$	$S_t=0$
P^{OX} (P^{2+})	($2\ Fe^{2+}$, $2\ Fe^{2.5+}$), ($3\ Fe^{2+}$, $1\ Fe^{3+}$)	$[Fe_4S_4]^+$ $S_1=1/2$ $[Fe_4S_4]^+$ $S_2=7/2$	$S_t=3$ or 4

structure of reference 29a as a search model.[28] To date, this is the highest-resolution X-ray structure for any nitrogenase protein component. The dithionite-reduced P^N state is reported as PDB code 1QGU, the phenosafranin-oxidized state P^{OX} is reported as 1QH1, and an as-crystallized $P^{semi\text{-}OX}$ state is reported as 1QH8. The P^N state structure of Kp1 is largely in agreement with that of Av1 reported in reference 24, (see Table 6.2), except for the Fe_6–S_{center} bond distance (2.92 Å in Av1 and 2.47 Å in Kp1). For the P^{OX} state of Kp1 (1QH1) the central sulfur atom is coordinated to only four iron atoms with Fe5 and Fe6 having moved 1.45 and 1.36 Å further away from the central sulfur. (Compare to Fe_5–S_{center} and Fe_6–S_{center} bond differences of +1.38 and 0.94 Å in Table 6.2 from reference 24.) The reference 28 authors cannot confirm coupled electron and proton transfer mediated by P-cluster cysβ188–O^{γ}–serβ188 protonations or deprotonations mentioned as a possibility for Av1 by reference 24 authors. They do speculate that serβ90 of Kp1 may be implicated in this role. Mutagenic studies on the serβ90 (serβ92 in Av1) could confirm or deny this postulate.

6.5.3 Details of the M Center

Nitrogenase's M center is buried approximately 10 Å below the surface of the MoFe–protein α-subunit and is the site of substrate binding and reduction. The FeMoco (M center) is paramagnetic, having spin system $S = 3/2$, and can undergo a one-electron oxidation to a diamagnetic state.[8] EPR spectra of the dithionite-reduced FeMoco cluster yield a characteristic rhombic signal with g factors close to 4.3, 3.7, and 2.01 arising from the lowest transition of the $S = 3/2$ system. By analogy to the model compound $[Mo_2Fe_6S_8(SEt)_3(SR)_6]^{3-}$ with $S = 3/2$, the M center might contain (1) Mo(III), 5 Fe(III), and 2 Fe(II) ions or (2) Mo(IV), 4 Fe(III), and 3 Fe(II) ions.[38] These model compounds are discussed in Section 6.6. Iron spins may be ferro- or antiferromagnetically coupled to produce the total spin state. This factor plus the capability of iron and molybdenum to take on differing oxidation states and electron pairings into high-spin or low-spin configurations combine to make the exact nature of FeMoco metal centers ambiguous.

Structural models for the metal centers in MoFe–protein have been proposed based on X-ray crystallographic studies.[29] The original 2.2-Å resolution difference map for the FeMo cofactor (FeMoco or M center) placed two MFe_3S_3 (M = Mo or Fe) clusters linked by three nonprotein ligands: two bridging nonprotein sulfides and a third ligand, lighter than sulfur and possibly oxygen or nitrogen. The structure, deposited as PDB code 1MIN, has been declared obsolete and replaced by PDB code 3MIN structural data of reference 24 which places as sulfur atom at position "Y." The M center schematic diagram is shown in Figure 6.8 as reprinted from reference 29b.

In the diagram shown in Figure 6.9, a sulfur atom has been inserted at the cluster's center in agreement with current thinking.[24] The M center is attached to protein side chains at only two locations, cysteine 275 (at Fe1) and histidine 442 (at Mo), and is unusual in having three-coordinate irons rather than the normal four-coordinate tetrahedral iron coordination sphere.[29b] The molybdenum forms

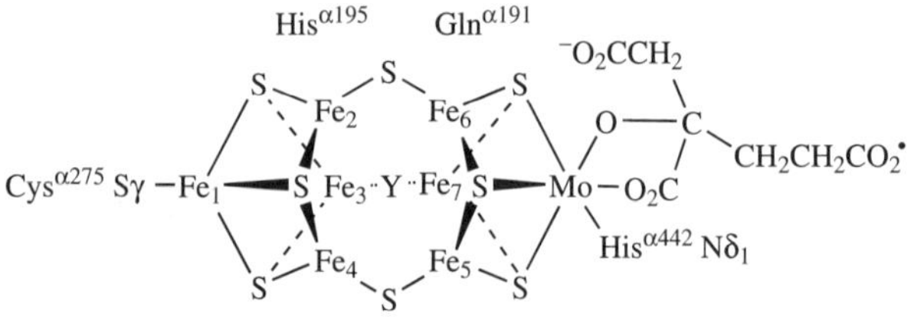

Figure 6.8 The M center (FeMoco) of MoFe–protein as reprinted from Figure 2A of reference 29b. "Y" atom assigned as sulfur in reference 24. (Reprinted with permission from Kim, J.; Rees, D. C. *Science* 1992, **257**, 1677–1682. Copyright 1992, American Association for the Advancement of Science.)

a distorted octahedral coordination sphere coordinating three S^{2-} ligands and a bidentate homocitrate molecule in addition to the his442 nitrogen. Fe–Fe separations within bridged iron sites are about 2.7–2.8 Å, whereas nonbridged iron sites (cross cluster) have a separation of ~3.8 Å. The cavity between the two cluster fragments has an ~4.0-Å diameter. This cavity is thought to be too small to accommodate a dinitrogen molecule, and the exact site of N_2 binding in nitrogenase has not been established to date. Two possible N_2 coordination modes are discussed in Section 6.6.2. These are a symmetric (side-on) mode[43,48] and an asymmetric (end-on) mode.[49,50] Homocitrate, an essential component of FeMoco, is coordinated to the molybdenum center by hydroxyl and carboxyl oxygen ligands. Other possible ligating side chains are those of hisα195 and glnα191. The X-ray crystallographic results are consistent with results of analytical (1Mo:7Fe:9S:1 homocitrate) and spectroscopic (EXAFS, ENDOR and Mössbauer) studies. A schematic diagram of the M center is shown in Figure 6.9.

More recent crystallographic data at 2.0-Å resolution were collected on two crystals, one believed to have the P-cluster and M center in oxidized states

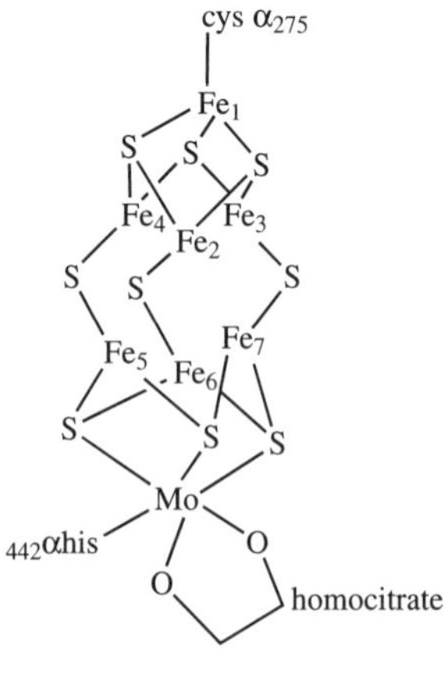

Figure 6.9 Schematic diagram of the M center of MoFe–protein.

Table 6.4 Bond Lengths for M^N and M^{OX} in Å

Bond	M^N	M^{OX}	Difference $M^{OX}-M^N$
Fe_1–Fe_2	2.62	2.69	+0.07
Fe_2–Fe_3	2.58	2.68	+0.10
Fe_3–Fe_4	2.56	2.65	+0.09
Fe_4–Fe_5	2.55	2.55	0.00
Fe_5–Fe_6	2.57	2.60	+0.03
Fe_6–Fe_7	2.46	2.47	+0.01
Fe_7–Mo	2.63	2.54	−0.09

Source: Reference 24.

(P^{OX}/M^{OX}) and the other, after dithionite reduction of the crystal, having the P-cluster and M center in reduced states (P^N/M^N).[24] Insignificant changes in the M center coordination sphere are noted; however, P-cluster changes are significant as described in Section 6.5.2. As can be seen in Table 6.4, Fe–Fe bond lengths are shorter in the M^N reduced form; however, differences are quite small compared to those noted for P-clusters in Table 6.2. Also, the Fe–Mo bond distance is longer in M^N.

Concurrently with the X-ray crystallographic studies, extended X-ray absorption fine structure (EXAFS) studies confirmed many of the bond distances proposed for nitrogenase's FeMoco cluster. The EXAFS data of reference 25 indicate short Fe–Fe distances of 2.61, 2.58, and 2.54 Å for M^+, M (resting state), and M^- forms, respectively. The authors believe that the short M center bond lengths indicate Fe–Fe bonds in this cluster. In another study using dithionite-reduced MoFe–protein Fe–S, Fe–Fe, Fe–Mo distances of 2.32, 2.64, and 2.73 Å, respectively, were found in the 1 to 3 Å region and Fe–Fe, Fe–S and Fe–Fe distances of 3.8, 4.3, and 4.7 Å, respectively, were found in the 3 to 5 Å region.[30]

The authors of reference 28 find a strikingly similar FeMoco environment in Kp1 as compared to that of Av1 discussed above, although the Kp1 interatomic distances were generally slightly longer.

Hypotheses for dinitrogen binding and reduction at nitrogenase's M center include:

1. A possible increased coordination number for Mo by adding hydrides and/or η^2-dihydrogen ligands for transfer to dinitrogen
2. Homocitrate acting as acid/base catalyst for protonating bound substrates or metal centers to produce hydrides
3. Homocitrate acting as a leaving group modulating the reduction potential of Mo and opening up coordination sites for hydrides

Studies on the extracted FeMo cofactor has allowed researchers to answer questions about the intrinsic reactivity associated with free clusters.[31] The reference 31 authors, for instance, studied an NifV$^-$ mutant nitrogenase from *Klebsiella pneumoniae* containing FeMoco in which the homocitrate ligand had been replaced by citrate ($MoFe_7S_9$ citrate), comparing its activity to an extracted wild-type FeMoco ($MoFe_7S_9$ R-homocitrate). Their analytical tool was to follow the kinetics of reaction of mutant and wild-type extracted FeMoco with the phenylsulfide anion (PhS^-). The PhS^- moiety reacts with Fe_1 (the point of attachment of the cysα275 ligand to FeMoco before its extraction from MoFe–protein) at different rates depending on the attachment to the molybdenum site (the point of attachment of the hisα442 ligand to FeMoco before its extraction from MoFe–protein). No differences in reactivity with PhS^- was noted between wild-type and mutant when the extracted cofactors were complexed with CN^-, N_3^-, or H^+. However, when imidazole was bound, the kinetics of reaction of PhS^- with the two cofactors was quite different. The authors propose that the wild-type R-homocitrate ligand (but not the mutant's citrate) can hydrogen-bond to the imidazole ligand on Mo, that this perturbs the electron distribution within the cluster core, and that the reactivity with PhS^- is therefore affected. Molecular modeling calculations, using X-ray crystallographic data for MoFe–protein, indicated that R-homocitrate is uniquely capable of facilitating the binding of dinitrogen by allowing the substrate access to Mo after the dissociation of the Mo–carboxylate bond. At the same time, the dissociated $-CH_2CH_2CO_2^-$ arm of R-homocitrate may hydrogen-bond to the imidazole group and/or other amino acid side chains in the vicinity.

Fitting together of nitrogenase's overall structure with the intimate mechanism for dinitrogen reduction remains the subject of continuing research. Reference 22 presents a well-explained and illustrated discussion of current knowledge of the subject. Figure 1b of reference 22, for instance, shows a stereodiagram of one-half of the overall Fe–protein/MoFe–protein complex. Updates to nitrogenase's X-ray crystallographic structure will be found by reference to the Protein Data Bank (PDB) at http://www.rcsb.org/pdb/.

6.6 NITROGENASE MODEL COMPOUNDS

Leigh[32] has discussed chemical systems for dinitrogen activation and reduction describing these as the following:

1. Using protons from water (and an energy input) for reaction with metal-coordinated dinitrogen (the metalloenzyme nitrogenase)
2. Reacting dinitrogen with dihydrogen on a catalytic iron surface (the Haber–Bosch process)
3. Splitting of dinitrogen to form nitrido (N^{3-}) complexes on a three-coordinate molybdenum(III) complex (process described below)[33]
4. Development of systems using vanadium(II) that fix nitrogen in an aqueous environment[34]
5. Reactivity of dinitrogen in a complex with electrophiles, including the proton and organic free radicals[35]

6. Use of transition metal complexes such as $[RuCl(Ph_2PCH_2CH_2CH_2PPh_2)_2]^+$, which will reversibly add dihydrogen and, by heterolytic cleavage of this hydrogen molecule, subsequently protonate dinitrogen on a tungsten complex such as $[W(N_2)_2(PMe_2Ph)_4]$, finally yielding ammonia (process described below).[36]

Some of these processes have been illustrated in Figure 6.2.

Chemists considering the design of model compounds for bioinorganic systems usually take one of two approaches:

1. *The Structural Model.* Design of a molecule that mimics the assumed structure of the enzyme active site. Such a model is considered totally successful if it carries out the desired function.
2. *The Functional Model.* Synthesis of a chemically similar or dissimilar molecule that will mimic the desired function of the enzyme be it dioxygen transport, electron transport, isomerization, or small molecule reduction.

6.6.1 Structural Models

Synthetic strategies for mimicking the metal sulfur clusters present in nitrogenase include (a) "spontaneous self-assembly" in which large molecules are generated from mononuclear starting materials or (b) "fragment condensation" where more complex polynuclear starting materials are used.[37] Holm and co-workers have used the spontaneous self-assembly approach to generate single cubanes from $[MS_4]^{x-}$, NaSR, and $FeCl_{2,3}$ and double cubanes of formula $(Et_4N)_3[M_2Fe_6S_8(SEt)_9]$ (M = Mo, Re) from $[MS_4]^{x-}$, $FeCl_2$, NaSEt, and Et_4NCl in acetonitrile.[38] These self-assembly reactions have been mentioned in Section 6.3. The Holm group's Mo–Fe–S cluster was one of the first to approximate the core composition and Mo–Fe distance distribution of the nitrogenase M center. The cluster $MoFe_4S_6(PEt_3)_4$, assembled from $MoCl_3(THF)_3$, $(Me_3Si)_2S$, and $FeCl_2(PEt_3)_2$ in THF, contains $[Fe_4(\mu_3\text{-}S)_3(\mu_2\text{-}S)_3]$, a 10-atom cluster present in the nitrogenase M center.[39] Fragment condensation was used by the authors of reference 37 to synthesize $[MoFe_6S_6(CO)_{16}]^{2-}$ from the fragments $[Fe_2S_2(CO)_6]^{2-}$ and $[Mo(CO)_4I_3]^-$. None of the clusters mentioned so far mimic either the total structure or any function of those present in MoFe–protein.

More recently, structural models have been synthesized that at least partially mimic the structure of MoFe–protein metal sulfur clusters. Holm and co-workers have reported a compositional analog of FeMoco in the form of a sulfido-bridged double cubane core $[Fe_4S_4\text{–}S\text{–}MoFe_3S_4]$.[40] The bridged μ_2-sulfide clusters are in their oxidized states ($[Fe_4S_4]^{2+}$ and $[MoFe_3S_4]^{3+}$), whereas those found in the as-isolated forms of FeMoco and P^N are substantially reduced. The rhomb-bridged double cubane $[(Cl_4cat)_2Mo_2Fe_6S_8(PEt_3)_6]$ has reduced cores ($[Fe_4S_4]^+$ and $[MoFe_3S_4]^{2+}$) stabilized by phosphine ligation.[41]

Starting with cluster $[(Cl_4cat)_2Mo_2Fe_6S_8(PEt_3)_6]$, where Cl_4cat is the tetrachlorocatecholate dianion, the cluster $[Mo_6Fe_{20}S_{30}]^{8-}$ was synthesized and characterized

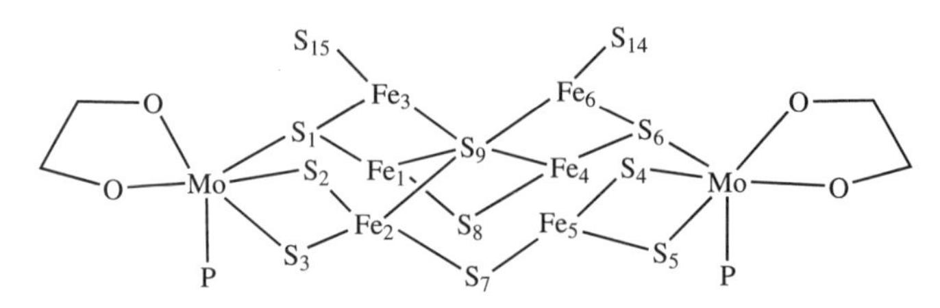

Figure 6.10 Upper/Lower Fragment 1, $Mo_2Fe_6S_9$. (Adapted with permission from Figure 2 of from Osterloh, F.; Sanakis, Y.; Staples, R., Münck, E.; Holm, R. H. *Angew. Chem. Int. Ed.*, 1999, **38**, 2066–2070. Copyright 1999, Wiley-VCH, STM.)

by X-ray crystallography.[42] As of the date of its synthesis, the cluster was the highest nuclearity Fe–S or heterometallic Fe–S cluster known to exist. The initial cluster has undergone fracture and rearrangement, and the sulfur atoms have divided into bridging modes $2 \times \mu_2 + 24 \times \mu_3 + 2 \times \mu_5 + 2 \times \mu_6$. The structure can be divided into three fragments. The upper and lower fragments $Mo_2Fe_6S_9$ are symmetry-related and have the structure shown in Figure 6.10 as Fragment 1. This fragment provides the closest synthetic approach (as of this date) to the structure of the $Fe_8S_7(SR)_2$ P-cluster of nitrogenase MoFe–protein if one replaces Mo(Cl_4cat)-(PEt_3) with Fe^{II}(SR) and replaces S(7, 8, 14, 15) with RS^-. The six Fe–S(9)–μ_6-S distances range from 2.381 to 2.429 Å in the fragment (mean 2.402 Å), whereas the central sulfur atom in the P^N cluster has five comparable Fe–S distances and one Fe–S distance of 2.9 Å. Fe–Fe distances in the fragment are in the same range as that found for the P^N cluster by EXAFS and by X-ray crystallography.[24] In comparison with the 4.2 K Mössbauer isomer shift for the complete cluster ($\delta = 0.52\ mm\ s^{-1}$) to that of Fe^{III}–rubredoxin ($\delta = 0.25\ mm\ s^{-1}$) and Fe^{II}–rubredoxin ($\delta = 0.70\ mm\ s^{-1}$), the authors calculate an average iron oxidation state of +2.4. Whereas the P^N state shows iron sites with ΔE_Q values of approximately $3.0\ mm\ s^{-1}$ (two sites), $1.3\ mm\ s^{-1}$ (one site), and $0.9\ mm\ s^{-1}$ (five sites), ΔE_Q values for the complete cluster described here range from 0.75 to 1.1 $mm\ s^{-1}$ for six inequivalent iron sites.

The central fragment $Mo_2Fe_8S_{12}$, with idealized C_{2h} symmetry is connected to the upper and lower fragments by six μ_3–S–Fe bridges involving S(8, 14, 15) and their symmetry-related counterparts S (8a, 14a, 15a) (see Figure 6.11).

In 1990 Coucouvanis and co-workers reported synthesis of a heterometallic double-cubane doubly-bridged molybdenum–iron cluster with formula $\{[MoFe_3S_4Cl_2(Cl_4cat)]_2(\mu_2{-}S)_2\}^{6-}$, ($Cl_4$cat is the tetrachlorocatecholate dianion).[43] Later it was found that similar clusters containing MFe_3S_4 cuboidal cores (M = Mo, V) will reduce the nitrogenase substrates hydrazine, acetylene, and protons to yield ammonia, ethylene, and dihydrogen.[44] Cobaltocene was used as the outer-sphere reductant and lutidine hydrochloride as the source of protons in these reactions. The clusters do *not* reduce dinitrogen. Further development of these model systems have yielded the clusters $(Cl_4cat)MoFe_3S_3(PEt_3)_2(CO)_6$ and $(Cl_4cat)Mo(O)Fe_3S_3(PEt_3)_3(CO)_5$.[45] The $MoFe_3S_3$ cores represent distorted versions of

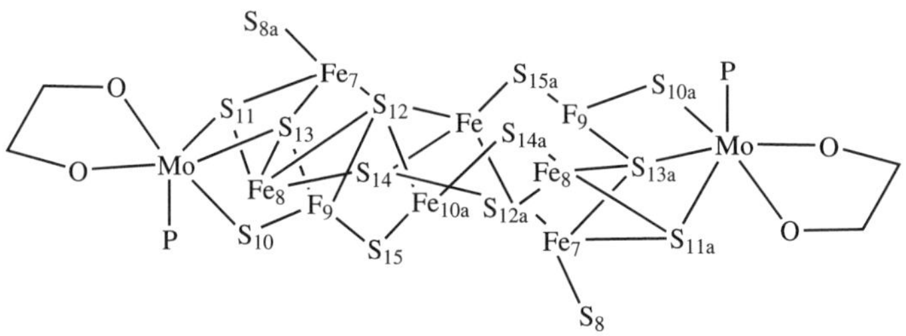

Figure 6.11 Central fragment 2 $Mo_2Fe_8S_{12}$. (Adapted with permission from Figure 3 of Osterloh, F.; Sanakis, Y.; Staples, R., Münck, E.; Holm, R. H. *Angew. Chem. Int. Ed.*, 1999, **38**, 2066–2070. Copyright 1999, Wiley-VCH, STM.)

the FeMoco (M center) of nitrogenase, although iron ions are four- and five-coordinate rather than three-coordinate as found in the nitrogenase M center. $(Cl_4cat)Mo(O)Fe_3S_3(PEt_3)_3(CO)_5$ contains an Mo–Fe bond length of 2.597 Å and other M–M bond distances of 2.570–2.691 Å in comparison to a mean Fe–Fe bond length of 2.56 Å (values ranging from 2.46 to 2.74 Å are reported in reference 24) for the nitrogenase cofactors.

More recently, this group has reported synthesis of (Cl_4cat)(py = pyridine)Mo-$Fe_3S_3(CO)_4(P^nPr_3)_3$ (Fe–Fe = 2.59 Å (2.565- to 2.609-Å range), Mo–Fe = 2.73 Å and $(Cl_4cat)(py)MoFe_3S_3(CO)_6(PEt_3)_2$ (Fe–Fe = 3.62, 2.68, 2.70 Å, Mo–Fe = 2.76 Å.[46] The authors explain the M–M bond distances as typical of extensive electron sharing between the metal atoms, indicating that the addition of two electrons in the latter structure increases one Fe–Fe bond length by >1 Å. They explain that the electron-deficient FeMoco is stabilized considerably by Fe–Fe bonding (M atoms in FeMoco fall considerably short of 18 electrons) as well as by donation of more electrons by the μ_3-S^{2-} and μ_2-S^{2-} ligands. The ability of bridging sulfide ligands to shift electrons into and out of the cofactor bonding scheme and the making/breaking of M–M bonds are cited as factors facilitating structural changes taking place during nitrogenase's catalytic cycle.

The group of Richard Shrock has synthesized iron–molybdenum dinitrogen complexes.[47] One of these is shown in Figure 6.12. As indicated in an X-ray crystallographic study, the central high-spin Fe(III) ion occurs in an unusual trigonal planar environment. The oxidation state and anisotropic magnetic environment for the iron center was confirmed by Mössbauer, EPR, and magnetic susceptibility measurements. The $[N_3N]Mo(N_2)$ unit may be viewed as a bulky ligand in which the trimethylsilane (TMS) groups preclude higher coordination numbers. The three Mo–N–N linkages are essentially linear as are two of the N–N–Fe linkages. One of the Mo–N–N–Fe linkages is significantly bent, perhaps as a consequence of steric crowding. At the time of its publication in 1999, this complex was unique as a structurally characterized iron–molybdenum dinitrogen complex relevant to the structure of MoFe–protein's FeMo cofactor.

It appears from the preceding discussion that producing metal–sulfur clusters similar in *both* structure and function to those found in native nitrogenases is not a

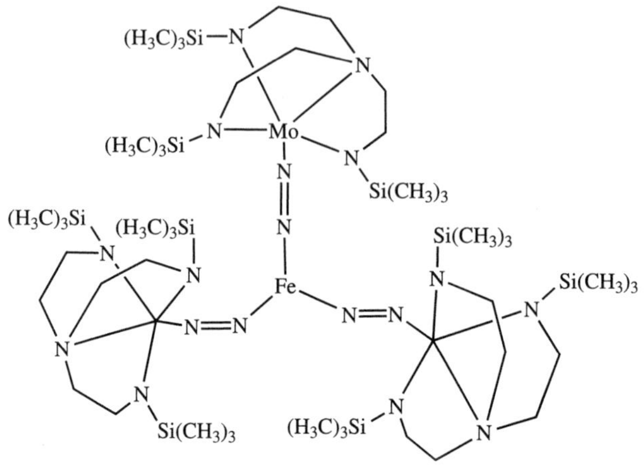

Figure 6.12 Structurally characterized iron–molybdenum dinitrogen complex. (Adapted from Figure 1 of reference 47.)

viable option for modeling nitrogenase activity at this time. This difficult problem remains an active research area for many groups interested in synthesis of large metal–sulfur clusters.

6.6.2 Functional Models

Functional models do not attempt to mirror the structure of active nitrogenase cofactors but rather to mimic the function—breaking the N_2 triple bond and producing NH_3. The work described in this section treats various parts of the reaction sequence producing intermediates diazene (N_2H_2) or hydrazine (N_2H_4). Some, but not all, produce the desired end product: ammonia. None of the functional models studied to date duplicate all parts of the reaction sequence beginning with dinitrogen and ending with ammonia.

In earlier work, not involving mechanistic studies, Laplaza and Cummins[33] achieved a major result in breaking the N_2 triple bond at $-35°C$ and 1 atm pressure. The reaction took place in hydrocarbon solvent using the molybdenum(III) complex $[Mo(NRAr)_3]$, where R is $C(CD)_2CH_3$ and Ar is 3,5-$C_6H_3(CH_3)_2$. The reaction yielded two moles of N^{3-}-ion-bonded molybdenum(VI) product NMo-$(NRAr)_3$ according to the reaction scheme shown in Figure 6.13. Both regeneration of the molybdenum(III) starting material and production of ammonia from the nitride would be necessary to make this a viable approach to the synthesis of ammonia.

As they report in reference 36, Nishibayashi and co-workers succeeded in producing ammonia through the use of ruthenium and tungsten complexes. $[RuCl(\eta^2\text{-}H_2)(Ph_2PCH_2CH_2CH_2PPh_2)_2]^+$ (**2**) is reversibly generated from

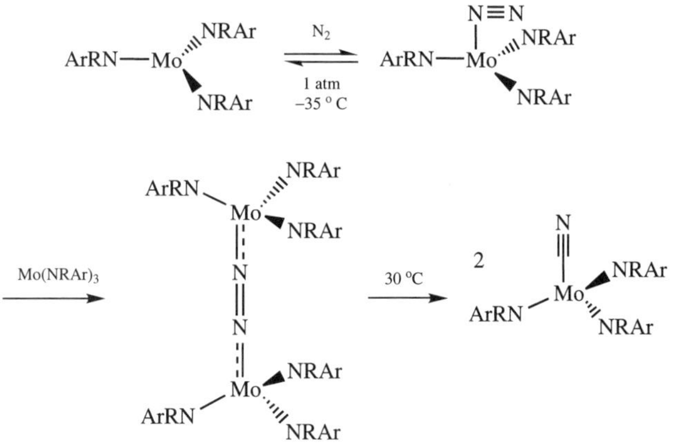
ArRN—Mo
NRAr
NRAr
N2
1 atm
−35 °C
N≡N
ArRN—Mo
NRAr
NRAr
Mo(NRAr)3
ArRN
Mo
NRAr
NRAr
N
N
ArRN
Mo
NRAr
NRAr
30 °C
2
N
ArRN
Mo
NRAr
NRAr

Figure 6.13 Activation and cleavage of the N_2 triple bond on molybdenum as discussed in reference 33.

$[RuCl(Ph_2PCH_2CH_2CH_2PPh_2)_2]^+$ (**1**) The hydrogen–hydrogen bond of $[RuCl(\eta^2\text{-}H_2)(Ph_2PCH_2CH_2CH_2PPh_2)_2]^+$ is cleaved heterolytically in the presence of a tungsten complex such as $[W(N_2)_2(PMe_2Ph)_4]$. In a typical reaction with $X = PF_6^-$, an equilibrium mixture of (**1**) and (**2**), and 10 equivalents of Ru complex to one equivalent of W complex, a 55% total yield of ammonia was achieved. The reaction scheme is shown in Figure 6.14.

Many researchers have considered models for possible intermediates in the nitrogenase reaction. Two possible dinitrogen attachments to the FeMoco factor of MoFe–protein have been put forward. Symmetric, edge- or side-on modes discussed by Dance[48] would lead to a reaction sequence such as is shown in reaction 6.11. In contrast, the asymmetric end-on terminal mode discussed in the work of Nicolai Lehnert[50] may be favored thermodynamically and by molecular orbital calculations. Reaction sequence 6.13 below illustrates one scenario for the asymmetric model.

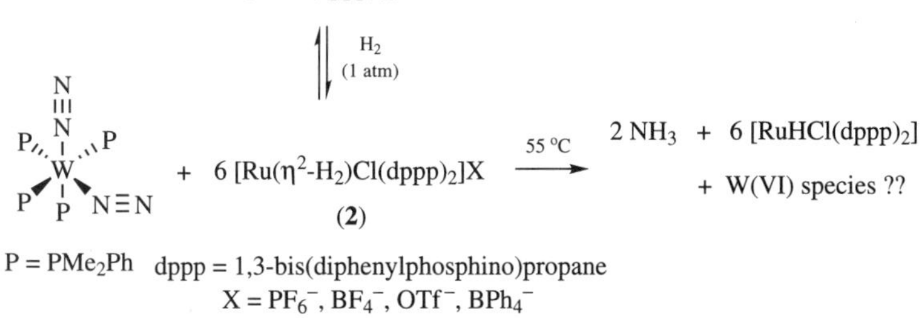
6 [RuCl(dppp)2]X (1)
H2
(1 atm)
N
N
P
P
W
P
P
N≡N
+ 6 [Ru(η2-H2)Cl(dppp)2]X
(2)
55 °C
2 NH3 + 6 [RuHCl(dppp)2]
+ W(VI) species ??
P = PMe2Ph dppp = 1,3-bis(diphenylphosphino)propane
X = PF6−, BF4−, OTf−, BPh4−

Figure 6.14 Production of NH_3 on tungsten with H_2 provided by a ruthenium complex. Reaction as described in reference 36.

$$\text{M–N}_2\text{–M} \xrightarrow{2e^-,\ 2\text{H}^+} \text{M–N}_2\text{H}_2\text{–M} \xrightarrow{2e^-,\ 2\text{H}^+} \text{M–N}_2\text{H}_4\text{–M} \xrightarrow{2e^-,\ 2\text{H}^+} 2\text{M–NH}_3 \quad (6.11)$$

Reduction from N_2 to diazene (N_2H_2) is assumed to be the most difficult step. In reference 14b, Sellmann and Sutter have estimated that an energy input of +523 kJ/mol would be required to produce diazene according to reaction 6.12:

$$\text{N}_2 \xrightarrow{2e^-,\ 2\text{H}^+} \text{N}_2\text{H}_2 \qquad +523\,\text{kJ/mol} \quad (6.12)$$

Additionally, diazene is unstable in its free state but is stabilized by coordination to metal complexes such as the iron–sulfur clusters synthesized by Sutter and Sellman. Nicolai Lehnert and Dieter Sellman have investigated the electronic structure (self-consistent field Xα scattered-wave (SCF-Xα-SW) and vibrational properties (resonance Raman and IR spectroscopy) of the systems $[\{Fe'N_HS'_4\}_2(N_2H_2)]$ ($'N_HS'_4$ = 2,2′-bis(2-mercaptophenylthio)diethylamine(2^-)) and $[\{Fe\text{'}S_4\text{'}(PPr_3)\}_2(N_2H_2)]$ ('S_4' = 1,2′-bis(2-mercaptophenylthio)ethane(2^-)).[49] Mössbauer spectroscopy indicated that the Fe–diazene systems are Fe(II) low-spin. X-ray crystallography indicated that diazene, (N_2H_2), is coordinated end-on in a trans-μ-1,2 mode as shown in Figure 6.15. The authors propose that N_2H_2 is first reduced to $N_2H_2^{2-}$ as Fe(II) is oxidized to Fe(III). Protonation of $N_2H_2^{2-}$ is followed by a change to Fe(III) high-spin. As one Fe–N bond breaks, Fe–N_2H_3 is protonated to N_2H_4. Reduction may stop here unless the remaining Fe–N bond breaks.

Lehnert and Tuczek further studied end-on terminal coordination by density functional theory (DFT) calculations on the compounds $[Mo(N_2)_2(dppe)_2]$, $[MoF(NNH)(dppe)_2]$, and $[MoF(NNH_2)(dppe)_2]^+$, where dppe = 1,2-bis(diphenylphosphino)ethane.[50] They proposed a reaction scheme, shown in reaction 6.13, for asymmetric dinitrogen reduction and protonation. The end-on model favored by Lehnert in reference 50, as shown in reaction 6.13, appears to be a less thermodynamically unfavorable pathway, at least to reach the M–NNH_3 intermediate. Step 1 produces a metal-attached diazenido ion (NNH^-), step 2 produces a hydrazido ion (NNH_2^{2-}), and step 3 produces a hydrazidium ion (NNH_3^+).

$$\text{M–N}_2 \xrightarrow{e^-,\text{H}^+} \text{M–NNH} \xrightarrow{e^-,\text{H}^+} \text{M–NNH}_2 \xrightarrow{e^-,\text{H}^+} \text{M–NNH}_3 \xrightarrow{-\text{NH}_3} \text{M–N} \xrightarrow{3e^-,\,3\text{H}^+} \text{M–NH}_3 \quad (6.13)$$

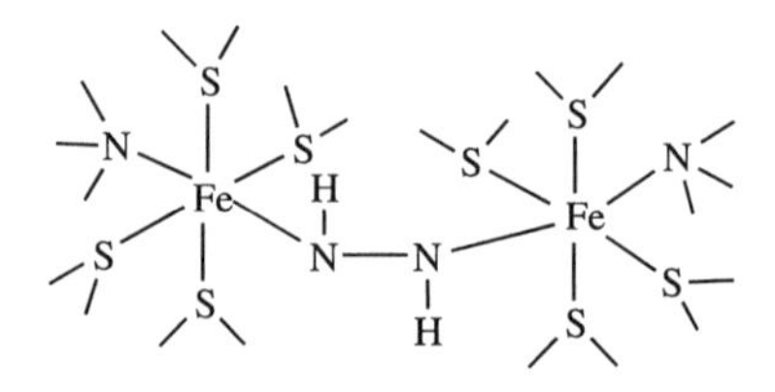

Figure 6.15 Iron–sulfur–diazene complexes as studied in reference 49.

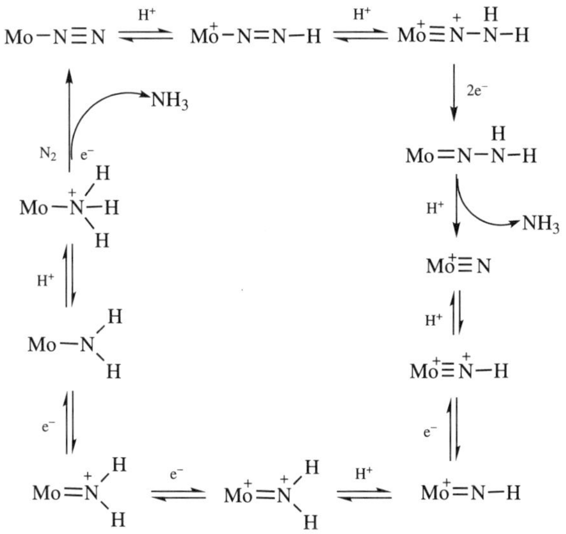

$$Mo{-}N{\equiv}N \underset{}{\overset{H^+}{\rightleftharpoons}} \overset{+}{Mo}{-}N{=}N{-}H \overset{H^+}{\rightleftharpoons} \overset{+}{Mo}{\equiv}\overset{+}{N}{-}\overset{H}{N}{-}H$$

$$\overset{+}{Mo}{\equiv}\overset{+}{N}{-}\overset{H}{N}{-}H \xrightarrow{2e^-} Mo{=}N{-}\overset{H}{N}{-}H \xrightarrow[-NH_3]{H^+} \overset{+}{Mo}{\equiv}N \overset{H^+}{\rightleftharpoons} \overset{+}{Mo}{\equiv}\overset{+}{N}{-}H \overset{e^-}{\rightleftharpoons} \overset{+}{Mo}{=}N{-}H$$

$$\overset{+}{Mo}{=}N{-}H \overset{H^+}{\rightleftharpoons} \overset{+}{Mo}{=}\overset{+}{N}H_2 \overset{e^-}{\rightleftharpoons} Mo{=}\overset{+}{N}H_2 \overset{e^-}{\rightleftharpoons} Mo{-}NH_2 \overset{H^+}{\rightleftharpoons} Mo{-}\overset{+}{N}H_3 \xrightarrow[-NH_3]{N_2,\ e^-} Mo{-}N{\equiv}N$$

Figure 6.16 Proposed nitrogenase catalytic cycle. (Adapted with permission from Scheme 1 of Pickett, C. J. *J. Biol. Inorg. Chem.*, 1996, **1**, 601–606. Copyright 1996, Society of Biological Inorganic Chemistry.)

One possible reaction scheme, an elaboration of the so-called Chatt cycle, results in the nitrogenase cycle shown in Figure 6.16.[51] The cycle shown operates between molybdenum oxidation states Mo(0) and Mo(IV), but others have been proposed that operate between Mo(II) and Mo(VI) levels.

The symmetric edge- or side-on N_2 attachment mode appears to favor initial complexation at Fe–Fe edges or faces in the M center. This mode is less able to explain how, or if, partially reduced and protonated intermediates migrate to the molybdenum atom for generation of ammonia. The asymmetric N_2 attachment mode appears to involve the molybdenum atom of the M center but not the iron atoms except as possible electron providers.

6.7 CONCLUSIONS

Chemists and biochemists have studied the complex enzyme nitrogenase for all of modern scientific times. Many models for the enzyme's efficient reduction and protonation of dinitrogen to the useful product ammonia have been put forward. Many different research groups have based these models on analytical and instrumental observations. Crystallization of the enzyme's subunits and subsequent X-ray crystallographic structures in the 1990s yielded an intimate portrait informing all aspects of research on nitrogenase. In spite of the many structural and analytical successes, aspects of the enzyme's structure and function remain controversial or unclear up to the present time.

It is hoped that a fuller understanding of nitrogenase's catalytic reduction of dinitrogen to ammonia will lead to more efficient catalysis for industrial ammonia production than that of the Haber–Bosch process. While studies of the enzyme itself contribute to this goal, chemists continue to design and synthesize model compounds for the same purpose. To date, chemical model compounds have succeeded in performing some, but not all, reaction sequences of nitrogenase's catalytic cycle. Molecular orbital calculations on simplified chemical models have helped also in determining probable reaction pathways. Chemists will continue to synthesize model compounds and study these, and the enzyme itself, not only to understand its complexities but also to duplicate its simple reaction for human purposes.

For updating the information presented in this chapter, a literature search on the keyword "nitrogenase" modified with "structure," "X ray," "Mössbauer," "iron sulfur cluster," or "model compound" will generate citations referring to the newest research results. A search of the Protein Data Bank (PDB) at the website address http://www.rcsb.org/pdb/ will yield the latest updates on X ray, NMR, and other submitted structural data.

REFERENCES

1. Bolin, J. T.; Campobasso, N. S.; Muchmore, W.; Minor, W.; Morgan, T. V.; Mortenson, L. E. In Palacios, R.; Mora, J.; Newton, W. E., eds. *New Horizons in Nitrogen Fixation*, Kluwer Academic Publishers, Dordrecht, 1993, p. 89.
2. Howard, J. B.; Rees, D. C. *Chem. Rev.*, 1996, **96**, 2965–2982.
3. Hay, R. W. *Bio-Inorganic Chemistry*, Ellis Horwood Limited, Halsted Press, New York, 1984.
4. Orme-Johnson, W. H. *Science*, 1992, **257**, 1639–1640.
5. Leigh, G. J. *Science*, 1995, **268**, 827–828.
6. Georgiadis, M. M.; Komiya, H.; Chakrabarti, P.; Woo, D.; Kornuc, J. J.; Rees, D.C., *Science*, 1992, **257**, 1653–1659. (PDB: 1NIP)
7. Seefeldt, L. C.; Dean, D. R. *Acc. Chem. Res.*, 1997, **30**, 260–266.
8. Thorneley, R. N. F.; Lowe, D. J. *J. Biol. Inorg. Chem.*, 1996, **1**, 576–580.
9. Cowan, J. A. *Inorganic Biochemistry, An Introduction*, 2nd ed., Wiley-VCH, New York, 1997.
10. Jang, S. B.; Seefeldt, L. C.; Peters, J. W. *Biochemistry*, 2000, **39**, 641–648 (PDB: 1FP6).
11. (a) Lanzilotta, W. N.; Seefeldt, L. C. *Biochemistry*, 1997, **36**, 12976–12983. (b) Ryle, M. J.; Lanzilotta, W. N. *Biochemistry*, 1996, **35**, 9424–9434.
12. Chan, J. M.; Christiansen, J. *Biochemistry*, 1999, **38**, 5779.
13. Duyvis, M.; Wassnick, H; Haaker, H. *Biochemistry*, 1998, **37**, 17495.
14. (a) Sellman, D., Sutter, J. *J. Biol. Inorg. Chem.*, 1996, **1**, 587–593. (b) Sellman, D., Sutter, J. *Acc. Chem. Res.*, 1997, **30**, 460–469.
15. Beinert, H.; Holm, R. H.; Münck, E. *Science*, 1997, **277**, 653–659.
16. Yoo, S. J.; Angove, H. C.; Burgess, B., Hendrich, M. P., Münck, E. *J. Am. Chem. Soc.*, 1999, **121**, 2534.

17. Hagen, K. S.; Reynolds, J. G.; Holm, R. H. *J. Am. Chem. Soc.*, 1981, **103**, 4054–4063.
18. Lippard, S. J.; Berg, J. M. *Principles of Bioinorganic Chemistry*, University Science Books, Mill Valley, CA, 1994.
19. Rawls, R. L. *C&E News,* 2000, **Nov. 20**, 43–51.
20. Eady, R. R. *Chem. Rev.*, 1996, **96**, 3013–3030.
21. Schlessman, J. L., Woo, D., Joshua-Tor, L., Howard, J. B., Rees, D. C. *J. Mol. Biol.* 1998, **280**, 669–685 (PDB: 2NIP and 1CP2).
22. Schindelin, H.; Kisker, C.; Schlessman, J. L.; Howard, J. B.; Rees, D. C. *Nature*, 1997, **387**, 370–376 (PDB: 1N2C).
23. (a) Rees, D. C.; Kim, J. *Nature*, 1992, **360**, 553–560. (b) Kim, J.; Rees, D. C. *Science*, 1992, **257**, 1677–1682.
24. Peters, J. W.; Stowell, M. H. B.; Soltis, S. M.; Finnegan, M. G.; Johnson, M. K.; Rees, D. C. *Biochemistry*, 1997, **36**, 1181–1187 (PDB: 3MIN).
25. Christiansen, J.; Tittsworth, R. C.; Hales, B. J.; Cramer, S. P. *J. Am. Chem. Soc.*, 1995, **117**, 10017–10024.
26. Mouesca, J.-M.; Noodleman, L.; Case, D. A. *Inorg. Chem.*, 1994, **33**, 4819–4830.
27. Surerus, K. K.; Hendrich, M. P.; Christie, P. D.; Rottgardt, D.; Orme-Johnson, W. H.; Münck, E. *J. Am. Chem. Soc.*, 1992, **114**, 8579–8590.
28. Mayer, S. M.; Lawson, D. M.; Gormal, C. A.; Roe, S. M.; Smith, B. E. *J. Mol. Biol.*, 1999, **292**, 871–891 (PDB: 1QGU, 1QH1, 1QH8).
29. (a) Chan, M. K.; Kim, J.; Rees, D. C., *Science*, 1993, **260**, 792–794 (PDB: 1MIN). (b) Kim, J.; Rees, D. C. *Science*, 1992, **257**, 1677–1682.
30. Filipponi, A.; Gavini, N.; Burgess, B.; Hedman, B.; DiCicco, A.; Natoli, C. R.; Hodgson, K. O. *J. Am. Chem. Soc.*, 1994, **116**, 2418.
31. Grönberg, K. L. C.; Gormal, C. A.; Durrant, M. C.; Smith, B. E.; Henderson, R. A. *J. Am. Chem. Soc.*, 1998, **120**, 10613–10621.
32. Leigh, G. J. *Science*, 1998, **279**, 506–507.
33. Laplaza, C. E.; Cummins, C. C. *Science*, 1995, **268**, 861–862.
34. Bazhenova, T. A.; Shilov, A. E. *Coord. Chem. Rev.*, 1995, **144**, 69.
35. Leigh, G. J. *Acc. Chem. Res.*, 1992, **25**, 177.
36. Nishibayashi, Y.; Iwai, S.; Hidai, M. *Science*, 1998, **279**, 540–544.
37. Eldredge, P. A.; Bose, K. S.; Barber, D. E.; Bryan, R. F.; Sinn, E.; Rheingold, A.; Averill, B. A. *Inorg. Chem.*, 1991, **30**, 2365–2373.
38. Cen, W.; Lee, S. C.; Li, J.; MacDonnell, F. M.; Holm, R. H. *J. Am. Chem. Soc.*, 1993, **115**, 9515–9523.
39. Cen, W.; MacDonnell, F. M.; Scott, M. J.; Holm, R. H. *Inorg. Chem.*, 1994, **33**, 5809–5818.
40. (a) Huang, J.; Mukerjee, S.; Segal, B. M.; Akashi, H.; Zhou, J.; Holm, R. H. *J. Am. Chem. Soc.*, 1997, **119**, 8662-8674. (b) Huang, J.; Holm, R. H. *Inorg. Chem.*, 1998, **37**, 2247–2254.
41. (a) Tyson, M. A.; Demadis, K. D.; Coucouvanis, D. *Inorg. Chem.*, 1995, **35**, 4519–4520. (b) Goh, C; Segal, B. M.; Huang, J.; Long, J. R.; Holm, R. H. *J. Am. Chem. Soc.*, 1996, **117**, 11844–11853.
42. Osterloh, F.; Sanakis, Y.; Staples, R., Münck, E.; Holm, R. H. *Angew. Chem. Int. Ed.*, 1999, **38**, 2066–2070.

43. Coucouvanis, D. *Acc. Chem. Res.*, 1991, **24**, 1–8.

44. Coucouvanis, D. *J. Biol. Inorg. Chem.*, 1996, **1**, 594–600.

45. Tyson, M. A.; Coucouvanis, D. *Inorg. Chem.*, 1997, **36**, 3808–3809.

46. Han, J.; Beck, K.; Ockwig, N.; Coucouvanis, D. *J. Am. Chem. Soc.*, 1999, **121**, 10448–10449.

47. O'Donoghue, M. B.; Davis, W. M.; Schrock, R. R.; Reiff, W. M. *Inorg. Chem.*, 1999, **38**, 243–252.

48. Dance, I. *J. Biol. Inorg. Chem.*, 1996, **1**, 581–586.

49. (a) Lehnert, N.; Wiesler, B. E.; Tuczek, F.; Hennige, A.; Sellmann, D. *J. Am. Chem. Soc.*, 1997, **119**, 8869–8878. (b) Lehnert, N.; Wiesler, B. E.; Tuczek, F.; Hennige, A.; Sellmann, D. *J. Am. Chem. Soc.*, 1997, **119**, 8879–8888.

50. Lehnert, N.; Tuczek, F. *Inorg. Chem.*, 1999, **38**, 1671–1682.

51. Pickett, C. J. *J. Biol. Inorg. Chem.*, 1996, **1**, 601–606.

7

METALS IN MEDICINE

7.1 INTRODUCTION

As a discipline, medicinal inorganic chemistry has only existed for about the last 35–40 years, since the serendipitous discovery of the antitumor activity of cisplatin, *cis*-$[Pt(NH_3)_2Cl_2]$.[1] Orvig and Abrams, in their introduction to an issue of *Chemical Reviews* devoted to Medicinal Inorganic Chemistry, cited here as reference 1, define the discipline as specifying a known chemical compound to have a specific activity. Furthermore, to meet the criteria of the discipline, studies should have been carried out to elucidate the compound's mechanism of medicinal action and to optimize and improve the compound's physiological activity. The rapid growth of understanding in bioinorganic chemistry has seen a parallel growth of interest in the medicinal properties of various metals whether or not they are found in biological species. The reference 1 authors describe the contributions of Paul Ehrlich, who first proposed a structure–activity relationship (SAR) for the inorganic compound arsphenamine (Salvarsan or Ehrlich 606) at the beginning of the twentieth century. Ehrlich is considered the founder of chemotherapy, which he defined as the use of drugs to injure an invading organism without injuring the host. He first formulated a chemotherapeutic index and the "magic bullet" concept—a specific compound can be designed to treat and cure a specific disease.

7.1.1 Inorganic Medicinal Chemistry

Inorganic or metal-containing medicinal compounds may contain either (a) chemical elements essential to life forms—iron salts used in the treatment of anemia—or (b) nonessential/toxic elements that carry out specific medicinal purposes—platinum-containing compounds as antitumor agents or technetium and gadolinium complexes as medical diagnostic tools. Chapter 1 defines and illustrates the meaning of chemical essentiality in Section 1.2. Any compound or element is subject to the limitations of the so-called Bertrand diagram (see Figure 1.1 of Chapter 1), which indicates the relationship among deficiency, optimum physiological response, and toxicity for the species in question. Every medicinal compound or element will have a different response curve defining its limits of deficiency, optimal response, and toxicity.

7.1.2 Metal Toxicity and Homeostasis

The essentiality of chemical elements and compounds was introduced in Chapter 1. Figure 1.1 illustrates survival, deficiency, optimal, toxic, and lethal behavior for two elements, fluorine and selenium. Although many elements and compounds are required at some dosage for an organism's survival, all elements and compounds may be deleterious if taken in overly large doses. Some elements and compounds are required in a certain range of concentration, while some appear to be toxic at minimal dosage levels. Two examples of the latter are elemental lead and mercury as well as their compounds. While these elements have always been present in our atmosphere, they are especially dangerous because they form long-lasting compounds with organic species and accumulate low in the food chain. Methyl mercury compounds accumulating in fish, for instance, caused a terrible wasting disease in Japan before the toxic effects of CH_3Hg^+ contamination of food was recognized. Tetraethyl lead, $Pb(C_2H_5)_4$, was used as a octane-increasing gasoline additive until dangerous accumulations of lead in the atmosphere were traced to its usage. Air pollution by lead and lead compounds remains a great problem in countries where leaded gasoline is still used extensively.

Toxic elements and compounds may not be equally dangerous for all organisms at all levels. The radioactive elements discussed in this chapter have great value in that they can be used to screen for disease and to kill tumors at given body locations. On the other hand, radioactivity is known to create mutations that may cause diseases such as leukemia and other cancers. Lead, mercury, and thallium appear to be dangerous no matter how they appear in biological species. While chromium is an essential element required for normal carbohydrate and lipid metabolism and its deficiency can lead to adult onset diabetes, Cr(VI) uptake through anion channel transport into cells via CrO_4^{2-} causes several toxic effects. Following reduction inside cells, Cr(III) forms adducts involving the phosphate backbone of DNA, the N7 atom of guanine, amino acids such as cysteine, glutathione, and larger peptides and protein molecules.

Table 7.1, as excerpted from reference 2, summarizes the toxic effects of some nonbiological metals on humans and animals. Further discussion of each metal is found in Chapter 22 of reference 2.

In addition to essentiality, chemical elements and compounds found in biological systems obey homeostatic criteria. Homeostasis, the maintenance of chemical elements and compounds at optimal physiological levels, is a complex interactive system involving many elements and the biological energy supply. Interactions

Table 7.1 Toxic Effects of Some Nonbiological Metals

Metal (Class)	Effect of Excess	Comments
Aluminum (hard, Al^{3+})	Implicated in Alzheimers disease	May interact with phosphates, may cross-link proteins.
Cadmium (soft, Cd^{2+})	Renal toxicity	Blocks sulfhydryl groups in enzymes and competes with zinc. Stimulates metallothionein synthesis and interferes with Cu(II) and Zn(II) metabolism.
Mercury (soft, Hg_2^{2+}, Hg^{2+})	Damage to central nervous system, neuropsychiatric disorders	CH_3Hg^+ compounds are lipid-soluble.
Lead (soft, Pb^{2+})	Injuries to peripheral nervous system, disturbs heme synthesis and affects kidneys	Pb^{2+} may replace Ca^{2+} with loss of functional and structural integrity. Reacts with sulfhydryl groups, replaces Zn^{2+} in δ-aminolevulinic acid dehydratase.
Thallium (soft, Tl^+)	Poisonous to nervous systems, enters cells via K^+ channels	Although similar to K^+, Tl^+ binds more tightly to N and S ligands.

Source: Reference 2.

between deficiencies and excesses of elements and compounds interfere with homeostasis and may cause disease states in organisms. How the complex processes of homeostasis are managed within systems has been the subject of much research and writing. The book *The Biological Chemistry of the Elements: The Inorganic Chemistry of Life*, written by J. J. R. Frausto da Silva and R. J. P. Williams, provides a comprehensive survey of the field.[2] Alteration of homeostatic concentrations of metals within the human body or removal of oversupply of toxic metals has been accomplished medically by the addition of metal chelators such as ethylenediaminetetracetate (EDTA). Metal chelators, while useful, have many disadvantages, one of which is the undesired complexation and removal of other essential metals. Researchers have therefore tried to understand the processes by which essential metal cofactors such as copper or zinc ions, toxic if found in inappropriate locations or in excess concentration, are acquired by intracellular proteins such as superoxide dismutase. With this understanding, it is believed that one might specifically target and remove only the offending metal ions. The subject, discussed in much more detail in Section 7.4, has been reviewed recently by O'Halloran and Culotta in reference 119. Findings so far have elucidated that the concentration of "free" copper or zinc ions maintained by homeostasis is too low to allow apoproteins to acquire these metals within a cell. In fact, proteins called metallochaperones are needed to carry out delivery of metal ions to their destinations. Once metallochaperones are understood, it may be possible to design compounds that alter intracellular metal ion availability, ultimately controlling biological phenomena such as the proliferation of cancer cells or disorders of metal metabolism.

Introducing metal ions into a biological system may be carried out for therapeutic or diagnostic purposes, although these purposes overlap in many cases. In this chapter, therapeutic and diagnostic applications will be treated in separate sections, although the reader will note the overlaps especially in the technetium and gadolinium imaging agent sections. In 1991, Peter Sadler[3] noted that most elements of the periodic table, up to and including bismuth with an atomic number of 83, have potential uses as drugs or diagnostic agents. Table 7.2, adapted from references 2 and 4, outlines some of these purposes.[4] Inorganic compounds have found usage in chemotherapeutic agents such as

1. Anticancer agents like *cis*-$[Pt(NH_3)_2Cl_2]$
2. The gold-containing antiarthritic drug Auranofin
3. Metal-mediated antibiotics like bleomycin, which requires iron or other metals for activity
4. Technetium-99m and other short-lived isotopes used as radiopharmaceuticals in disease diagnosis and treatment
5. Magnetic resonance imaging (MRI)-enhancing gadolinium compounds
6. Antibacterials, antivirals, antiparasitics, and radiosensitizing agents

Medicinal inorganic chemistry is a multidisciplinary field combining elements of chemistry (synthesis, reactivity), pharmacology (pharmacokinetics, toxicology), biochemistry (targets, structure, conformational changes), and medicinal chemistry (therapeutics, pharmacodynamics, structure–activity relationships (SAR)). The fields are joined by the necessity of bringing both a deep understanding of inorganic chemistry and state-of-the-art research in biology to bear on problems affecting the use of metal-containing drugs. Medicinal inorganic chemists can be characterized as being focused on basic research while attempting to maintain a clear vision of their research's potential future medical application. An attempt has been made to touch on each of the named fields in discussing the individual topics of this chapter. The discussion of platinum anticancer agents, vanadium compounds as insulin substitutes for diabetics, 99m-technetium reagents used in diagnostic and treatment capabilities, and MRI contrast reagents containing gadolinium are but a few of the topics that could have been chosen. Articles describing other important topics are reviewed in references 5–9. These include topics such as bleomycin–metal complex antibiotics,[5] the inorganic pharmacology of lithium,[6] medicinal and biologically active bismuth compounds,[7] gold antiarthritic compounds,[8] and metal complexes as photo- and radiosensitizers.[9]

Table 7.2 Some Examples of Inorganic Elements and Compounds with Medicinal Purposes

Element	Example of a Product Name	Active Compound in the Product	Medicinal Usage
Li	Camcolit	Li_2CO_3	Manic depression
N	Laughing gas	N_2O (nitrous oxide)	Anesthetic
F		SnF_2	Tooth protectant
Mg	Magnesia	MgO	Antacid, laxative
Fe		Fe(II) fumarate, succinate	Dietary iron supplement
Co	Cobaltamin S	Coenzyme vitamin B_{12}	Dietary vitamin supplement
Zn	Calamine	ZnO	Skin ointment
Zn		Zn undecanoate	Antifungal (athlete's foot)
Br		NaBr	Sedative
Tc	TechneScan PYP	^{99m}Tc-pyrophosphate	Bone scanning
Sb	Triostam	NaSb(V) gluconate	Antileishmanial (antiprotozoal)
I		I_2	Antiinfective, disinfectant
Ba	Baridol	$BaSO_4$	X-ray contrast medium
Gd	Magnevist™	$[Gd(III)(DTPA)(H_2O)]^{2-}$ DTPA = diethylenetriamine pentaacetic acid	MRI contrast agent
Pt	Cisplatin, platinol, cisDDP	cis-$[Pt(NH_3)_2Cl_2]$	Anticancer agent
Pt	Carboplatin	$[Pt(NH_3)_2(CBDCA)]$ CBDCA = cyclobutanedicarboxylic acid	Anticancer agent
Au	Auranofin	$Au(I)(PEt_3)$ (acetylthioglucose)	Antiarthritic
Bi	De-Nol	$K_3[Bi(III)(citrate)_2]$	Antacid, antiulcer

7.2 THERAPEUTIC COMPOUNDS

7.2.1 Superoxide Dismutase Mimics

The enzyme superoxide dismutase (SOD) occurs in three forms in mammalian systems: (1) CuZnSOD (SOD1) found in the cytosol, (2) MnSOD (SOD2) found in mitochondria, and (3) CuZnSOD found in extracellular space (SOD3). Additionally, many bacterial SOD enzymes contain iron. SOD1 has been discussed in detail

Section 5.2.3 in Chapter 5. Superoxide dismutase enzymes catalyze dismutation of the superoxide anion radical ($O_2^{-\bullet}$) according to the summary reactions in equation 7.1:

$$\begin{aligned} O_2^{-\bullet} + M^{n^+} &\rightarrow O_2 + M^n \\ HO_2^{\bullet} + M^n + H^+ &\rightarrow H_2O_2 + M^{n^+} \end{aligned} \tag{7.1}$$

The superoxide anion ($O_2^{-\bullet}$) exhibits numerous physiological toxic effects including endothelial cell damage, increased microvascular permeability, formation of chemotactic factors such as leukotriene B_4, recruitment of neutrophils at sites of inflammation, lipid peroxidation and oxidation, release of cytokines, DNA single-strand damage, and formation of peroxynitrite anion ($ONOO^-$), a potent cytotoxic and proinflammatory molecule generated according to equation 7.2[10]:

$$O_2^{-\bullet} + NO \rightarrow NOO_2^- \tag{7.2}$$

Superoxide radical anion and the peroxynitrite anion formed in its reaction with NO cause reperfusion injury in ischemic tissue after myocardial infarction (heart attack) or stroke. Ischemic (i.e., O_2-deprived) tissue suffers increased superoxide concentration and lowered pH ($\sim$5), followed by cell death. When this tissue is reperfused with fresh, oxygenated blood, a large concentration of $HO_2^{\bullet}$ is generated. Superoxide anion and HO_2 radical have been implicated in sufferers of neurological diseases such as Parkinson's disease and in inflammations suffered by arthritis patients. Medical researchers have attempted to design low-molecular-weight SOD mimics (synzymes) that would mimic the natural SOD enzyme in removing $O_2^{-\bullet}$ and the perhydroxyl radical, $HO_2^{\bullet}$, as well as preventing formation of $ONOO^-$ in the ischemic–reperfusion injury tissue.[11]

Many transition metal complexes have been considered as synzymes for superoxide anion dismutation and activity as SOD mimics. The stability and toxicity of any metal complex intended for pharmaceutical application is of paramount concern, and the complex must also be determined to be truly catalytic for superoxide ion dismutation. Because the catalytic activity of SOD1, for instance, is essentially diffusion-controlled with rates of $2 \times 10^9\,M^{-1}\,s^{-1}$, fast analytic techniques must be used to directly measure the decay of superoxide anion in testing complexes as SOD mimics. One needs to distinguish between the uncatalyzed stoichiometric decay of the superoxide anion (second-order kinetic behavior) and true catalytic SOD dismutation (first-order behavior with $[O_2^{-\bullet}] \gg [\text{synzyme}]$ and many turnovers of SOD mimic catalytic behavior). Indirect detection methods such as those in which a steady-state concentration of superoxide anion is generated from a xanthine/xanthine oxidase system will not measure catalytic synzyme behavior but instead will evaluate the potential SOD mimic as a stoichiometric superoxide scavenger. Two methodologies, stopped-flow kinetic analysis and pulse radiolysis, are fast methods that will measure SOD mimic catalytic behavior. These methods are briefly described in reference 11 and in Section 3.7.2 of Chapter 3.

The SOD mimic synzyme must not react with hydrogen peroxide, H_2O_2, a product of superoxide dismutation, and should reduce or eliminate the production of peroxynitrite anion via equation 7.2. The potential metal complex synzyme must behave as well in vivo as in vitro and, as stated above, have excellent stability and nontoxic behavior. These criteria have led to the choice of manganese as the metal ion in potential drugs rather than copper or iron ions because aqueous Mn(II) ions (assuming some decomposition of the metal complex in the body) are the least toxic of the three metal ion choices, the least likely to react with H_2O_2 to generate hydroxyl radicals (Fenton chemistry), or the least likely to undergo redox reactions to generate nitrogen dioxide, NO_2. In summary, one wishes the synzyme to possess a high catalytic rate (ideally diffusion-controlled), optimal biodistribution into the desired target tissue, and low kinetic lability (both chemical and metabolic). In addition, the ideal complex's rate of reaction and rate of excretion should far exceed its rate of decomposition.

Drug design for these agents, and any other, must optimize partitioning of the drug to the desired tissue. For instance, one desires a lipophilic reagent for treatment of Parkinson's disease because the drug must cross the blood–brain barrier. The measurement most often cited as a measure of a drug or diagnostic agent's lipophilicity or hydrophilicity is the partition coefficient log P, where $P = [\text{agent}]_{n\text{-octanol}}/[\text{agent}]_{\text{water}}$. Two Mn(III) complexes of salicyldialdehyde ethylenediimine (salen), EUK-8 and EUK-134 developed by Eukarion, Inc. (Figure 7.1), and a third compound, EUK-189 having $-OCH_2CH_3$ substituents, show efficacy as SOD mimics in animal models of neurological disorders. The agents show increased SOD mimetic capability as their log P values become more positive—that is, as they become more lipophilic.[11b] They also exhibit protection of pulmonary function in respiratory disease and protection of renal ischemia–reperfusion injury. In stopped-flow kinetic analysis, these agents, shown in Figure 7.1, show some catalytic activity as SOD dismutases.[11a]

Mn(II) ions complexed by porphyrinato(2^-) ligands have shown catalytic superoxide anion dismutation. One SOD mimic, M40403, complexes Mn(II) via a macrocyclic ligand, 1,4,7,10,13-pentaazacyclopentadecane, containing added bis(cyclohexyl) and pyridyl functionalities. M40403 carries the systematic name [manganese(II) dichloro{(4*R*,9*R*,14*R*,19*R*)-3,10,13,20,26-pentaazatetracyclo[20.3.1.0(4,9)0(14,19)]hexacosa-1(26),-22(23),24-triene}]. The molecule is shown in

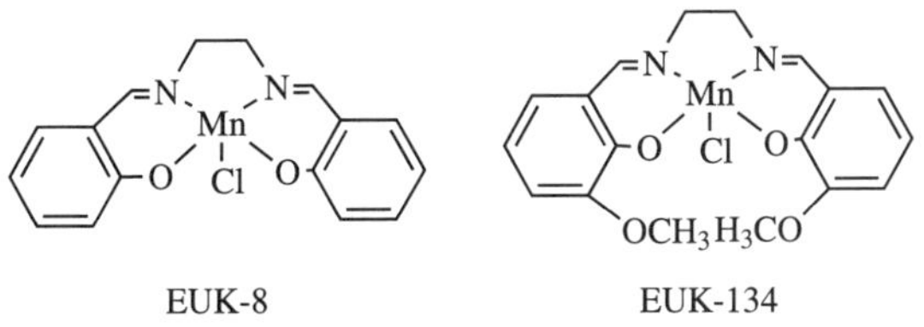

EUK-8 EUK-134

Figure 7.1 Superoxide dismutase mimetic Mn(III) complexes of salicyldialdehyde ethylenediimine.

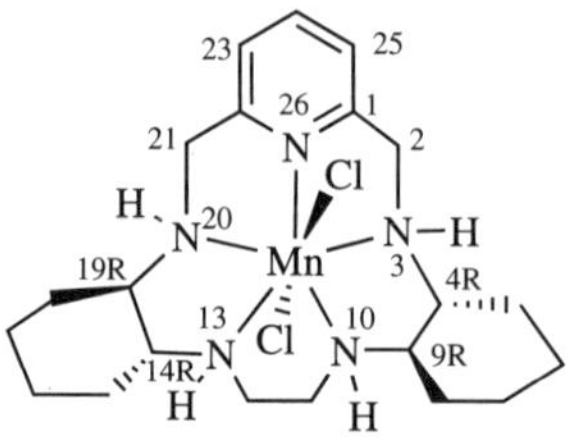

M40403

Figure 7.2 M40403 Mn(II) macrocycle tested as a catalytic SOD mimic.

Figure 7.2.[12] The MW = 484.4 complex catalyzes $O_2^{-\bullet}$ dismutation at a rate of $1.2 \times 10^7\,M^{-1}\,s^{-1}$ comparable to that of the MnSODs at pH ~6. The complex is thermodynamically stable (log $K > 17$) and does not decompose for up to 10 hours in whole rat blood at 37°C. After intravenous injection, M40403 distributes into the heart, lungs, brain, liver, and kidneys, retaining intact chemical identity, and is excreted intact in urine and feces. M40403 does not react with NO, H_2O_2, or $ONOO^-$. This stable SOD mimic appears to have therapeutic activity in animal models of inflammation and ischemia. It may prevent generation of $ONOO^-$ via equation 7.2 by elimination of $O_2^{-\bullet}$ by so-far undetermined mechanisms. Other similar Mn(II) macrocycles discussed in reference 11 exhibit two possible mechanistic pathways to SOD dismutation involving rate-determining steps of Mn(II) to Mn(III) oxidation. The faster major pathway is proton-dependent, involving oxidation of aquo Mn(II) complex to Mn(III) hydroxo via an outer-sphere electron transfer—that is, H-atom transfer. The slower, lesser path involves an inner-sphere path in which a vacant coordination site on Mn(II) must be generated so that superoxide can bind to Mn(II). Molecular modeling studies correlate the catalytic rate with the ability of the macrocycle to accommodate a particular folded geometry that facilitates a six-coordinate nonplanar geometry about Mn(II)—that is, a pseudo-octahedral complex.[13] More recently, Riley and co-workers investigated the stereoisomers of 2,21-dimethyl-substituted derivatives of M40403.[14] As predicted from molecular modeling studies, the 2*S*,21*S*-dimethyl derivative had a catalytic SOD dismutation rate of $1.6 \times 10^9\ M^{-1}s^{-1}$, faster than that of the native mitochondrial MnSOD enzyme. The stereochemistry of the 2*S*,21*S*-dimethyl derivative was confirmed by X-ray crystallography, which also indicated a seven-coordinate pentagonal bipyramidal geometry around Mn with an approximately planar macrocylic ring. Conversely, the 2*R*,21*R*-dimethyl derivative had low SOD activity, again in agreement with modeling studies. The meso (or *cis*-*R*,*S*) dimethyl complex has SOD activity comparable to unsubstituted M40403. The authors believe that the 2*S*,21*S*-dimethyl complex generates the highest activity because in this stereochemistry the Mn(II) complex is "preorganized" to switch to octahedral geometry upon oxidation to Mn(III) during catalysis. In other words, the best catalyst in this series will be one in which the Mn(II) center exists in a geometry that promotes fast electron transfer—that is, a pseudo-octahedral geometry in the Mn(II) catalyst that is capable of fast conversion to the octahedral geometry preferred by Mn(III).

In late 2001 the pharmaceutical company MetaPhore Pharmaceuticals initiated a Phase I study of M40403 to test its safety and tolerability and to determine the pharmacokinetics of the compound in normal, healthy human subjects, as a precursor to Phase II trials in which M40403 will be tested as a co-therapy with interleukin-2 (IL-2) for advanced skin and end-stage kidney cancers. Preclinical studies in animals indicated that M40403 significantly improved the effectiveness of IL-2, an approved treatment for inoperable metastatic melanoma and metastatic renal-cell carcinoma. IL-2 immunotherapy works by activating natural killer (NK) cells that have the ability to recognize and destroy many types of tumors. Its use is limited, however, by potentially life-threatening side effects, including extremely low blood pressure (hypotension), particularly at the high-dosage level indicated for end-stage cancers. By reducing the level of superoxide, the SOD enzyme mimetics, including M40403, have been shown to reverse hypotension and effectively restore blood pressure in animal studies. The studies also showed that M40403 enhanced the direct antitumor properties of IL-2 therapy. This may be explained by other research revealing an excess of superoxide in tissue surrounding cancerous sites, paired with deficient amounts of SOD in the cancer cells themselves. In these states, superoxide also inhibits the activity of NK cells, thus hampering the effectiveness of immunotherapies such as IL-2. By selectively catalyzing the removal of excessive superoxide in cancer states, the enzyme mimetic appears to work synergistically with IL-2.

7.2.2 Vanadium-Based Diabetes Drugs

7.2.2.1 Introduction. Insulin injections have been used in the treatment of diabetes mellitus, beginning in the 1920s. Insulin cannot be given orally because it is a protein destroyed by the digestion process. Oral therapies now exist for treatment of diabetics; however, these drugs do not mimic insulin signaling. Rather these drugs (1) stimulate insulin release (sulfonylureas), (2) potentiate insulin action (thiazolidinediones), or (3) lower liver glucose production (biguanides).[15]

Vanadium, an essential trace metal for a broad range of organisms including humans, has been postulated to be a cofactor for a number of enzymatic processes, including those involved in insulin signaling and production. Salts of this element have been known for a decade to inhibit the action of a number of phosphatases in in vitro situations. In particular, vanadate has been used frequently in laboratory settings with isolated tissues and cell cultures as a tool in biochemical studies of insulin action mechanisms and experimental insulin-resistant states. For in vitro testing of insulin mimetics, two enzymes systems, phosphotyrosine phosphatases (PTPases) and tyrosine kinases, have been found that play key regulatory roles in insulin receptor binding.[16] They are sensitive to vanadium inhibition and stimulation, respectively, and are used as markers for vanadium-containing insulin mimics. The vanadium ion has shown insulin mimetic properties in preparations of muscle, liver, and adipose tissue as well as in whole animals with various forms of diabetes.

The most widely accepted in vivo test animal is the streptozotocin (STZ)-induced diabetic rat. STZ is an antibiotic that attacks the insulin-secreting pancreatic β-cells in a dose responsive fashion. Nondiabetic laboratory animals appear to show much less response to vanadium compound administration. Despite the lack of consensus on the precise biochemical mechanism of action for vanadate, it is clear that a broad array of cellular and physiologic processes are modified in an insulin mimetic pattern by administration of vanadate compounds. The magnitude and universality (i.e., across tissues, across species, across diabetes type or model) of these effects are still being debated. In addition, vanadium can exist in a number of oxidation states (both cationic and anionic), and the state most relevant to insulin action has not been established. In addition to its insulin mimetic effect, vanadium has been shown to lower growth of human prostate cancer cells in tissue cultures and to reduce bone cancer and liver cancer in animals. These widespread effects on cancer and diabetes, along with the protective effect seen with another trace mineral, selenium, on certain cancers, suggest that trace minerals are likely to come under more scrutiny for potential health benefits and toxicity. Unfortunately, vanadium's effects are not all positive. Vanadium works by blocking dozens of enzymes, including ribonucleases, mutases, kinases, and synthases, and this indiscriminate blocking action may have negative as well as positive effects.

Vanadium and vanadium compounds have been studied in humans for their essentiality as well as their therapeutic effects. Vanadium may act as a cofactor for enzymes involved in blood sugar metabolism, lipid and cholesterol metabolism, bone and tooth development, fertility, thyroid function, hormone production, and neurotransmitter metabolism. Although vanadium deficiency has not been characterized in humans, deficiency in animals causes the following: infertility; reduction in red blood cell production, leading to anemia; iron metabolism defects; and poor bone, tooth, and cartilage formation. It is possible that vanadium deficiency in humans may lead to high cholesterol and triglyceride levels and increase susceptibility to heart disease and cancer. A daily intake of 10–100 μg of vanadium is probably safe and adequate. Some foods that are rich in vanadium are radishes, parsley, dill, and wheat grains. Toxic effects of vanadium include nerve damage, blood vessel damage, kidney failure, liver damage, stunted growth, loss of appetite, and diarrhea. Excess vanadium in humans has been suggested as a factor in bipolar disorder. Since 1980 when research first showed that vanadium compounds could lower blood sugars, animal experiments have shown that vanadium can mimic the effects of insulin and reduce blood sugar levels from high to normal. In human studies it has been shown to have some ability to lower cholesterol levels and blood pressure. These benefits are seen with low doses. There have been limited clinical trials with vanadium salts in patients with Type II diabetes, indicating that vanadium may have therapeutic potential in the treatment of this disease. Type II diabetes, or mature onset diabetes, is known medically as non-insulin-dependent diabetes mellitus (NIDDM). The main effect of this disease is the development of insulin resistance. Vanadium compounds appear to enhance the effects of insulin in individuals with NIDDM, lowering glucose, lipids and cholesterol levels. Because elevated insulin levels in NIDDM is associated with hypertension, vanadium compounds help in this manner also. Vanadium compounds have also been shown to be useful in patients with insulin-dependent diabetes mellitus (IDDM), where they operate to improve sensitivity to insulin. Unfortunately, no one has been "cured" while very serious concerns have been raised about vanadium's potential toxic effects. After oral intake, effects of vanadium ingestion are seen weeks to months later due to its accumulation in tissues like the kidneys and bone.

A November 1997 American Chemical Society symposium focused attention on the importance of understanding the interrelationships among the aqueous chemistry, the biochemistry, and the therapeutic utilization of vanadium compounds.[17] In addition to coverage of basic vanadium chemistry and biochemistry, examples of the topics included in the symposium were (1) insulin-mimetic action of selected vanadium compounds, (2) synthetic models for vanadium haloperoxidases, (3) the vanadium-containing nitrogenase system of *Azotobacter vinelandii*, (4) a possible role for amavadine in some Amanita fungi, and (5) vanadium salts in the treatment of human diabetes mellitus. Aspects of the essentiality, toxicology, and pharmacology were discussed along with insulin-mimetic compounds and their influences in cell cultures.

7.2.2.2 Examples of Vanadium Compounds Tested as Insulin Mimetic Agents. Vanadium, a group 5 transition metal, may exist in many oxidation states: −3, −1, 0, +1, +2, +3, +4, and +5. The thermodynamically and kinetically possible oxidation states under physiological conditions are +5, +4, and +3. Compounds that have been tested as insulin mimetics, both as inorganic salts and coordination complexes having organic ligands, are the V(IV) and V(V) species. The vanadyl ion, VO^{2+}, has been administered to animal and human patients as the $VOSO_4$ salt. This inorganic vanadium salt, long known for its antidiabetic properties, has been the subject of preliminary clinical trials. However, it is often poorly absorbed by the body and can have gastrointestinal side effects. Improved solubility, capability for oral administration and insulin mimetic activity, was found originally for the bidentate chelate prepared by reacting $VOSO_4$ with maltol to form bis(maltolato) oxovanadium(IV), BMOV, whose structure is shown in Figure 7.3A.[16] BMOV has square-pyramidal geometry about the central V(IV) ion and demonstrates the one unpaired electron characteristic of the vanadyl unit.

In one comparison study between inorganic vanadium salts and organovanadium complexes,[18] $VOSO_4$ doses of 25, 50, and 100 mg V daily to human patients produced no relationship between plasma glucose levels and vanadium in blood serum, the same result being found in animal studies. With administration of BMOV in animals, low plasma glucose did correlate with high blood vanadium levels. To determine if V binding to serum proteins could diminish biologically active serum V, binding of both agents to human serum albumin (HSA), human apoTransferrin (apoHTf), and pig immunoglobin (IgG) was studied by electron paramagnetic resonance (EPR) spectroscopy. Both $VOSO_4$ and BMOV bound to HSA and apoHTf, forming different V–protein complexes, while neither V compound bound to the IgG. The agents also showed differences when levels of plasma glucose and blood V in diabetic rodents were compared, as well as

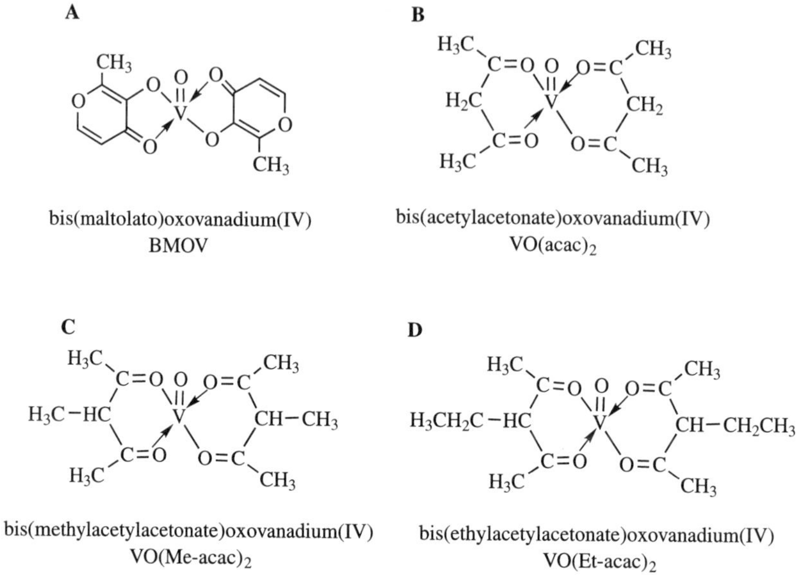

A

bis(maltolato)oxovanadium(IV)
BMOV

B

bis(acetylacetonate)oxovanadium(IV)
$VO(acac)_2$

C

bis(methylacetylacetonate)oxovanadium(IV)
$VO(Me\text{-}acac)_2$

D

bis(ethylacetylacetonate)oxovanadium(IV)
$VO(Et\text{-}acac)_2$

Figure 7.3 Structures of vanadium(IV) insulin mimetic agents. (A) BMOV; (B) $VO(acac)_2$; (C) $VO(Me\text{-}acac)_2$; (D) $VO(Et\text{-}acac)_2$.

differences in the formation of V–protein complexes with abundant serum proteins. These data suggested to the authors that binding of V compounds to ligands in blood, such as proteins, may affect the available pool of V for biological effects.

Five-coordinate bis(acetylacetonate)oxovanadium(IV), $VO(acac)_2$ (Figure 7.3B), along with derivative complexes $VO(Et\text{-}acac)_2$ (Figure 7.3D) and $VO(Me\text{-}acac)_2$ (Figure 7.3C), have been shown to have long-term in vivo insulin-mimetic effects in streptozotocin-induced Wistar rats, a commonly studied animal model.[19] Three organovanadium complexes, $VO(acac)_2$, $VO(Et\text{-}acac)_2$, and BMOV, were compared to the inorganic salt $VOSO_4$ in oral administration of 125 mg vanadium element in drinking fluids for up to 3 months. All organovanadium complexes induced a faster and larger fall in glycemia ($VO(acac)_2$ being the most potent) than $VOSO_4$. Rats treated with $VO(acac)_2$ exhibited the highest levels of plasma or tissue vanadium, most likely due to a greater intestinal absorption. There appeared to be no relationship between plasma and tissue vanadium levels and serum glucose or glucose metabolic processes, leading the authors to believe that $VO(acac)_2$ exhibited greater insulin-like properties than the inorganic salt. In further studies, the concentrations of EPR-observable V(IV) species for $VO(acac)_2$ and derivatives in water solution at 20°C were determined and apportioned on the basis of computer simulations of the spectra.[20] Three time-, pH-, temperature-, and salt-dependent species were found to be the *trans*-$VO(acac)_2 \cdot H_2O$ and *cis*-$VO(acac)_2 \cdot H_2O$ adducts as well as a hydrolysis product containing one V and one acetylacetonate group. Reaction rates for conversion of species was in the order

$VO(acac)_2 < VO(Et\text{-}acac)_2 \approx VO(Me\text{-}acac)_2 \approx$ BMOV. Conversion rates for all hydrolysis products were faster than for the original species. Both EPR and visible spectroscopic studies of solutions prepared for administration to diabetic rats ocumented both a salt effect on the species formed and formation of a new halogen-containing complex. The authors concluded that vanadium compound efficacy with respect to long-term lowering of plasma glucose levels in diabetic rats traced the concentration of the hydrolysis product in the administration solution.

Peroxovanadates having V(V) have been tested for insulin-mimetic properties. In vitro, they have been shown to stimulate insulin receptor tyrosinase kinase (IRTK or IRK) activity in heptoma cells and to inhibit phosphotyrosine phosphatase (PTPase) activity in rat liver endosomes.[21] These agents were found to be potent inhibitors of PTPases. (Figure 1 of reference 15 shows a schematic diagram indicating the sequence of insulin receptor tyrosinase kinase (IRTK) activation by insulin, IRTK activation, and autophosphorylation along with inhibition of PTPase-catalyzed dephosphorylation of IRTK through the activity of vanadium complexes.) Although stable in the solid state, the peroxovanadates are subject to decomposition in aqueous solution. Because oral administration is not possible, the compounds were administered intravenously to normal rats and found to lower glucose concentrations in a dose-dependent manner.[22] In insulin-deprived BB rats, peroxovanadium compounds were administered intravenously or subcutaneously and found to lower plasma glucose levels in relation to the degree of PTPase inhibition observed in vivo.[23] The compound potassium oxodiperoxo(1,10-phenanthroline)vanadate(V) trihydrate, $[bpV(phen)]^-$ is heptacoordinate with the geometry about the V atom being pentagonal bipyramidal as shown in Figure 7.4.[24] A monoperoxovanadate(V) complex, oxoperoxopicolinatovanadium(V) dihydrate [mpV(pic)], shown in Figure 7.4, was found to achieve a 20% decrease in plasma glucose in STZ-diabetic rats when administered by intraperitoneal or subcutaneous injection.[25]

A proprietary organo-V(IV) compound, KP-102, produced by Kinetek Pharmaceuticals, Inc., is currently in phase II clinical trials. KP-102, as an orally active insulin sensitizer, appears to work by acting through the intracellular signal transduction system, rather than eliciting insulin release or affecting the insulin

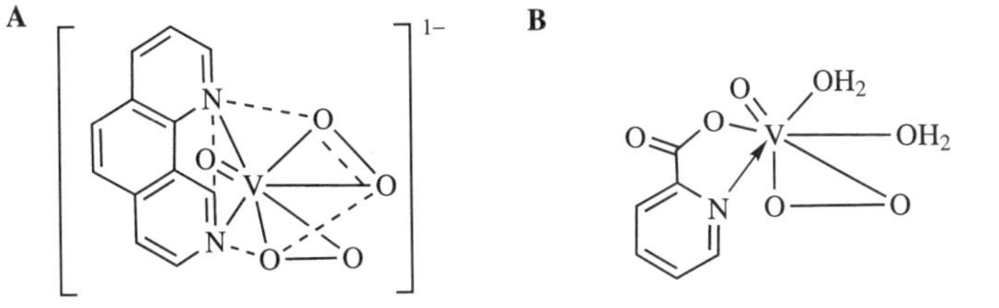

A

oxodiperoxo(1,10-phenanthroline)vanadate(V)
$[bpV(phen)]^-$

B

oxoperoxopicolinatovanadium(V)
[mpV(pic)]

pentagonal bipyramidal geometry

Figure 7.4 Vanadium(V) insulin mimetic agents. (A) $[bpV(phen)]^-$; (B) [mpV(pic)].

receptor. According to U.S. Patent No. 5,866,563, entitled "Vanadium Compositions," KP-102 appears to work by acting through the intracellular signal transduction system, rather than by eliciting insulin release or affecting the insulin receptor.

Recently, several vanadium(III) complexes have been evaluated for their insulin mimetic properties.[26] Usually, V(III) complexes adopt six-coordinate octahedral or pseudo-octahedral geometry and are hydrolytically stable. However, they undergo rapid oxidation to V(IV) and V(V) at pH values >3 and therefore have not been considered suitable as insulin-enhancing agents. In this research, complexes with maltol (ma), ethyl maltol, kojic acid, and several other ligands were synthesized and tested for hydrolytic and oxidative stability. The compound $V(ma)_3$ was found to be hydrolytically stable and the most stable to oxidation of any compound tested in this series. In STZ-induced diabetic rats, $V(ma)_3$ exhibited glucose lowering to <9 mM within 24 hours of intraperitoneal (i.p.) injection, a behavior similar to that seen for BMOV, $VO(ma)_2$, discussed previously. The authors concluded that V(III) compounds can be comparable to V(IV) in insulin-enhancing capability and are continuing to study speciation and biodistribution of the V(III) compounds.

7.2.2.3 The Role of Chromium. The role of essential chromium(III) ion has recently been tied to the maintenance of proper carbohydrate and lipid metabolism at a molecular level even though it was known in the 1950s that rats deficient in chromium developed an inability to remove glucose efficiently from their bloodstreams.[27] Two problems exist in determining reasons for Cr(III) essentiality: (1) Cr(III) is inert to substitution, and (2) the concentration of Cr(III) is so low that reliable analytical techniques for its detection were not known until the 1980s. Chromodulin, an ~1500-Da oligopeptide comprised of only glycine, cysteine, glutamate, and aspartate residues, has been found to bind four equivalents of Cr(III) ions in a neutral tetranuclear arrangement.[28] The chromium-containing chromodulin complex potentiates the effects of insulin on the conversion of glucose to carbon dioxide and water. More recently it was found that chromodulin slightly activates PTPases and more significantly carries out the stimulation of insulin receptor tyrosine kinase (IRTK) activity.[29] Studies of apochromodulin and other common biological transition metals complexed to chromodulin indicated that chromium (III) is essential to chromodulin's physiological function. The proposed mechanism is as follows (and is illustrated in Figure 7.5):

1. In response to increases in blood sugar levels, insulin is released into the bloodstream.
2. Insulin binds to the external α subunit of the transmembrane protein insulin receptor.
3. The receptor autophosphorylates tyrosine residues on the internal portion (β subunit), turning the receptor into active kinase.
4. Meanwhile chromodulin, stored in apo form in the cell cytosol and nucleus, is metallated by transfer of Cr(III) from the blood to insulin-dependent cells, perhaps by the metal transport protein transferrin.

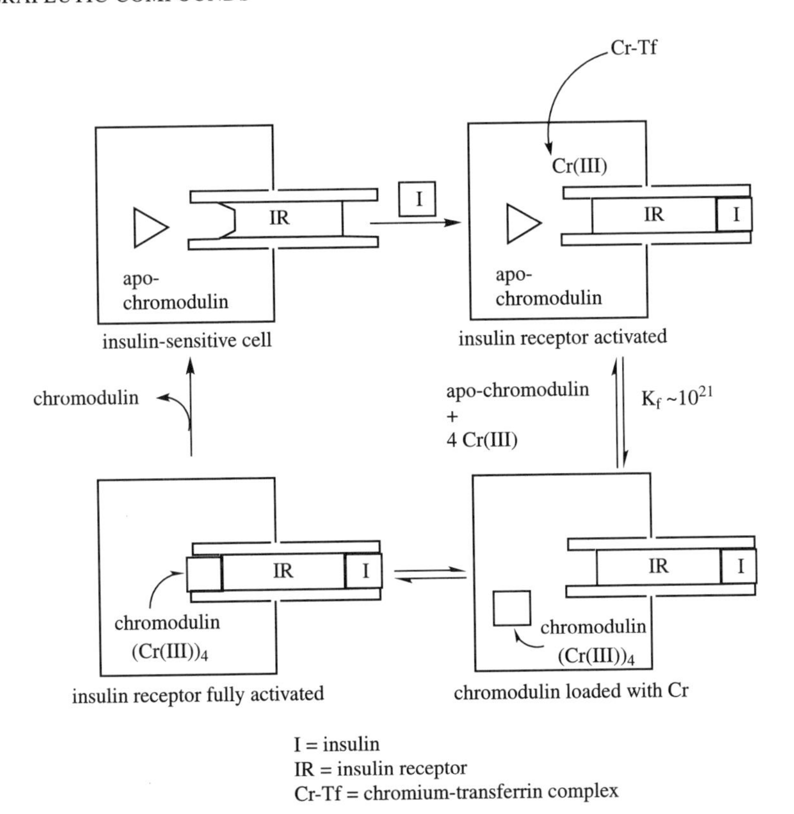

Figure 7.5 Schematic diagram illustrating proposed mechanism for chromodulin-activated insulin receptor kinase activity. (Adapted with permission from Figure 1 of Vincent, J. B. *J. Nutr.*, 2000, **130**, 715–718.)

5. The holochromodulin binds to the insulin-stimulated insulin receptor, helping to maintain its active conformation and increasing insulin signaling.

It is known that part of this process involves the 80-kDa blood serum protein transferrin that tightly binds and transports two ferric iron ions. Because the iron binding uses only ~30% of transferrin's metal binding capacity, it has long been thought to bind and transfer other metal ions (including perhaps chromium) in vivo, although this has not been demonstrated by experiment.

7.2.3 Platinum-Containing Anticancer Agents

7.2.3.1 **Cis- *and* trans-*Dichlorodiammineplatinum(II)*.** The compounds *cis*- and *trans*-dichlorodiammineplatinum(II) were known to chemists even before the geometric characterization work of Alfred Werner in the nineteenth century.[30] In

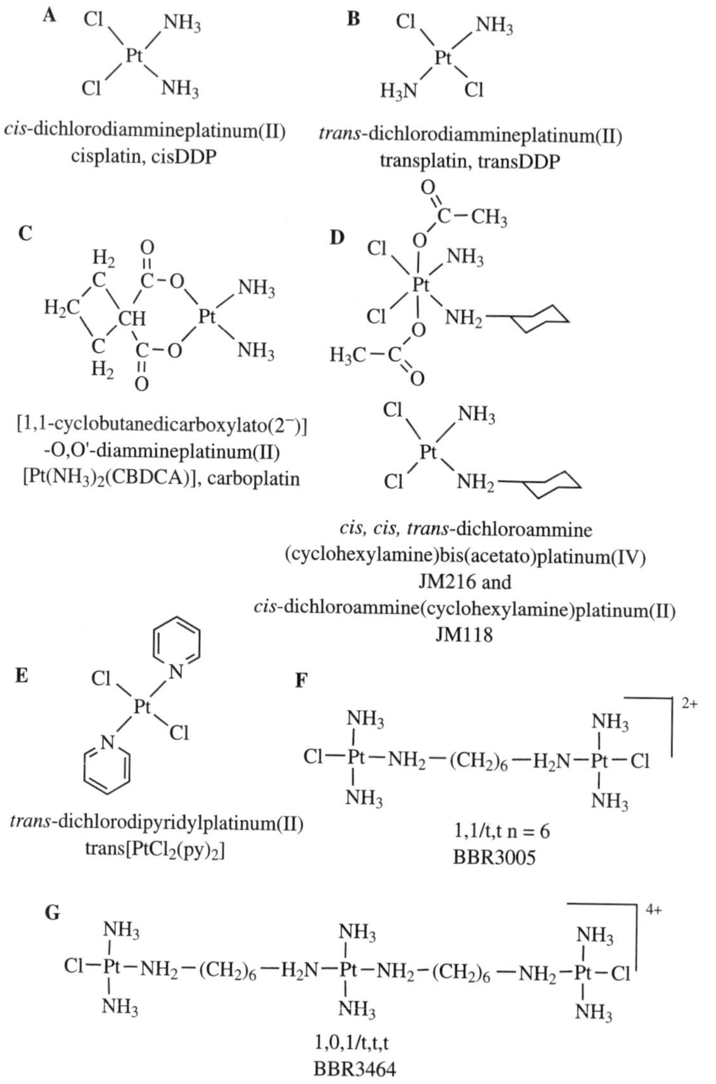

A *cis*-dichlorodiammineplatinum(II)
cisplatin, cisDDP

B *trans*-dichlorodiammineplatinum(II)
transplatin, transDDP

C [1,1-cyclobutanedicarboxylato(2⁻)]
-O,O'-diammineplatinum(II)
$[Pt(NH_3)_2(CBDCA)]$, carboplatin

D *cis, cis, trans*-dichloroammine
(cyclohexylamine)bis(acetato)platinum(IV)
JM216 and
cis-dichloroammine(cyclohexylamine)platinum(II)
JM118

E *trans*-dichlorodipyridylplatinum(II)
trans$[PtCl_2(py)_2]$

F 1,1/t,t n = 6
BBR3005

G 1,0,1/t,t,t
BBR3464

Figure 7.6 Platinum antitumor compounds: (A) *cis*-$[Pt(NH_3)_2Cl_2]$, cisDDP; (B) *trans*-$[Pt(NH_3)_2Cl_2]$, transDDP; (C) $[Pt(NH_3)_2(CBDCA)]$, carboplatin; (D) JM216, Pt(IV) and JM118, Pt(II); (E) *trans*-$[PtCl_2(py)_2]$; (F) BBR3005, 1,1/*t*,*t* $n = 6$; (G) BBR3464, 1,0,1/ *t*, *t*, *t*.

the mid-1960s, Barnett Rosenberg and his co-workers serendipitously discovered that *cis*-dichlorodiammineplatinum(II) (*cis*-$[Pt(NH_3)_2Cl_2]$, cisDDP, cisplatin) exhibited antitumor activity but that the trans isomer, *trans*-$[Pt(NH_3)_2Cl_2]$, did not.[31] See Figure 7.6A and 7.6B. Rosenberg, a biophysicist, had been studying the growth of bacterial cells in solution in the presence of electrical fields. He observed elongation of the *Escherichia coli* bacterial cells, a phenomenon known as filamentous growth, rather than cell division.[32] Rosenberg also discovered that the supposedly inert platinum electrodes of the electrical circuit were reacting with ammonium and chloride ions in the nutrient broth in the electrical cell forming a number of platinum(II) and platinum(IV) compounds. These platinum compounds, *cis*-$[Pt(NH_3)_2Cl_2]$ and *cis*-$[Pt(NH_3)_2Cl_4]$ in particular, were found to cause filamentous growth in cells, absent any electrical current. Faced with his discoveries, Rosenberg made a spectacular "leap of logic" postulating that platinum compounds causing filamentous growth rather than cell division could be antitumor agents. His reasoning was based on the well-known fact that one attribute of malignant tumors is rampant and uncontrolled cell division. As history now reports, his hypothesis was proven correct. In 1972 *cis*-$[Pt(NH_3)_2Cl_2]$ was first tested in clinical trials in human patients,[33] leading, in 1979, to its approval by the U.S. Food and Drug Administration for anticancer treatment. The drug's main targets are testicular, ovarian, bladder, and head and neck tumors. At the current time, *cis*-$[Pt(NH_3)_2Cl_2]$ is one of the most widely used anticancer agents and is the only anticancer agent known that can cure a malignancy. This result has been achieved for testicular cancer when the disease is discovered in its early stages.

Because *cis*-$[Pt(NH_3)_2Cl_2]$ exhibits severe renal toxicity and has narrow applicability to few tumor types, the search has continued for less toxic and wider spectrum platinum-containing anticancer agents. A second drug, $[Pt(NH_3)_2(CBDCA)]$ (CBDCA = the cyclobutane-1,1-dicarboxylate ligand), carboplatin (Figure 7.6C), has been brought into clinical usage. Its activity spectrum is similar to that of *cis*-$[Pt(NH_3)_2Cl_2]$, but carboplatin is less toxic to the kidneys (less nephrotoxic). Thousands of other platinum complexes have been investigated for anticancer activity, and it is now known that platinum(IV) compounds (Figure 7.6D),[34] platinum(II) compounds of trans geometry (Figure 7.6E),[35] and dinuclear platinum(II) and trinuclear platinum(II) compounds (Figure 7.6F,G)[36] show antitumor activity.

7.2.3.2 Mechanism of cisDDP Antitumor Activity. Thousands of research papers have made platinum antitumor compounds the most studied inorganic drugs in history and possibly the most studied drugs of any type. An excellent review of the field up to 1987 was published in *Chemical Reviews*.[37] In 1999 Bernhard Lippert edited a volume detailing the history and behavior of platinum anticancer agents.[38] At the beginning of its history as an anticancer agent, it was found that *cis*-$[Pt(NH_3)_2Cl_2]$ (or cisDDP as it will be abbreviated here) inhibits DNA replication.[39] Mechanistic studies to determine cisDDP's physiological activity have focused therefore on its reactions with nucleic acids. Before any reaction can take place between cisDDP and DNA, the platinum drug must enter the cell. This is believed to take place through passive or active diffusion, although the process is not well understood. It is known that in the bloodstream, cisDDP retains its chloride ions because Cl^- concentrations are high (~0.1 M). Inside the cell, chloride concentration is much lower (~0.004 M) and equilibria are established for the hydrolysis reactions shown in Figure 7.7. Because the chloride ion is a better leaving group than the ammonia ligand, which forms a more thermodynamically stable covalent

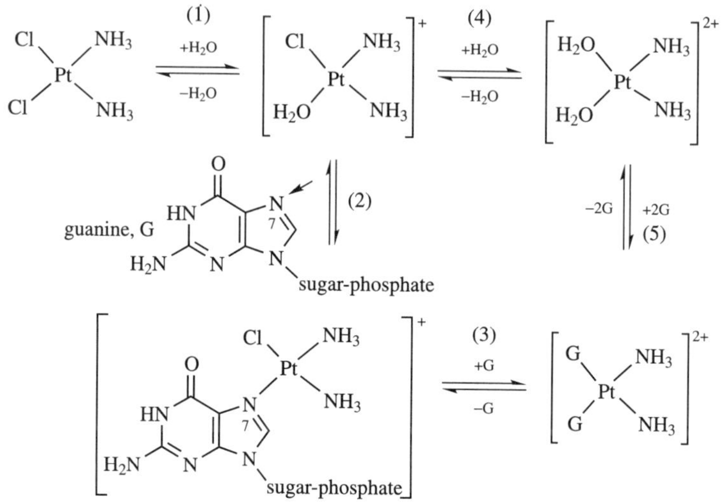

Figure 7.7 Physiological behavior of cisDDP in vivo. Aquation reactions followed by guanine N7 ligand attachment.

bond with metal ions, the ammine ligands remain while chloride ions are displaced by water. This aqueous chemistry for platinum(II) compounds, more complex than shown in Figure 7.7 because various OH^- as well as H_2O ligand combinations are possible at physiological pH, has been exhaustively studied and reviewed.[40] Water ligands in turn are replaced by nucleobase ligands from DNA strands; once formed, these nucleobase nitrogen attachments are thermodynamically stable. The preferred point of attachment along DNA strands has been found to be at the guanine N7 position. The most likely sequence of events is the hydrolysis of one chloride ligand followed by reaction with a dsDNA guanine ligand, hydrolysis of the second chloride ion, and attachment of a second dsDNA guanine (reaction sequence (1), (2), (3) in Figure 7.7).

Many other questions arise in determining the reactions of cisDDP with biological molecules. Is the dsDNA attachment intrastrand (same DNA strand) or interstrand (crosslink from a base on one strand to a base on the complementary strand)? Are other nucleobases involved? It is now known that 65% of DNA–cisDDP cross-links are intrastrand 1,2-d(GpG) cross-links with lesser contributions from 1,2-d(ApG), 1,3-d(GpXpG) (where X is any nucleobase), and interstrand crosslinks.[41] Much other evidence has confirmed the cisDDP cross-link to N7 atoms of adjacent guanines on the same strand of DNA. NMR studies of cisDDP cross-links with di- and oligonucleotides in the 1980s supported intrastrand cross-linking at adjacent guanines.[42] Two-dimensional NMR analysis by Jan Reedijk's group in 1985 indicated that chelation of cisDDP to two neighboring guanines indeed does occur, that base pairing on dsDNA remains essentially intact, and that the DNA bends or kinks by an angle of 40–50°.[43] NMR and X-ray crystallographic studies have continued to the present time, and more will be reported on these studies in the following paragraphs. Another question is: What kinds of cisDDP–protein interactions arise in vitro or in vivo? Because Pt^{2+} is a soft metal species, does it react with soft sulfur ligands? The answer is most definitely "yes," as has been shown for many biological sulfur–platinum systems by the group of Jan Reedijk. In fact, thioether sulfur donors such as methionine or perhaps oxidized glutathione may be additional intermediates in hydrolysis reactions before the thermodynamic end product with nucleobases is reached.[44]

Another known protein–platinum interaction occurs with the high mobility group (HMG) proteins that recognize bends in DNA brought about by cisDDP-d(GpG) or cisDDP-d(ApG) intrastrand cross-links. Kinked or bent DNA is recognized by HMG proteins which, upon binding to the cisDDP cross-link, significantly increase the bending of the helix. This behavior has been postulated to mediate the antitumor activity of cisDDP. Possibly the HMG–Pt–DNA lesion prevents binding of DNA-repair proteins, thus preventing replication and causing cell death (apoptosis).

To facilitate understanding of interactions between platinum antitumor compounds and nucleic acids, many X-ray crystallographic and NMR solution structures of platinum complex–oligonucleotide and later platinum complex–DNA cross-links have been published over the last 15 years. In 1987 Reedijk's group published the X-ray structure of cisDDP with the trinucleotide CpGpG, *cis*-$[Pt(NH_3)_2(CGG\text{-}N^7,N^7)]$.[45] The structural data indicated that the bis-N^7 chelate was formed by distortion of the planar guanine rings away from planarity with each other. The platinum(II) coordination sphere remained almost identical to normal square-planar geometry. Dimers and tetrameric units in the crystal were kept together by base stacking and Watson–Crick-type hydrogen bonding interactions. After many failed attempts by many researchers, Takahara and Lippard succeeded, in 1995, in determining the X-ray crystal structure of a double-stranded dodecamer oligonucleotide–platinum(II) unit.[46] The oligonucleotide, with* indicating the point of platination, was d(CCTCTG*G*TCTCC)/d(GGAGACCAGAGG). The basic structure appeared kinked as predicted with about a 45° bending of the helix. The platinum(II) ion lay about 1 Å out of the guanine plane, and the DNA conformation at the platinum's 5′ side changed from the expected B DNA to A DNA. Data for this structure are deposited in the Protein Data Bank (PDB) with code name 1AIO. In 1998 a high-resolution 2D NMR solution structure of a DNA dodecamer duplex, d(CCTCTGGTCTCC/d(GGAGACCAGAGG), containing the cisplatin d(GpG) 1,2-intrastrand cross-link was published.[47] The *cis*-$[Pt(NH_3)_2d(GpG\text{-}N7(1), N7(2))]$ lesion caused the adjacent guanine bases to roll toward one another by 49°, leading to an overall helix bend angle of 78°. These features appeared more exaggerated than those observed in the X-ray crystal structure determined for the same platinated duplex described in reference 46. A common property of the solution and crystal structures is the widening and flattening of the minor groove opposite the platinum adduct, yielding geometric parameters resembling those found in A-form DNA. This deformation is especially noteworthy for the solution structure because its sugar puckers are primarily those of B DNA. (See Section 2.3.2 in Chapter 2 for a discussion of sugar ring puckers.)

The unwinding of the helix at the site of platination is 25°. The curvature and shape of the platinated duplex are remarkably similar to those observed in DNA duplexes complexed by the HMG-domain proteins revealing that cisplatin binding alters DNA in such a manner as to facilitate HMG-domain protein recognition as mentioned in the previous paragraph.

Although the major adducts between platinum antitumor agents and DNA appear to be intrastrand 1,2-d(GpG) or d(ApG) cross-links, much evidence exists for other types of lesions. Interstrand GG adducts were reported in the 1980s,[37,41] and more recently two cisDDP-GG interstrand cross-link structures were published. The crystal structure of a (5′-d(CpCpTpCpGpCpTpCpTpC)-3/5′-d(GpApGpApGp CpGpApGpG)-3′)-cisDDP interstrand cross-link between G5-N^7 and G15-N^7 (PDB code: 1A2E) was reported in 1999.[48] The group of P. B. Hopkins and B. R. Reid at University of Washington, Seattle reported an NMR solution structure for the cisDDP-d(CpApTpApGpCpTpApTpG) duplex oligonucleotide which indicated that the double helix was locally reversed to a left-handed form with the helix unwound and bent toward the minor groove.[49] DNA–protein cross-links have also been reported.[50] Platinated DNA is a substrate for repair enzymes[51] which may remove the lesion; however, all adducts are not removed with the same efficiency or at all (see discussions of HMG-protein-platinated DNA later in this section). It is not known which adduct is primarily responsible for the cytotoxicity (antitumor activity) of cisDDP; furthermore, other platinum compounds may cause entirely different cytotoxic lesions. Although research has focused on cisDDP's ability to modify genomic DNA (gDNA) in the cell nucleus, mitochondrial DNA (mtDNA) may also be a cisDDP target. The mtDNA–cisDDP adducts have been found to be persistent perhaps because mitochondria do not perform nucleotide excision repair.[52] One may infer then that mtDNA–cisDDP lesions contribute to anticancer activity.

An interesting target area for platinum antitumor drugs are the so-called telomeric DNA regions occurring at the ends of eukaryotic chromosomes and containing guanine-rich regions.[52] For instance, the human telomere sequence is 5′-TTAGGG-3′. Telomeres protect the ends of chromosomes from degradation and ensure that genetic information is correctly passed to the next DNA generation during cell division. Telomerases, on the other hand, are proteins that catalyze telomere addition to chromosome ends, and they are implicated in the growth of malignant tumors. Also, because telomerase plays a role in cellular resistance to apoptosis, which is the primary mode of cell death induced by several drugs, telomerase could be involved in determining the chemosensitivity profile of tumor cells. In a study of telomere loss in cisDDP-treated HeLa cells, researchers found that low doses of cisDDP caused telomere loss sufficient to cause lethal damage in $\sim$61% of the cells.[53] Reaction of cisDDP with the protein component of telomerase may also inhibit telomere formation, and although little information is available currently, this constitutes a fertile field for further investigation. One in vitro study has attempted to correlate the relationship between telomerase activity, telomere length, and chemosensitivity to effective antitumor agents in a panel of human melanoma and ovarian cancer cell lines. It was found that telomeres were generally longer in melanoma than in ovarian carcinoma cell lines. No precise relation was found between telomerase activity and cellular sensitivity to different DNA damaging agents including doxorubicin (an organic anthracycline antibiotic anti-cancer agent sometimes used in combination protocols with platinum drugs), cisDDP, or BBR 3464 (discussed in Section 7.2.3.4). In contrast, longer telomeres were associated with resistance to the drugs, even though the association reached statistical significance only for cisDDP. Because platinum compounds may have affinity for telomere sequences, it is conceivable that the interaction is relevant for drug sensitivity and drug resistance status depending on telomere length.[54] High mobility group (HMG) domain proteins have been mentioned previously as representing a large family of proteins that bind specifically to cisDDP–DNA adducts. Experiments performed using cell extracts identified a ~91,000-Da protein that bound *cis*-GG and *cis*-AG adducts but not *cis*-GNG adducts or transDDP modified DNA.[55] Many other experiments have identified the HMG domain as a DNA binding motif consisting of approximately 80 amino acids.[56] Proteins in the HMG family may contain multiple HMG domains and specifically recognize DNA structures or sequences. They bind in the minor DNA groove and bend DNA upon binding.[52] The work concerning HMG domain binding to cisDDP-modified DNA has been reviewed.[57] Some of these HMG domain proteins binding to cisDDP-DNA adducts include HMG1, HMG2, and a transcription factor known as Ixr1 (Ord1), which has a 15-bp binding site centered around the cisDDP lesion. The NMR solution structures of HMG1 domains A and B have been solved.[58] Both domains A and B have three α-helices forming the shape of an L with an ~80° angle between the arms. Figure 19 of reference 52 shows a representation of the HMG1 domain B from reference 58b. The L-shaped proteins bind in the minor groove severely bending and unwinding the DNA. In a recent study, it has been found that domain A of the structure-specific HMG-domain protein HMG1 binds to the widened minor groove of a 16–bp DNA duplex containing a site-specific cisDDP-[d(GpG)-N7(1), N7(2)] adduct. The protein binds to the DNA through the concave surface of the L-shape, bending DNA by ~61°. The strongly kinked DNA is centered at a hydrophobic notch created at the platinum–DNA cross-link. HMG protein binding extends exclusively to the 3′ side of the platinated strand. An HMG protein phenylalanine residue at position 37 intercalates into a hydrophobic notch created at the platinum cross-linked d(GpG) site. Binding of the domain is dramatically reduced in a mutant in which alanine is substituted for phenylalanine at position 37.[59] The X-ray crystallographic data are deposited in the Protein Data Bank (PDB) with code name 1CKT. A representation of the structure is shown in Figure 20 of reference 52. One hypothesis for a mechanism relating HMG domain binding to cisDDP sensitivity is that cisDDP-DNA adducts hijack proteins away from their normal binding sites, disrupting their normal cellular function. Because many HMG domain proteins function as transcription factors, their removal could disrupt cellular processes in tumor cells. The other hypothesis, repair shielding, suggests that HMG domain proteins block cisDDP–DNA adducts from damage recognition needed for repair functions. If cisDDP–DNA adducts are not excised and the site is not repaired, cell death will eventually occur. The authors of reference 52 speculate that HMG domain proteins could be engineered to modulate the response of cells to cisDDP. They suggest a combination of gene and chemotherapy with the goal of HMG domain protein sensitization of the tumor cell to cisDDP binding.

More recently, Brabec and co-workers have studied how BBR3464–DNA intrastrand cross-links are recognized by HMG1 protein and removed from DNA during in vitro nucleotide excision repair reactions.[60] (BBR3464's structure is shown in Figure 7.6G. The compound is further discussed in Section 7.2.3.4.) Intrastrand BBR3464–DNA cross-links create local conformational distortion but no stable DNA bending. The intrastrand cross-links are not recognized by HMG1 proteins, and they are removed by nucleotide excision repair. The authors believe these facts indicate that intrastrand cross-links are *not* responsible for BBR3464's antitumor properties.

7.2.3.3 ***Drug Resistance and DNA Repair Mechanisms.*** Resistance of malignancies to cisDDP at the onset of anticancer treatment (innate or intrinsic resistance) or upon subsequent treatment (acquired resistance, ADR) limits its curative potential.[61] However, cisDDP is not affected by multidrug resistance (MDR), making it useful in complementary therapy in MDR-dependent drug-resistant cancers. Complementary therapy can include X-ray or hormone treatment or combination chemotherapies with other anticancer drugs as well as medications to alleviate side effects of chemotherapy. Cellular resistance to platinum anticancer drugs is multifactorial and consists of mechanisms that limit the formation of lethal platinum–DNA adducts, altered or reduced drug transport, and inactivation mechanisms that enable a cell to repair or tolerate platinum–DNA damage once it occurs. These mechanisms, which may occur through the overexpression or inactivation of certain genes, have been identified primarily through the use of cell lines selected for cisDDP resistance in vitro. Such model systems may exhibit high levels of resistance, thus facilitating the identification of genes involved in resistance and providing a system to study potential resistance modulators—compounds that would increase the sensitivity of cancer cells to platinum drugs. Numerous mechanisms potentially contributing to clinical cisDDP resistance have been identified, including (1) changes in membrane permeability, (2) detoxification pathways, and (3) the ability to remove cytotoxic lesions from DNA. Changes triggered by cisDDP selection in cells that exhibit resistance involve a secondary layer of complexity that may include alterations in (1) oncogene and protein kinase signal transduction pathways, (2) growth factor and hormone responsiveness, (3) chromosome structure and gene expression, (4) ion transport, and (5) nutrient transport and utilization. It is likely that all of these changes are part of an interconnected, multipart response to cisDDP.

Three main activities have been implicated in platinum drug resistance. These include changes in intracellular accumulation of the drug as described in the previous paragraph, increased production of intracellular thiols that modulate heavy metal toxicity, and increased capability of cells to repair cisDDP–DNA damage.[52] The last topic is discussed in following paragraphs. In general, chronic, long-term exposure to increasing concentrations of cisDDP seems to lead to permanent elevations in the levels of the nucleophiles glutathione (GSH) and metallothionein. These molecules should be good ligands for platinum compounds because they contain many soft sulfur ligands attractive to the soft Pt^{2+} metal center. An excellent review of Pt–S interactions that must play a key role in platinum drug distribution in the body, in metabolism mechanisms of platinum drugs, in their cytotoxic effects, and in acquired drug resistance (ADR) has been written by Jan Reedijk.[62] The response of two types of cancer cells (lung and ovarian carcinomas) exhibiting ADR indicates the complications arising from attempts to quantify the problem. In resistant human small cell lung carcinoma, the amount of GSH and the total amount of sulfhydryl compounds increased in comparison to the cisDDP-sensitive cell line.[63] The ovarian cell line results are much more variable and nonconsistent. The level of GSH was found to fluctuate in some ovarian cell lines over time,[64] while some resistant lines had similar GSH levels as the nonresistant line. In this case, blockage of a key enzyme in GSH synthesis did not increase sensitivity to cisDDP.[65] One promising avenue for circumventing cisplatin resistance is the development of non-cross-resistant platinum analogs. This is a major reason for continued testing of new types of platinum-containing antitumor compounds such as reported in Section 7.2.3.4, which describes the compound BBR3464.

A number of cisDDP-resistant cell lines with known resistance mechanisms are used for evaluation of new platinum-containing complexes. Some of these include 41McisR and HX/155cisR that are resistant due to reduced platinum accumulation,[66] GCT27cisR and CH1cisR that are resistant because of enhanced removal or increased tolerance to Pt–DNA adducts,[67] and A2780cisR and SKOV-3 that are resistant due to detoxification via elevated glutathione levels, decreased uptake, and increased DNA repair.[68] Glutathione has been implicated in resistance by reducing drug accumulation through the multidrug resistance-associated protein (MRP),[69] by reacting with drugs to form inactive species,[70] and by enhancing DNA repair.[71]

The precise mechanism by which platinum–DNA damage results in cell death is unknown. In many cases, cells treated with cytotoxic levels of cisDDP display the biochemical and morphologic features associated with programmed cell death (apoptosis). The cell cycle may be involved in this process because proliferating (fast growing) cells have been shown to be more sensitive to cisDDP than nonproliferating cells. It is known that cells can replicate past Pt–DNA lesions, and this behavior may be predictive of acquired drug resistance (ADR). Post-replication repair is an important mechanism of DNA damage tolerance. It is best defined as replication of damaged DNA without the introduction of potentially lethal secondary lesions (gaps or discontinuities in the nascent DNA) and/or the repair of those secondary lesions following replication. A great deal of what has been called "postreplication repair" actually occurs during replication, rather than following it. For that reason the group of S. G. Chaney at the University of North Carolina uses the term "replicative bypass" as synonymous with postreplication repair and uses the term "translesion synthesis" to define replication through and beyond the damaged site on the template DNA by individual polymerases. DNA polymerase complexes may catalyze translesion synthesis past Pt–DNA adducts.

DNA damage recognition proteins may bind to the Pt–DNA adducts and interfere with translesion synthesis either by removing the newly synthesized DNA (as likely occurs with the mismatch repair (MMR) system) or by physically blocking translesion synthesis. High mobility group (HMG) proteins, also called HMG-box proteins, and their interactions with platinated DNA have been described in Section 7.2.3.2. Their connection to DNA repair mechanisms may be indicated by their strong affinity for cisDDP-modified DNA, which can be as much as 100 times greater than that for unmodified DNA. The affinity occurs because bent DNA duplex lesions, caused by Pt–DNA cross-links, resemble HMG's normal binding site. Therefore, it is thought that HMG-box proteins increase cisDDP cytotoxicity by binding onto Pt–DNA adducts and obstructing cellular excision repair.

Another factor related to drug resistance is the behavior of the p53 gene. This gene, encoding the p53 protein, exhibits a high frequency of mutation in human cancers. The p53 protein is a transcription factor involved in the regulation of many genes. Activation of the normal p53 gene due to DNA damage or other cellular stress can result in arrested cell cycles and apoptosis, helping to maintain gene stability. While disruption of normal p53 function occurs in about 50% of human cancers, testicular tumors, cured by early cisDDP intervention, generally do not contain mutated p53 genes. Loss of p53 function appears to confer cisDDP resistance in some human ovarian cell lines, presumably by interfering with the regulation of apoptosis. In general, overexpression of p53 seems to confer cisDDP resistance; and although the mechanism is unclear, DNA repair capabilities or enhanced cellular tolerance to DNA damage are probably involved. An excellent review of recent knowledge on the topics of drug resistance and repair mechanisms, entitled "Structure Recognition and Processing of Cisplatin–DNA Adducts," has appeared in *Chemical Reviews*.[52]

7.2.3.4 A New Nonclassical Platinum Antitumor Agent. Nicholas Farrell has written a review of the "nonclassical platinum antitumor agents" that describes a series of platinum compounds whose activities in vivo may differ considerably from the classical case discussed so far.[35b] Complexes of formula *trans*-[$PtCl_2(py_2)$], where py = pyridine (Figure 7.6E), are more cytotoxic than their *cis* isomers and indeed than cisDDP itself.[72,35a] Dinuclear platinum(II) and trinuclear platinum(II) compounds (Figure 7.6F, G)[36] are active in murine and human cell lines both sensitive and resistant to cisDDP. In binding to DNA, these compounds form quite different DNA lesions. For example, the compound BBR3305, $1,1/t,t$ $n = 6$ (Figure 7.6F), may form DNA (Pt, Pt) interstrand cross-links by binding of one Pt atom to each strand of DNA[73] or may form DNA (Pt, Pt) intrastrand cross-links by binding of the two Pt atoms to the same strand.[74]

The compound BBR3464 [{*trans*-$PtCl(NH_3)_2$}$_2$μ-*trans*-$Pt(NH_3)_2${$H_2N(CH_2)_6NH_2$}$_2$]$^{4+}$, whose structure is shown in Figure 7.6G, was in Phase II clinical trials in late 2001. More than 200 patients were involved in three trials determining the drug's effectiveness in ovarian, lung, and gastric cancers. Previously, in a Phase I studies of 47 cancer patients, the new class of drugs showed activity against pancreatic cancer, melanoma, and lung cancer. The drug also appeared to be active

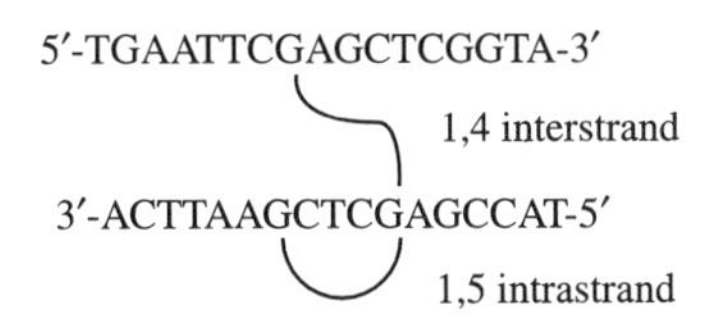

Figure 7.8 Modeled 1,4 interstrand and 1,5 intrastrand cross-links formed by platinum bifunctional compound BBR3464.

in cisDDP-resistant ovarian cancer. BBR3464 shows cytotoxicity at 10-fold lower concentrations than does cisDDP and shows sensitivity in cisDDP-resistant cell lines. It is also active in cell lines characterized as mutant p53, and therefore it is thought to be able to bypass the p53 pathway for tumors in which p53 causes tumorogenicity.

Several mechanistic differences noted between BBR3464 and cisDDP binding to DNA may help explain the differences in activity.[75] Because of its high positive charge, BBR3464 binds significantly more quickly to DNA than does cisDDP. DNA unwinding properties of BBR3464 (an unwinding angle of approximately 14°) is similar to that of other bifunctional platinum(II) compounds and is different from that for monofunctionally binding platinum(II) compounds such as $[Pt(NH_3)_3Cl]$, which exhibits an unwinding angle of approximately 6°. Assays in vitro have shown 20% interstrand DNA cross-linking for BBR3464, higher than for cisDDP (6%) but lower than for BBR3005 (Figure 7.6F), which exhibits 70–90% interstrand cross-linking. Long-range intrastrand cross-links are also postulated for BBR3464 as shown in Figure 7.8. The guanine N7 position appears to be a favored binding site as found for most other platinum compounds. Both interstrand and intrastrand cross-linking hypotheses were tested by molecular modeling on the oligonucleotide sequence shown in Figure 7.8.[75]

The 1,4 interstrand cross-link was modeled with the linking platinum species within the major DNA groove (energy = −2751.4 kJ/mol) and outside the major groove (energy = −2751.8 kJ/mol). The 1,5 intrastrand cross-link was found to have the lowest relative energy at −2894.1 kJ/mol. The modeling indicates that the long-range interstrand and intrastrand cross-links are similar in energy perhaps because the linker platinum compound is itself flexible. Figure 8 of reference 75 illustrates these models. The authors conclude that the altered DNA-binding mode is an important factor in its altered cytotoxicity profile when compared to cisDDP. Both BBR3005 and BBR3464 exhibit greatly enhanced physiological accumulation and DNA binding compared to cisDDP. This may be one reason for lower dosage requirements for the drugs in vivo. It has also been found that BBR3464 has enhanced binding to both single-stranded (ss) DNA and RNA, compared to cisDDP.[76] Continued clinical trials testing this "nonclassical" platinum antitumor agent may result in Federal Food and Drug Administration (FDA) approval for the drug's administration to cancer patients in normal clinical settings, a process described in the final section of this chapter.

7.2.3.5 Other Platinum-Containing Anticancer Compounds. Thousands of platinum compounds have been tested for antitumor activity; less than 30 of these, such as BBR3464 discussed above, have reached clinical trials in patients. Several advantages must be displayed by any drug in comparison to cisDDP: (1) The drug must be less toxic to the body while maintaining the same cytotoxicity; (2) the drug should have a wider or different spectrum of activity in treating malignancies; (3) the drug should overcome cisDDP drug resistance and be effective in tumors that have become refractory (chemotherapy has no effect on the tumor); and (4) the drugs have an administrative method other than the intravenous delivery now practiced.

The first criteria, lowered toxicity, is displayed by carboplatin (Figure 7.6C). This drug, having the same spectrum of treatment as cisDDP, has replaced it in many drug protocols for treating ovarian cancers because it is much less toxic to the kidneys and can be given in much higher dosages. One reason for its lowered toxicity is believed to be the slower intracellular hydrolysis of the bidentate CBDCA ligand compared to the monodentate chloro ligands of cisDDP.

The second criteria, a different activity spectrum, is met by oxaliplatin (Figure 7.9A), the L isomer of [oxalato(*trans*-1,2-diaminocyclohexane)platinum (II)], oxaliplatin, [Pt(II)(oxalato)(DACH)]. This platinum agent is used for secondary treatment of metastatic colorectal cancer.[77] Oxaliplatin, like carboplatin, has a kinetically slower leaving group, and is also less nephrotoxic than cisDDP. The limiting toxicity of oxaliplatin is peripheral sensory neuropathy, also seen with cisDDP. The neuropathy affects the extremities and increases in incidence and

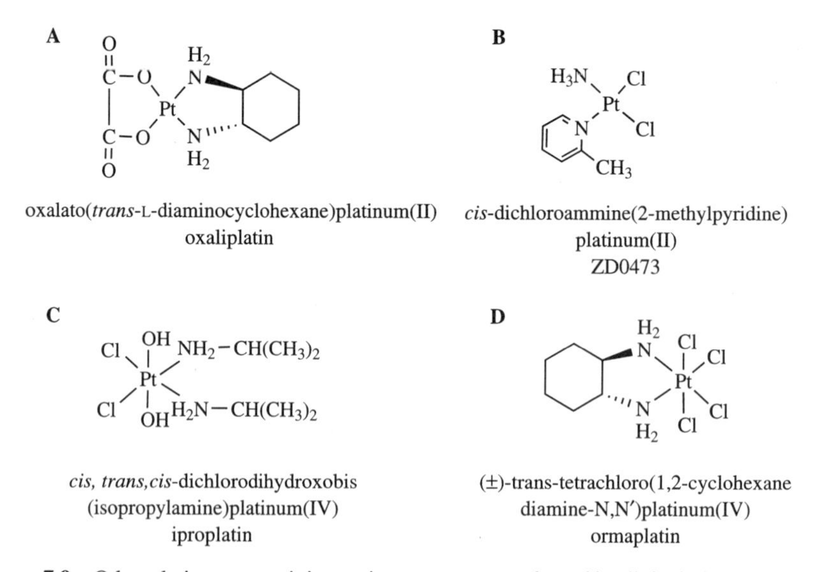

Figure 7.9 Other platinum-containing anticancer agents evaluated in clinical trials. (A) Oxaliplatin, Pt(II); (B) ZD0473, Pt(II); (C) iproplatin, Pt(IV); (D) ormaplatin, Pt(IV).

severity with repeated doses. The target and mechanism of action for a drug, along with the drug's mechanism of resistance, influence sensitivity profiles. The dissimilarity in profiles between oxaliplatin and cisDDP suggests that these two platinum compounds have a different target and mechanism of action, a different mechanism of resistance, or most likely both. Because oxaliplatin has a different spectrum of activity and low cross-resistance to cisDDP, it should be valuable in patients who have developed resistance to cisDDP.[78]

The third criteria, ability to overcome cisDDP resistance, has been demonstrated for oxaliplatin and BBR3464. A third drug, *cis*-[dichloroammine(2-methylpyridine) platinum(II)], ZD0473 (Figure 7.9B), was rationally designed to circumvent resistance by sterically hindering cellular detoxification by glutathione and other cellular thiols while still retaining the ability to form cytotoxic lesions with DNA.[79] X-ray crystallographic analysis indicates that the tilt of the pyridine ring places the 2-methyl group directly over the platinum square plane introducing steric hindrance to approach of an axial ligand and thereby reducing the rate of substitution reactions.[80] Testing in vitro has shown that ZD0473 is indeed slower in reacting with DNA as well as with sulfur-containing ligands such as metallothionein or glutathione. For instance, ZD0473 exhibited lower resistance than cisDDP in A2780cisR cells in which detoxification due to elevated glutathione levels takes place.[81] ZD0473 also circumvented resistance in 41McisR and CH1cisR cell lines where cisDDP resistance is due to enhanced removal or increased tolerance to Pt–DNA adducts.[80] In 2001, ZD0473 was being developed for both intravenous (i.v.) and oral administration, the i.v. drug was used in phase I clinical trials for the potential treatment of a range of solid tumors, including colorectal cancer, as well as in trials for patients with cisDDP-resistant ovarian tumors.

The fourth criteria, that of oral administration, is being developed and evaluated for ZD0473 as just mentioned. Another candidate in this field, JM216, *cis, trans, cis*-[dichlorodiacetatoammine(cyclohexylamine)platinum(IV)] (see Figure 7.6D), is an orally active platinum(IV) drug being evaluated in clinical trials. This platinum (IV) complex has demonstrated activity in cisplatin-resistant tumors and exhibits less nephrotoxicity and neurotoxicity than does cisDDP.

Both platinum(II) and platinum(IV) metabolites have been found in patients treated with JM216 including JM118, *cis*-[ammine(cyclohexylamine)dichloroplatinum(II)] (see Figure 7.6D) as a major metabolite. However, JM216, the platinum (IV) compound, has been found to be more antitumor active than its platinum(II) analog, JM118. JM216 showed a lack of cross-resistance in acquired cisDDP-resistant cell lines particularly those in which reduced platinum accumulation played a dominant role in resistance.[82] In a study of the reactions of the JM118 metabolite with calf thymus DNA, Hartwig and Lippard found that the major adduct (54%) was the intrastrand Pt-1,2 GG cross-link followed in frequency by interstrand or long-range intrastrand cross-links also involving guanosine (18%).[83] Fewer d(ApG) cross-links were found than for cisDDP. Additionally, the more common intrastrand Pt-1,2 GG cross-link had the cyclohexylamine ligand directed toward the 3′ end of the platinated strand. In either cyclohexylamine ligand orientation, the authors found that the *cis*-[Pt(NH_3)($C_6H_{11}NH_2$){d(GpG)-N7(1)-N7(2)}] isomers inhibited DNA replication less efficiently than did cisDDP. In a 1999 phase II clinical study, orally administered JM216 was found to be active in previously untreated patients with small cell lung carcinoma (SCLC) and to show milder toxic side effects than other platinum drugs.

Other platinum(IV) complexes evaluated previously include iproplatin, *cis*-dichloro-*trans*-dihydroxy-*cis*-bis(isopropylamine)platinum(IV) (see Figure 7.9C), and ormaplatin, also known as tetraplatin (Figure 7.9D). Development on ormaplatin was halted to due its severe neurotoxicity. Lack of superior performance stopped further testing in the case of iproplatin.[79] Platinum(IV) complexes have been called "inert" because they react much more slowly in ligand substitution reactions than do their platinum(II) analogs; however, they are also known to have their substitution reactions speeded up by the presence of even small amounts of platinum(II) compounds.[84] It is generally believed that platinum(IV) compounds are reduced to platinum(II) species extracellularly or at least prior to their reactions with DNA. Many studies have shown this to be possible; platinum(IV) metal centers are readily reduced by cellular components such as glutathione and ascorbic acid to form the platinum(II) analogs that bind more rapidly to DNA.

7.2.3.6 Conclusions. At this time, more is known about platinum-containing anticancer agents than about possibly any other known antitumor agent. In spite of this, many questions remain as to how platinum drugs carry out their antineoplastic (antitumor) activity. Scientists believe that rational design techniques will lead to better platinum–containing agents that have fewer toxic side effects, overcome all types of resistance, and are efficacious in different tumor types. The great success of platinum compounds, serendipitously discovered to have antitumor behavior, should be an inspiration to scientists who continue the quest to find the "magic bullet" cure for cancer, so long anticipated and so long in realization. Michael Clarke and co-workers have written a review of "Non-Platinum Chemotherapeutic Metallopharmaceuticals" describing gallium, ruthenium, rhodium titanium(IV), vanadium, and tin compounds that have been evaluated for antitumor activity.[85] Of these, the ruthenium compounds have been studied most extensively. Many of the properties of the ruthenium agents tested are similar to that of platinum compounds; that is, they bind to DNA (preferentially to guanine residues), their DNA-binding properties are affected by physiological sulfur ligands such as glutathione, and multiple ruthenium oxidation states and ligand systems appear to have anticancer activity.

7.3 DIAGNOSTIC AGENTS

7.3.1 Technetium Imaging Agents

7.3.1.1 Introduction. The element technetium exists in oxidation states ranging from -1 to $+7$ and consequently exhibits a wide range of coordination geometries utilizing many different ligands. Table 3 of reference 93 outlines examples of technetium oxidation states, coordination geometry and number, and magnetic

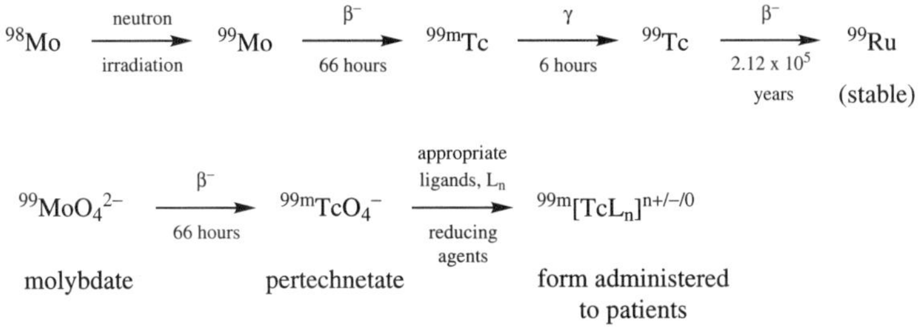

$$^{98}Mo \xrightarrow[\text{irradiation}]{\text{neutron}} {}^{99}Mo \xrightarrow[\text{66 hours}]{\beta^-} {}^{99m}Tc \xrightarrow[\text{6 hours}]{\gamma} {}^{99}Tc \xrightarrow[2.12 \times 10^5 \text{ years}]{\beta^-} {}^{99}Ru \text{ (stable)}$$

$$^{99}MoO_4^{2-} \xrightarrow[\text{66 hours}]{\beta^-} {}^{99m}TcO_4^- \xrightarrow[\text{reducing agents}]{\text{appropriate ligands, } L_n} {}^{99m}[TcL_n]^{n+/-/0}$$

molybdate — pertechnetate — form administered to patients

Figure 7.10 Technetium-99m generator. (Adapted from references 88 and 89.)

moments. Technetium has no stable isotopes; but one of its metastable isotopes, ^{99m}Tc or technetium-99m, has been a mainstay of diagnostic nuclear medicine since the 1970s.[86] In some chemical form, this isotope is used in more than 85% of diagnostic scans done in hospitals annually. Technetium-99m emits a 140-keV γ ray close to optimal for imaging cameras. Its half-life of six hours is sufficiently long to synthesize the needed pharmaceutical, inject it into patients, and perform the imaging studies, yet short enough to minimize radiation dosages in the patient. Technetium-99m is conveniently obtained from the ^{99}Mo–^{99m}Tc generator according to the reactions shown in Figure 7.10.

The medical technique utilizing the technetium-99m γ emitter has the name gamma scintigraphy because it requires emission of gamma radiation that is detected by a camera. The imaging center purchases a new generator weekly, eluting the $^{99m}TcO_4^-$ to be used in formulating the technetium-99m radiopharmaceuticals daily. Because the oxidation state of technetium-99m radiopharmaceuticals is almost always lower than the Tc(VII) state found in $^{99m}TcO_4^-$, reduction to the correct oxidation state must accompany the preparation of the radiopharmaceutical to be used in a patient. Some of these procedures are described in the "Technetium Chemistry" section of reference 93. Of the many possible technetium cores that will be further modified in designing the pharmaceutical, $[Tc{=}O]^{3+}$, containing Tc(V), is the most frequently used. The in vivo behavior of the technetium-99m radiopharmaceutical depends on the ligand field surrounding the metal and the stability, stereochemistry, charge, and lipophilicity of the complex. For example, negatively charged compounds tend to clear through the kidneys, many positively charged ions accumulate in the heart, and an overall neutral complex is required for crossing the blood–brain barrier.[87]

There are two classes of drugs: (1) "technetium essential," in which the technetium-99m is an integral part of the molecule and without which it would not be delivered to the target tissue; and (2) "technetium tagged," in which the technetium-99m labels the targeting moiety (an antibody, peptide, or hormone) and is carried along to the targeted tissue. Technetium-99m "tagged" pharmaceuticals come in two versions: integrated and bifunctional chelate modes. At this time, most technetium-99m radiopharmaceuticals are of the "technetium essential" variety, although many future drugs now under development will be of the "technetium tagged" type.

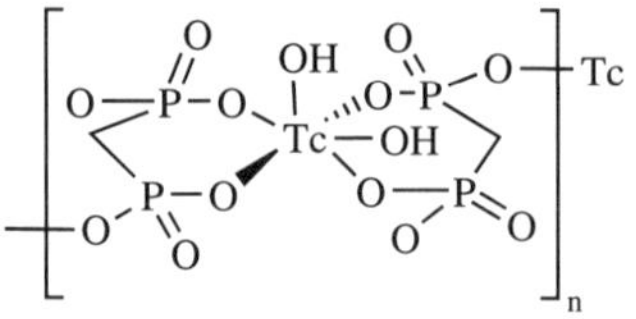

Figure 7.11 Technetium-99m(methylenediphosphonate), ^{99m}Tc-MDP. (Adapted from Figure 13 of reference 89.)

7.3.1.2 "Technetium Essential" Radiopharmaceuticals. In reference 88, Pinkerton et al. describe technetium-99m bioavailability and reagents used in thyroid, brain, kidney, liver function, heart, and bone imaging. Bone imaging agents were among the first technetium-99m agents developed.[88] Diphosphonate ligands, with their affinity for calcium in actively growing bone, guide the technetium-99m complexes to the target site.[89] The imaging reagent, technetium-99m(methylenediphosphonate), ^{99m}Tc-MDP, illustrated in Figure 7.11, was shown to be a 1-to-1 polymer in a structure reported by Deutsch and co-workers.[90] In this structure, each technetium ion is bound to two diphosphonate ligands and each diphosphonate is coordinated to two Tc centers. Although the precise oxidation state of the technetium is not known because the degree of protonation of the Tc–O–Tc bridge cannot be determined from the X-ray structure, it is believed to be Tc(IV).

Another technetium-99m reagent, originally developed for bone imaging, has been used to image myocardial infarcts (heart attacks). This agent, ^{99m}Tc pyrophosphate, accumulates in infarcted heart muscle through binding of the pyrophosphate ligand to calcium deposits in the infarct.[88] Other heart imaging agents include ^{99m}Tc-sestamibi (Cardiolite^{TM}), [(hexakis-2-methoxy-1,2-methylisopropyl)isonitrile technetium(I)]$^+$ (Figure 7.12A), and ^{99m}Tc-tetrofosmin (Myoview^{TM}), [(1,2-bis (bis-(ethoxyethyl)phosphino)ethane)-*trans*-dioxotechnetium(V)]$^+$ (Figure 7.12B).[86]

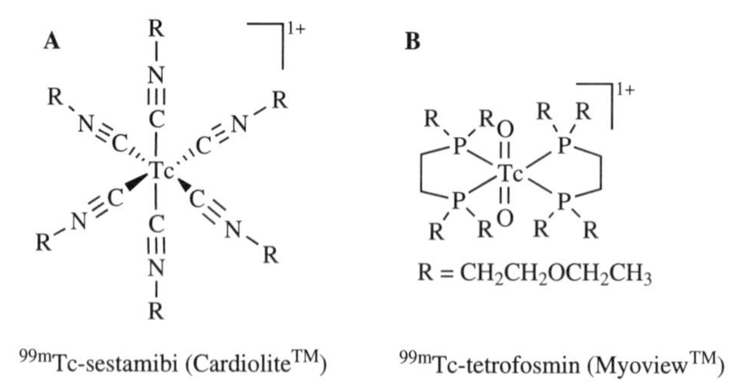

Figure 7.12 (A) ^{99m}Tc-sestamibi (Cardiolite^{TM}) and (B) ^{99m}Tc-tetrofosmin (Myoview^{TM}). (Adapted from Figure 1 of reference 86.)

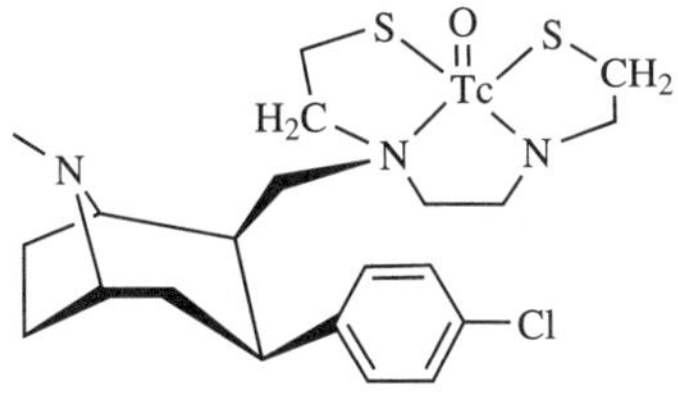

Figure 7.13 Technetium-99m TRODAT-1. (Adapted from Figure 2b of reference 86.)

Cardiolite was found to have good uptake in the target tissue (heart) and a desirable rapid clearance from blood, lung, and liver tissue. It has also been approved for breast cancer imaging. Myoview exhibits rapid lung and liver clearance on injection and good myocardial uptake.[86,89] The +1 cations are thought to be good heart imaging agents because heart tissue recognizes a variety of positively charged species (including the accumulation of Group I cations K^+, Cs^+, and Rb^+ in normal myocardial tissue).

Small molecule ^{99m}Tc receptor targeting agents also belong to the technetium essential class.[86] Examples of such agents include those targeting neuroreceptors for imaging and diagnostic applications in patients with Parkinson's disease, schizophrenia, Alzheimer's disease, epileptic seizures, and drug addiction. These agents must balance size (MW <600), lipophilicity (log $P = 1.5$–3), and receptor binding characteristics, and they must be neutral in order to cross the blood–brain barrier by diffusion. The dopamine transporter (DAT), implicated in Parkinson's disease and schizophrenia, appears to be a likely target, for instance by using tropane analogs that are derivatives of well-known DAT antagonists. One of these, [^{99m}Tc]TRODAT-1, shown in Figure 7.13, demonstrated localization in the basal ganglia consistent with DAT binding in humans. This agent was also shown to be safe for administration to humans and is undergoing further development. It is thought that estrogen, progesterone, and androgen receptors targeted by ^{99m}Tc agents may lead to faster and better diagnoses in breast cancer patients because patients with estrogen/progesterone-receptor-positive breast cancer respond better to hormonal therapeutic regimens. Although technetium-99m agents have been designed for this cancer diagnostic purpose, nonspecific binding of the steroid-derivatized technetium-99m chelates in vivo have so far limited their usefulness. It is known that multidrug resistance (MDR) occurs in tumors through overexpression of P-glycoprotein (Pgp), a transmembrane pump that transports cytotoxic materials out of cells and leads to treatment failure in cancer patients. Because ^{99m}Tc-sestamibi (Cardiolite™) is taken up by a variety of tumors[91] and also transported out of tumor cells overexpressing Pgp glycoprotein,[92] researchers have been encouraged to believe that diagnostic use of the agent may lead to greater understanding of multidrug resistance in cancer patients.

7.3.1.3 "Technetium Tagged" Radiopharmaceuticals. Small molecule ^{99m}Tc receptor targeting agents belong to the so-called integrated approach design methodology for ^{99m}Tc radiopharmaceuticals. In this motif the technetium chelate is a integral part of receptor binding to the target site. This approach may lead to the loss of receptor binding affinity as interjection of Tc–N, Tc–O, or Tc–S coordination bonds into a molecule containing only C–C or C–X bonds often has significant impact on the size and conformation of the targeting molecule. Additionally, the introduction of the $[Tc{=}O]^{3+}$ core leads to a change in lipophilicity of the receptor ligand.[93] Sometimes this problem can be overcome by the bifunctional approach, described in reference 93. The approach uses a high-binding-affinity receptor ligand as the targeting molecule and uses a bifunctional coupling agent (BFCA) for the attachment of the receptor ligand to the ^{99m}Tc radionuclide through a linker called a pharmacokinetic modifier (PKM). Thus the technetium-99m chelate is placed relatively distant from the binding receptor, minimizing the problems associated with the integrated approach. The targeting molecule, the linker, the BFCA, and a radionuclide comprise the four-part system, illustrated in Figure 7.14B as adapted from reference 93. The targeting molecules can be macromolecules such as antibodies, or small molecules including peptides and nonpeptide receptor ligands. Between the targeting molecule and the radionuclide is the BFCA, strongly coordinated to the metal ion and covalently attached to the targeting molecule, either directly or through the PKM. Much study, design, and testing of these systems is required to produce a usable diagnostic or treatment tool. The radiopharmaceutical must demonstrate biological efficacy—that is, high uptake in

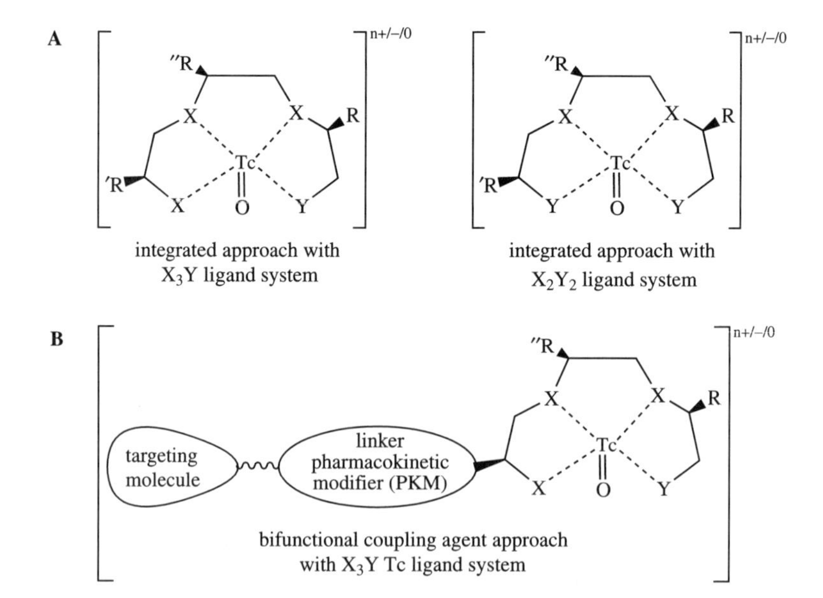

Figure 7.14 (A) Schematic representations for the integrated approach to technetium-99m radiopharmaceuticals. (B) Schematic representation for the bifunctional coupling agent approach for technetium-99m radiopharmaceuticals. (Adapted with permission from Figure 2 of Liu, S.; Edwards, D. S. *Chem. Rev.*, 1999, **99**(9), 2235–2268. Copyright 1999, American Chemical Society.)

the target tissue, high target to background ratio, high specificity and sensitivity for the targeted disease state, high radiochemical purity, and appropriately rapid clearance from the body.

Three examples of the bifunctional approach to technetium-99m radiopharmaceuticals will be discussed here: (1) a thrombus imaging agent P280 (AcuTec™), (2) an infection/inflammation imaging agent P483H, and (3) a tumor imaging agent P829 (Neotect™). These agents are being tested in clinical trials or have been approved for use in human patients.

Venous and arterial thrombus (blood clot) formation are potentially life-threatening events. In a rapidly growing thrombus, activated platelets express the GPIIb/IIIa receptor that recognizes peptides and proteins bearing the RGD (arg-gly-asp) tripeptide sequence. P280 (Figure 7.15) is an oligopeptide with two identical cyclic 13-amino-acid monomers. Each monomer contains an–(S-aminopropylcysteine)-gly-asp tripeptide sequence that mimics the RGD motif and gives P280 a high binding affinity for the GPIIb/IIIa receptor.[94] Added $[^{99m}Tc{=}O]^{3+}$ core is believed to bind at the N_2S_2 diaminedithiol chelating location, although no structural or compositional data are available. Other binding modes, such as purely S coordinate binding, are possible and have been observed for similar species. This agent was found to give excellent images of deep vein thrombosis (DVT).[94d]

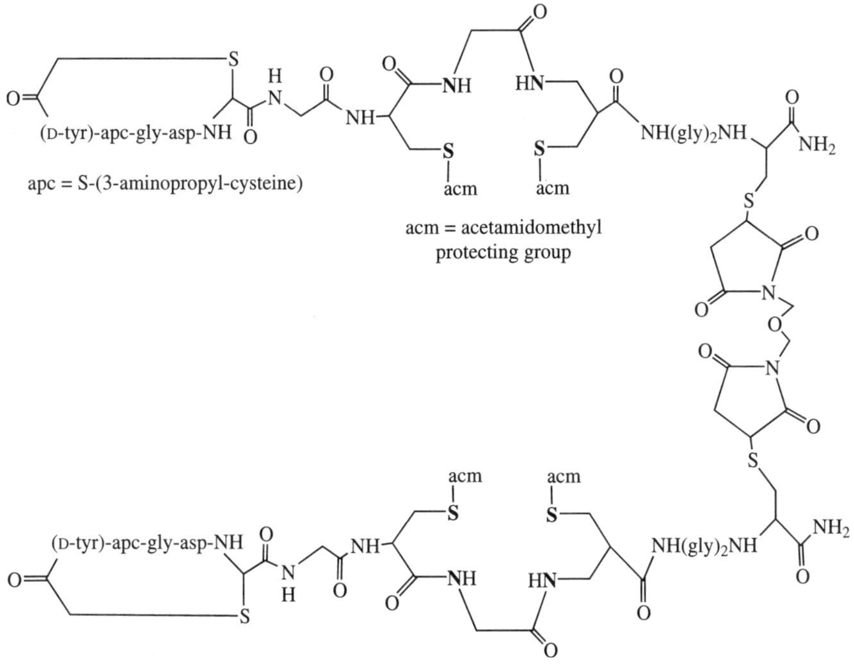

Figure 7.15 The P280 thrombus imaging agent peptide for labeling with technetium-99m. Possible Tc ligand atoms shown in bold type. (Adapted with permission from Figure 23 of Liu, S.; Edwards, D. S. *Chem. Rev.*, 1999, **99**(9), 2235–2268. Copyright 1999, American Chemical Society.)

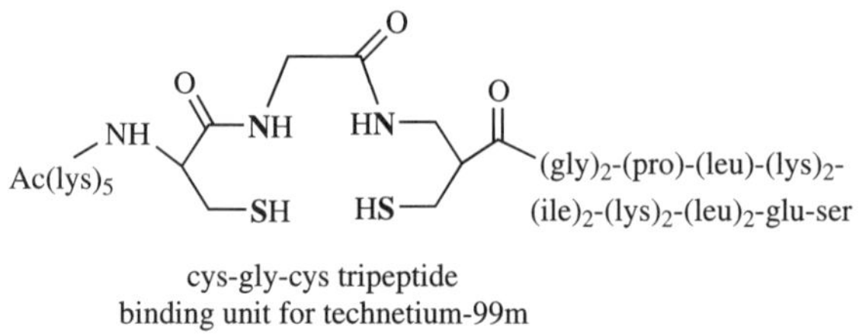

Ac(lys)$_5$ NH O NH HN O O SH HS (gly)$_2$-(pro)-(leu)-(lys)$_2$-(ile)$_2$-(lys)$_2$-(leu)$_2$-glu-ser

cys-gly-cys tripeptide
binding unit for technetium-99m

Figure 7.16 The 23-amino-acid peptide P483 for labeling with technetium-99m. Possible Tc ligand atoms in bold type. (Adapted with permission from Figure 24 of Liu, S.; Edwards, D. S. *Chem. Rev.*, 1999, **99**(9), 2235–2268. Copyright 1999, American Chemical Society.)

White blood cells (WBCs), particularly polymorphonuclear leukocytes (PMNLs) and monocytes, accumulate in high concentrations at sites of infection. A study carried out by reference 95 authors sought to evaluate the safety and efficacy of a leukocyte-avid peptide for the detection of infection, to determine the effects of peptide dose on performance, and to compare the peptide with in-vitro-labeled leukocytes.[95] A 23-amino-acid peptide, P483 (see Figure 7.16), containing the platelet factor-4 heparin-binding sequence, was labeled with 99mTc and complexed with heparin (P483H). The P483 peptide contains a cys-gly-cys tripeptide chelating unit for ^{99m}Tc labeling. The pentalysine sequence on the N-terminus is used to promote renal (kidney) clearance, leaving the abdominal area clear for imaging. It was found that ^{99m}Tc-P483H safely, rapidly, and accurately detected WBC infection, was comparable with in-vitro-labeled leukocyte imaging (^{111}In-labeled leukocyte scintigraphy), and therefore merits further investigation, ongoing at this time.

Radiolabeled receptor-based biomolecules are of great interest as diagnostic agents for primary and metastatic tumors. These agents—the example discussed is P829 (Neotect™)—have the potential to detect primary sites, identify metastatic lesions (cancerous cells at a remote location from the original site), guide surgical intervention, and predict efficacy of chemotherapeutic agents. They may also become therapeutic pharmaceuticals. A target peptide of interest is somatostatin, a tetradecapeptide that exhibits an inhibitory effect on secretion of numerous hormones. Somatostatin receptors (SSTR) are overexpressed (expressed at greater levels than that for normal tissue) on a number of human tumors and their metastases. This SSTR overexpression by many malignant tumors provides the basis for differentiating malignant tumors from other tissues by nuclear imaging using an SSTR-binding radiotracer. A somatostatin analog, P829 or Neotect™, labeled with ^{99m}Tc has been evaluated for this purpose. As shown in Figure 7.17, the peptide, also known as depreotide, is a synthetic 10-amino-acid peptide with the sequence cyclo(-homocys-*N*-Me-phe-tyr-D-trp-lys-val) $(1 \rightarrow 1')$-S-CH_2-CO-β-dap-lys-cys-lys-NH_2, TFA salt. Homocys is the amino acid homocysteine, and TFA is a trifluoroacetate counterion. The β-dap-lys-cys tripeptide sequence forms an N_3S monaminediamidethiol chelating unit for the technetium-99m. The substituted

cyclo(homocys-N-Me-phe-**tyr-D-trp-lys-val**)

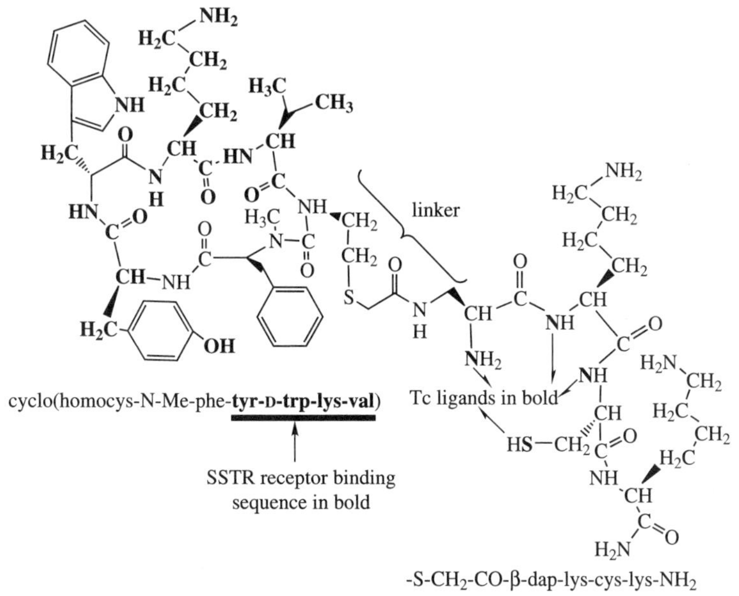

SSTR receptor binding sequence in bold

linker

Tc ligands in bold

-S-CH_2-CO-β-dap-lys-cys-lys-NH_2

Figure 7.17 The P829 peptide labeled with technetium-99m to prepare radiopharmaceutical 99mTc-P829, NeoTect™, as discussed in reference 96.

amino acid β-dap is an alanyl residue modified by a (mercaptoacetyl)amino group for connection to the SSTR binding peptide. The alanyl $-NH_2$ group provides one of the ligands of the N_3S monaminediamidethiol chelating unit. The peptide contains the tyr-(D-trp)-lys-val sequence that is responsible for SSTR binding. The peptide P829 was synthesized by *N*-alpha-Fmoc peptide chemistry (a method for peptide synthesis on a solid-phase support), purified by reversed-phase high-pressure liquid chromatography (HPLC) and characterized by fast-atom bombardment mass spectrometry (FAB-MS).

The peptides were labeled with 99mTc by ligand exchange from 99mTc-glucoheptonate, thus inserting the 99mTc ion at the chelating location shown in Figure 7.17.[96] The method is a two-step ligand exchange synthesis. First the $^{99m}TcO_4^-$ is reduced, in this case by added stannous chloride dihydrate ($SnCl_2{\bullet}2H_2O$), in the presence of a chelating agent such as glucoheptonate (GH). Following this the intermediate complex $[^{99m}TcO(GH)_2]^{n-}$ is reacted with the BFCA-peptide conjugate P829 to form the active agent ^{99m}TcOP829. The exact oxidation state of the technetium is unknown but is believed to be Tc(V). Technetium-99m-P829, also called (99m)Tc depreotide, prepared in >90% radiochemical yield, exhibited high-specific activity and high tumor uptake in in vivo experiments. Using this radiopharmaceutical, small-cell and non-small-cell lung cancerous lesions were evaluated in a successful phase III clinical trial. The radiopharmaceutical was found to be sensitive and accurate in diagnosis when compared to other standard medical evaluation methods.[97]

In addition to their frequent usage in diagnostic medicine, radiopharmaceuticals have long been used to treat cancerous tumors and their micrometastases. Because of their specificity in targeting and selective destruction of malignant cells, they are often described as "magic bullet" therapeutic agents first enunciated by Paul Ehrlich in the late nineteenth century.[98] Radionuclides that decay by β-particle emission are the most common type of radiopharmaceutical used in current medical practice. A great advance in their pharmacology resulted from the ability of scientists to produce monoclonal antibodies (MAbs) of predefined specificity in large quantities. Radioisotopes conjugated to MAbs have two great advantages over drugs for eradication of cancerous tumors: (1) α- or β-particles carried to the target tissue by the radiopharmaceutical will kill adjacent tumor cells whether or not they express the antigen targeted by the Mab; (2) radioisotopes are not subject to multidrug resistance (MDR). One example of the current technology is the synthesis of $^{186/188}$Re-labeled (both β-particle emitters) site-specific radiopharmaceuticals. The synthesis of these agents follows that of the technetium diagnostic tools just discussed because these metals exhibit similar chemistry. Several types of $^{186/188}$Re agents are being tested in clinical trials. In a direct labeling approach, $^{186/188}ReO^{3+}$ is attached directly to the MAbs, which have numerous coordination sites and thiol groups for the conjugation of the metal. Radioactivity clears the blood faster with the rhenium-labeled MAbs than does the BFCA system described next. One bifunctional coupling agent (BFCA) linked to MAbs, peptides, and other biomolecules is the MAG_2-GABA, mercaptoacetylglycylglycyl-γ-aminobutyric acid, peptide-linker system. The system, shown in Figure 7.18, provides an N_3S metal chelator and GABA linker from the metal center to the targeting molecule. The overall charge on the $^{99m}TcO^{3+}$ or $^{186/188}ReO^{3+}$ chelating framework is -1 because the coordinating thiol carries a -1 charge as do the three amido groups that each lose a proton at neutral pH upon complexation. Because the chemistry is so similar, technetium-99m (for diagnosis) and rhenium-186/188 metal centers can be combined in some developing drug treatment protocols for concurrent diagnosis and treatment protocols.

7.3.2 Gadolinium MRI Imaging Agents

7.3.2.1 Introduction. The water-tickling magnetic properties of the gadolinium (III) ion, with its seven unpaired electrons, have placed it in the forefront of medical

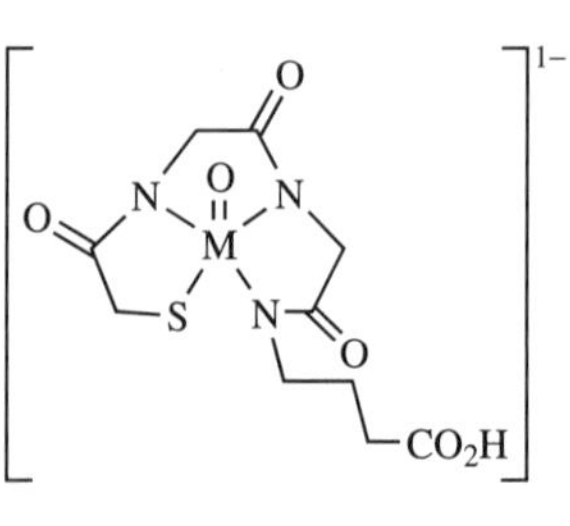

Figure 7.18 Metal–MAG_2–GABA complex for diagnostic and/or treatment protocols.

magnetic resonance imaging (MRI). Two other lanthanide ions, dysprosium(III) and holmium(III), have larger magnetic moments than Gd(III), but the symmetric S-state of Gd(III) leads to a slower (and more desirable for MRI) electronic relaxation rate favorable for water molecule excitation. Iron particles, manganese complexes, and hyperpolarized nuclei of noble gases are also used as MRI imaging agents but will not be discussed here.

7.3.2.2 Magnetic Imaging Considerations, Kinetics, and Thermodynamics of Complexes. Magnetic resonance imaging (MRI) signal intensity stems largely from the longitudinal relaxation rate of water protons, $1/T_1$, as well as from their transverse rate, $1/T_2$. See Section 3.5.5 in Chapter 3 for further information on NMR relaxation rates. Gadolinium(III) agents increase $1/T_1$ and $1/T_2$ by roughly similar amounts and are best visualized using T_1-weighted MRI images. Because improvements in MRI imaging techniques have favored T_1-weighted visualization methods, Gd(III) agents have correspondingly become more popular. The authors of reference 101 have reviewed relaxation theory essential to the understanding of Gd(III)-based MRI contrast agents, and the reader is referred there and to previous reviews for detailed information.[99,100] The factors are briefly described here. In order to maximize the observed longitudinal relaxation rate of water's hydrogen atoms, $(1/T_i)_{obs}$, the agent must have maximized relaxivity, r_i, as seen in the following equations. The subscripts d and p refer to diamagnetic and paramagnetic, respectively. The paramagnetic contribution is dependent on the concentration of the paramagnetic species, [Gd]. A plot of $(1/T_i)_{obs}$ versus [Gd] will yield the relaxivity value as the slope of the $y = mx + b$ line.

$$\frac{1}{(T_i)_{obs}} = \frac{1}{(T_i)_d} + \frac{1}{(T_i)_p} \qquad i = 1, 2 \qquad (7.3)$$
$$\frac{1}{(T_i)_{obs}} = \frac{1}{(T_i)_d} + r_i[\mathrm{Gd}]$$

The longitudinal relaxation rate, $1/T_1$, is further divided into inner-sphere (IS) relaxation (the first coordination sphere of the Gd(III) complex, i.e., a water molecule in the first ligand sphere) and outer-sphere (OS) relaxation. Inner-sphere relaxation effects, most important for Gd(III) imaging agents, will depend on factors affecting $1/T_1^{IS}$ and $1/T_2^{IS}$. These factors include the lifetime of the solvent molecule (H_2O) in the inner sphere of the complex (τ_m, also the reciprocal of the solvent exchange rate, k_{ex}), the correlation time (τ_{ci}) that defines the dipole–dipole relaxation for water's protons, and $\Delta\omega$, the chemical shift difference between the Gd(III) paramagnetic complex and a diamagnetic reference. The rotational correlation time (τ_R) is the dominant contributor to τ_{ci} for small Gd(III) chelates so that efforts to maximize relaxivity have focused on strategies to slow rotation of these molecules. In addition, relaxivity will be maximized when the correlation time (τ_{ci}) is the inverse of the proton Larmor frequency—in effect, maximizing relaxivity at low magnetic field strengths. Equations and graphs relating all the variables, as well as further amplification of the topic, are presented in reference 101. Outer-sphere relaxation may be broken down further into second-sphere relaxation (i.e., water molecules H-bonded to lone pairs of ligand carboxylate ions) and outer-sphere relaxation (i.e., bulk solvent water in the vicinity of the Gd(III) complex). Figure 36 of reference 101 illustrates the concepts. Inner-sphere effects are related to relaxation rates through a series of equations called the Solomon–Bloembergen–Morgan equations. A number of factors must be taken into account:

1. The effect of the magnetic field strength that is usually low for the MRI experiment (0.5–1.5 Tesla, ^{1}H frequencies of 20–60 MHz,)
2. Electronic relaxation rates that are also dependent on field strength
3. Hydration number, q, that is, the number of water molecules in the Gd(III) first coordination sphere
4. Rotation effects (slower rotation, i.e., larger molecule yields improved relaxivity)
5. Water exchange with bulk solvent

Researchers have accumulated a large body of thermodynamic and kinetic data to assess these effects, and many of these results are included in the tables of reference 101. Qualitatively, one concludes that for small molecule Gd(III) complexes—those of molecular weight <1000 Da—a high relaxivity, measured in $mM^{-1}s^{-1}$, will approach an upper limit of 5 $mM^{-1}s^{-1}$. Some data are collected in Table 7.3. Newer macromolecular conjugate–Gd(III) complex systems, also discussed below, may approach relaxivities five to six times larger per Gd(III) ion.

Solution equilibria for gadolinium imaging agents have been studied with consideration for pharmacokinetic, protein binding, elimination, and safety aspects of the drugs. The thermodynamic stability constant, K_{GdL} defined by equation 7.4 must be large for clinically viable agents. Some K_{GdL} data are listed in Table 7.3.

$$\begin{aligned} M + L &= ML \\ K_{ML} &= \frac{[ML]}{[M][L]} \end{aligned} \tag{7.4}$$

However, even with large K_{ML} values, the toxicity for a series of gadolinium was found to be very sizable with questionable LD_{50} values. The LD_{50} (lethal dose, 50%) value is typically expressed in milligrams of material per kilogram of subject body weight and indicates the quantity of material that, if administered to a population of subjects, will cause 50% of the subjects to perish. (In Table 7.3, LD_{50} values are expressed in mmol/kg.) The lower the LD_{50} value, the more toxic a substance is assumed to be. For example, the LD_{50} value for sucrose is 29,700 mg/kg (86.8 mmol/kg), orally in rats, while the value for the nerve gas sarin is 24 mg/kg (0.17 mmol/kg). The subject animal and the route by which the material was

Table 7.3 Properties of Some Gadolinium(III) MRI Imaging Agents

Chemical Formula	Brand Name (Figure No.)	MW (Da)	log K_{GdL} (log K_{sel})	LD_{50} (mmol/kg) in mice	Relaxivity, r_i ($mM^{-1}s^{-1}$)	^1H Frequency (MHz)
$[Gd(DTPA)(H_2O)]^{2-}$	Magnevist™ (as dimeglumine salt) (7.19A,B)	547	22.46 (7.04)	5.6	4.3	20
$[Gd(BOPTA)(H_2O)]^{2-}$	MultiHance™ (7.19D,E)		22.59		4.39	20
MS-325	AngioMARK™ (7.19F)				6.6	20
$[Gd(DTPA\text{-}BMA)(H_2O)]$	Omniscan™ (7.21A,B)		16.85 (9.04)	14.8	3.96	20
$[Gd(HP\text{-}DO3A)(H_2O)]$	ProHance™ (7.21C,D)	588	23.8 (6.95)	12	3.7	20
$[Gd_2(DO3A)_2L2]$	(7.24B)	1231			5.1	20
$[Gd_4(DO3A)_4L1]$	(7.24C)	2229			8.5	20
Gadomer-17	(7.25)	17,453[a] 24 Gd(III) ions			365 (mol) 15.2 (ion)	64.5
Gadomer-17	(7.25)	17,453[a] 24 Gd(III) ions			449 (mol) 18.7 (ion)	20

[a] Each Gadomer-17 molecule contains 24 Gd(III) ions. See Figure 7.25.

administered should be specified (e.g., oral, rat) in order for the LD_{50} value to be meaningful. Studies by various researchers led to the proposal of a selectivity factor, log K_{sel}, that takes into account the affinity of the gadolinium(III) chelating system (L) for other physiologically important ions such as Ca(II), Cu(II), and Zn(II) and the dependence of K_{ML} on pH. This consideration leads to calculation of K_{sel} based on equation 7.5:[100,101]

$$K_{sel} = K_{ML}(\alpha_{H}^{-1} + \alpha_{CaL}^{-1} + \alpha_{CuL}^{-1} + \alpha_{ZnL}^{-1})^{-1} \tag{7.5}$$

where

$$\begin{aligned}
\alpha_{H}^{-1} &= 1 + K_1[H^+] + K_1K_2[H^+]^2 + K_1K_2K_3[H^+]^3 + K_1K_2K_3K_n[H^+]^n \\
\alpha_{CaL}^{-1} &= K_{CaL}[Ca^{2+}] \\
\alpha_{CuL}^{-1} &= K_{CuL}[Cu^{2+}] \\
\alpha_{ZnL}^{-1} &= K_{ZnL}[Zn^{2+}]
\end{aligned}$$

While the K_{sel} and LD_{50} data appear to be related (higher K_{sel} = higher LD_{50}) as indicated in Table 7.3, this prediction has been found to be true only for those complexes that have sufficiently fast dissociation and substitution kinetics for transmetallation to occur during the time in which the Gd(III) chelate remains in vivo. In fact the search for kinetically inert as well as thermodynamically stable Gd(III) chelates has led to the development of agents containing a significantly more inert macrocyclic ligand, exemplified by the [Gd(HP-DO3A)(H_2O)] system described in Section 7.3.2.3.

7.3.2.3 Selected Drugs in Usage or in Trials. The gadolinium compound, gadopentetate dimeglumine, $[Gd(DTPA)(H_2O)]^{2-}$, Magnevist™ (Figure 7.19B), was first approved for use as a so-called contrast agent in 1988. DTPA = diethylenetriaminepentaacetic acid, dimeglumine = $(NMG)_2^{2+}$ as counterion, NMG = *N*-methylglucamine, the *N*-methylamine salt of D-glucose. Since then, 9 or 10 other gadolinium compounds have been approved for MRI use or are currently in clinical trials.[101] Several of these are shown in Figure 7.19 and described further below. Data on the complexes are listed in Table 7.3.

The choice of a proper ligand system for the toxic heavy metal Gd(III) ion has been of paramount importance, because for the time period in which the agent is in the body there must be no detectable complex dissociation. The success of the DTPA and similar ligand systems has depended on their strong Gd(III) complexation characteristics that achieve this goal. Assuming that the usually labile Gd(III) ion is rendered inert by chelation, one then regards the agent as its $[Gd(DTPA)(H_2O)]^{2-}$ complex. The complex, being hydrophilic, is unlikely to enter cells or be attacked by biological nucleophiles or electrophiles. There are no intercalation groups (usually planar aromatic substituents) to integrate themselves into DNA. The chelated metal ion is unlikely to bind to donor groups in proteins or enzymes.

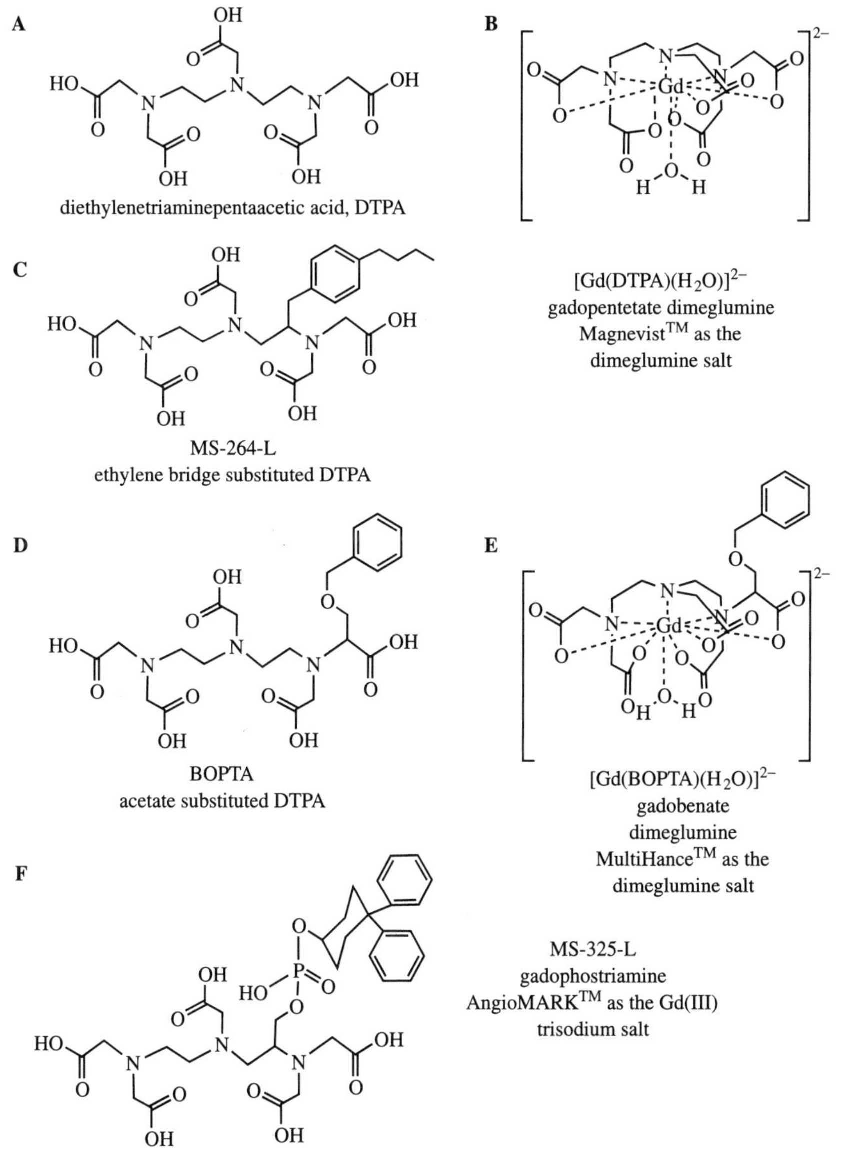

A

diethylenetriaminepentaacetic acid, DTPA

B

$[Gd(DTPA)(H_2O)]^{2-}$
gadopentetate dimeglumine
Magnevist™ as the
dimeglumine salt

C

MS-264-L
ethylene bridge substituted DTPA

D

BOPTA
acetate substituted DTPA

E

$[Gd(BOPTA)(H_2O)]^{2-}$
gadobenate
dimeglumine
MultiHance™ as the
dimeglumine salt

F

MS-325-L
gadophostriamine
AngioMARK™ as the Gd(III)
trisodium salt

Figure 7.19 Several ligands and gadolinium(III) complexes used in MRI imaging. (A) DTPA ligand; (B) $[Gd(DTPA)(H_2O)]^{2-}$, Magnevist™; (C) MS-264-L, ethylene-bridge-substituted DTPA; (D) BOPTA, acetate-substituted DTPA; (E) $[Gd(BOPTA)(H_2O)]^{2-}$, MultiHance™; (F) AngioMARK™. (Adapted with permission from Chart 1 of Caravan, P.; Ellison, J. J.; McMurry, T. J.; Lauffer, R. B. *Chem. Rev.*, 1999, **99**(9), 2293–2352. Copyright 1999, American Chemical Society.)

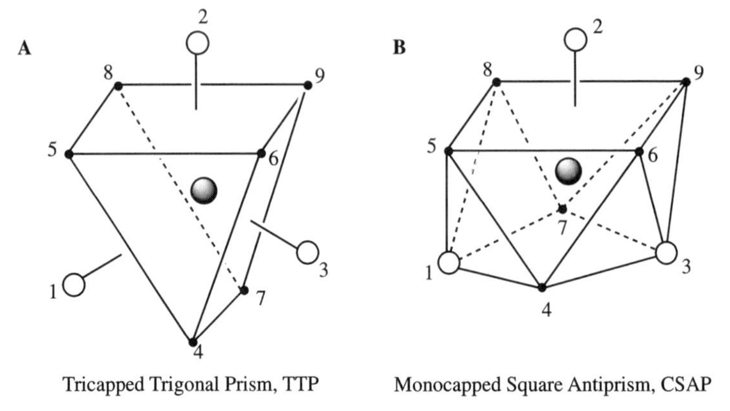

Figure 7.20 Nine-coordinate TTP (A) and CSAP (B) geometries exhibited by gadolinium (III) imaging agents. (Adapted with permission from Figure 1 of Caravan, P.; Ellison, J. J.; McMurry, T. J.; Lauffer, R. B. *Chem. Rev.*, 1999, **99**(9), 2293–2352. Copyright 1999, American Chemical Society.)

Consequently, most gadolinium imaging agents have extracellular applications, although newer agents are being tested for blood pool (AngioMARK™, Figure 7.19F, discussed below) or hepatobiliary [i.e., liver imaging (MultiHance™, Figure 7.19E, also discussed below)] applications. Currently used gadolinium(III)-based imaging agents are nine-coordinate with the chelating ligand occupying eight coordination sites and a water molecule the ninth site. Idealized geometries for nine-coordinate geometry are tricapped trigonal prism (TTP) and capped square antiprism (CSAP); however, the TTP geometry is favored for the Gd(III) complexes. An illustration of the TTP structure is shown in Figure 7.20A along with the CSAP geometry in Figure 7.20B.

Many X-ray crystallographic studies of gadolinium and other lanthanide (Ln) ion chelates have appeared, and these are reviewed extensively in reference 101. X-ray crystallographic structural data for $[Gd(DTPA)(H_2O)]^{2-}$ features a distorted TTP geometry with the following range of observed bond distances (all in Å): Gd–O_{water}, 2.490; Gd–$O_{carboxylate}$, 2.363–2.437; Gd–$N_{terminal}$, 2.629, 2.710; Gd–$N_{central}$, 2.582.[102] As is usually the case for the Ln–N bond length for the central nitrogen of the diethylenetriamine ligand, the Gd–$N_{central}$ is shorter than the Gd–$N_{terminal}$ bond length. Modifications of the DTPA ligand, as shown for the example of BOPTA (Figure 7.19D), MultiHance™ (Figure 7.19E), or MS-264-L (Figure 7.19C), feature substituents to the acetate arms or diethylenetriamine backbone of the ligand. The substituents cause further distortions to the solid-state structures as described in some detail in reference 101.

Figure 7.21 indicates the several neutral complexes used as MRI imaging agents. Neutrality is accomplished through bisamide substituents on DTPA [Gd(DTPA-BMA)(H_2O)], Omniscan™ (Figure 7.21A,B) or via the cyclic ligand HP-DO3A [Gd(HP-DO3A)(H_2O)], ProHance™ (Figure 7.21C,D).

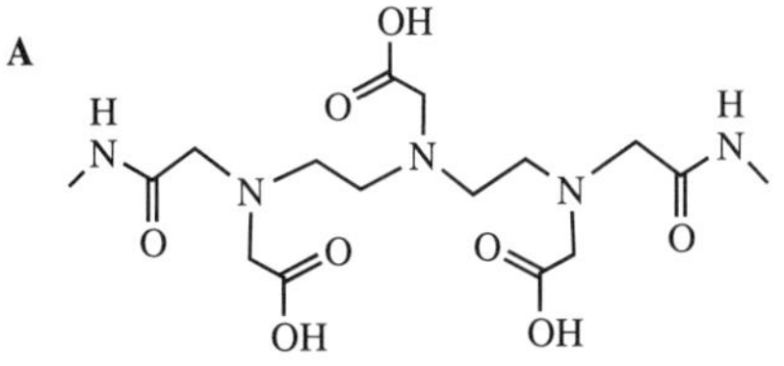

A

diethylenetriaminepentaacetic acid-bisamide
DTPA-BMA

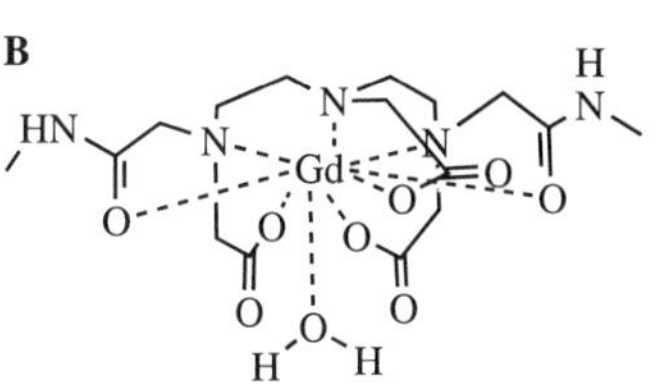

B

[Gd(DTPA-BMA)(H_2O)]
Omniscan™

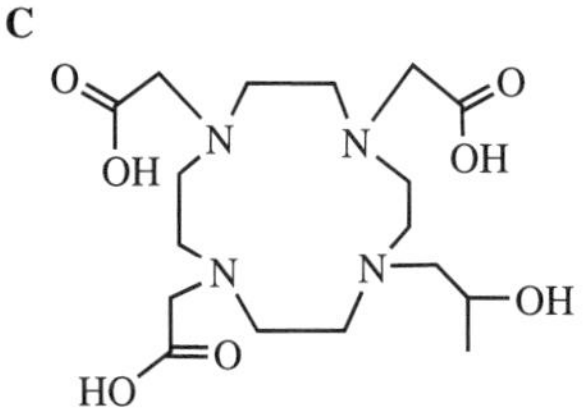

C

HP-DO3A

D

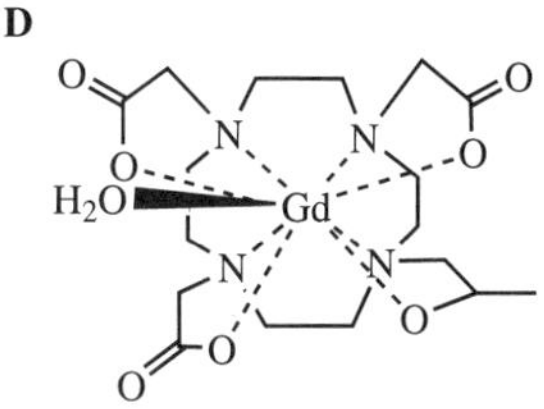

[Gd(HP-DO3A)(H_2O)]
ProHance™

Figure 7.21 Two neutral gadolinium(III) imaging agents. (Adapted with permission from Charts 4 and 6 of Caravan, P.; Ellison, J. J.; McMurry, T. J.; Lauffer, R. B. *Chem. Rev.*, 1999, **99**(9), 2293–2352. Copyright 1999, American Chemical Society.)

The crystal structure of [Dy(DTPA-BMA)(H_2O)][103] indicated a nonacoordinated metal ion with three nitrogen, three monodentate carboxylate oxygen, two monodentate amide oxygen, and one water oxygen atom ligands in a TTP arrangement. Four possible configurations for placement of the amide groups (*syn, cis, anti*, and *trans*) exist; the *trans* structure is found in the crystal structure (Figure 7.22B). The crystal structure of the complex [Gd(HP-DO3A)(H_2O)] also exhibited a nine-coordinate metal ion with four nitrogens, three monodentate carboxylate oxygens, and one hydroxyalkyl oxygen and a capping water molecule, this time in a CSAP-type arrangement, shown in Figure 7.20B.[104] Two independent molecules in the crystal structure exhibit different twist angles between the basal and capped planes as shown in Figures 13 and 14 of reference 101. The two diastereoisomers indicated by the crystal structural data interconvert by ring inversion, as shown in Figure 29 of reference 101.

In addition to X-ray crystallographic studies, two-dimensional NMR solution experiments (i.e., COSY, 1D-NOE, and NOESY, discussed in Sections 3.5.9 and 3.5.10) have been carried out on many lanthanide(III), Ln(III), chelate complexes to confirm that the structure of the MRI imaging agent, used in aqueous solution, will correspond to the solid-state X-ray crystallographic structure. Two-dimensional exchange spectroscopy (2D-EXSY) has been applied to lanthanide chelates to study the dynamics of conformational equilibria (how acetate arms chelate and how

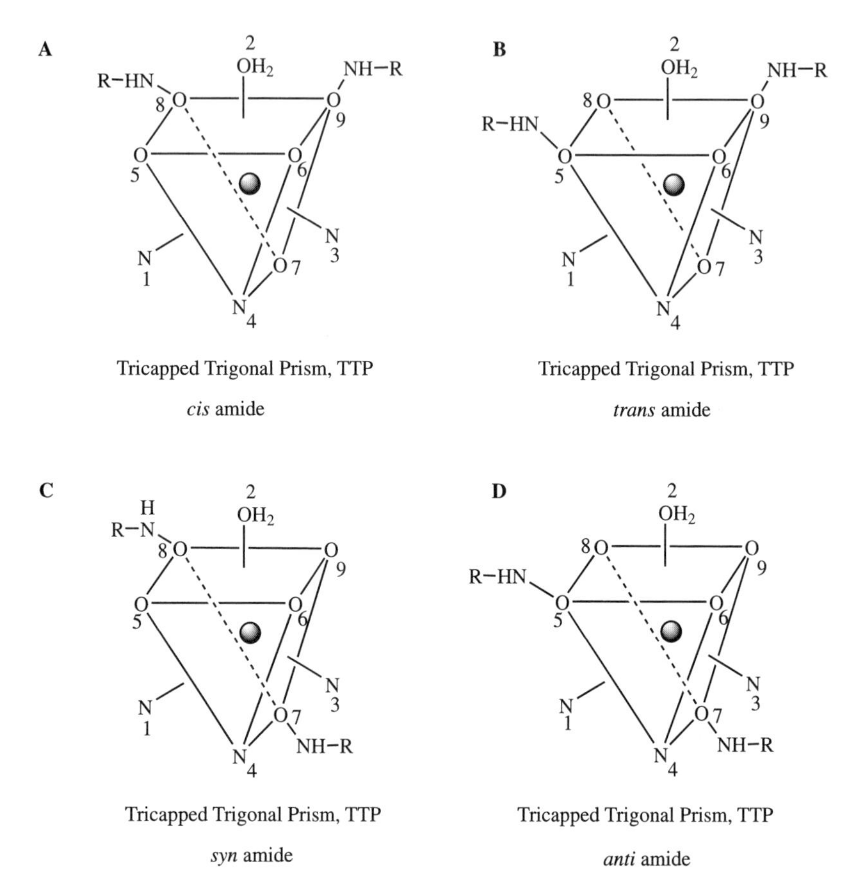

Figure 7.22 *cis, trans, syn, anti* isomers for TTP DTPA-bis amides. (Adapted with permission from Figure 7 of Caravan, P.; Ellison, J. J.; McMurry, T. J.; Lauffer, R. B. *Chem. Rev.*, 1999, **99**(9), 2293–2352. Copyright 1999, American Chemical Society.)

backbone ethylene protons flip between equatorial and axial positions). The studies have indicated nonadentate coordination (three nitrogens, five monodentate carboxylate oxygens with water ligation confirmed by other solution methods) consistent with X-ray crystallographic data. At higher temperatures, two existing chiral wrapping isomers, as shown in Figure 7.23, exchange rapidly in solution giving averaged signals for pairs of protons. The two acetate arms alternate coordination at position 7, and the axial ethylenediamine protons equilibrate with the equatorial conformations.[101]

As indicated in Figure 7.22, two terminal nitrogen atoms of the DTPA–BMA ligand become chiral upon chelation, resulting in four possible diastereomers (*cis, trans, syn, anti*). Two enantiomers (wrapping isomers) also exist for each diastereomer as shown in Figure 7.23, resulting in eight possible stereoisomers for the [Ln(III)(DTPA–BMA)] complexes. NMR studies indicate that all four diastereo-

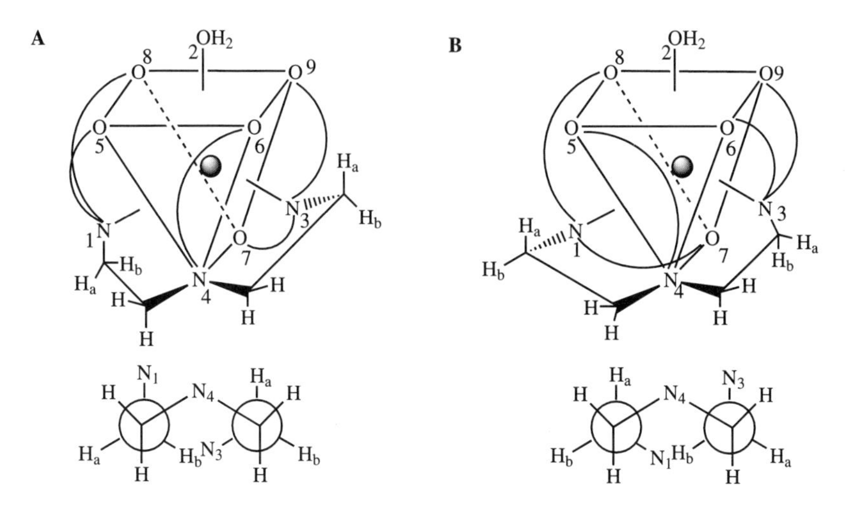

Figure 7.23 Wrapping isomers of Ln(DTPA) resulting in a pseudo-mirror plane. (Adapted with permission from Figure 28 of Caravan, P.; Ellison, J. J.; McMurry, T. J.; Lauffer, R. B. *Chem. Rev.*, 1999, **99**(9), 2293–2352. Copyright 1999, American Chemical Society.)

meric pairs of conformers are present in aqueous solution. Activation energies between two wrapping isomers were found for Nd(III) complexes, indicating that the barrier to exchange is determined by the eclipsing of the ethylene bridges in the transition state. A higher barrier to exchange was found for racemization of terminal nitrogen atoms because this requires partial decoordination (octadentate to pentadentate) of the ligand.[101]

A number of different isomers are possible for Ln(III) complexes of HP-DO3A, a cyclic ligand (Figure 7.21C). Replacement of an acetate arm with a hydroxypropyl group gives rise to diastereomeric differentiation of the four expected isomers. A chiral carbon in the hydroxypropyl arm yields eight stereoisomers (four pairs of enantiomers). The hydroxypropyl arm's introduction of an asymmetric site enabled the chemical shift separation of the acetate arms and ethylene groups in a ROESY analysis of [Y(HP-DO3A)(H_2O)].[105] Both rearrangement of the acetate groups and inversion of the macrocycle were observed. Exchange rate determinations showed that exchange of the ethylene groups in the ring is faster than exchange for the pendant arms and exchange for the hydroxypropyl arm is faster than for acetate arms. Multiple diastereomers are confirmed by the presence of multiple species, and ring inversions (as seen in the X-ray crystallographic study) are suggested by exchange processes involving the acetate arms.

In spite of the exchange processes evident for these lanthanide compounds, their overall stability is acceptable with respect to chelate decomposition. The search continues, however, for more stable, less toxic chelates. In this regard, research on new compounds for use as gadolinium imaging agents has concentrated on improving relaxivity values as well as Gd(III)ligand chelate stability, assessed by measuring the ligand affinity for Gd(III) in comparison to other M(II) ions,

A

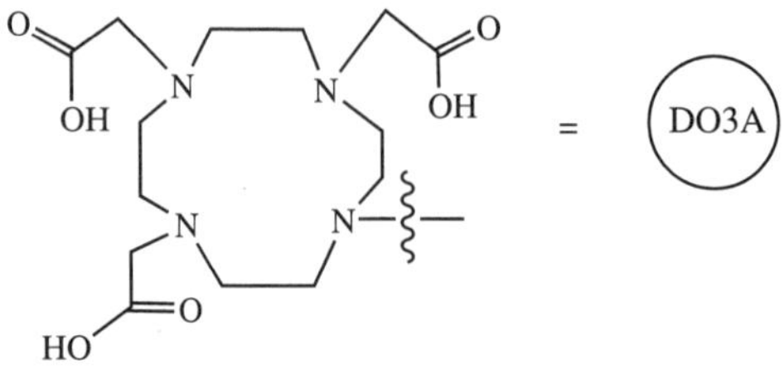

O OH N N OH O = DO3A

N N

HO O

B

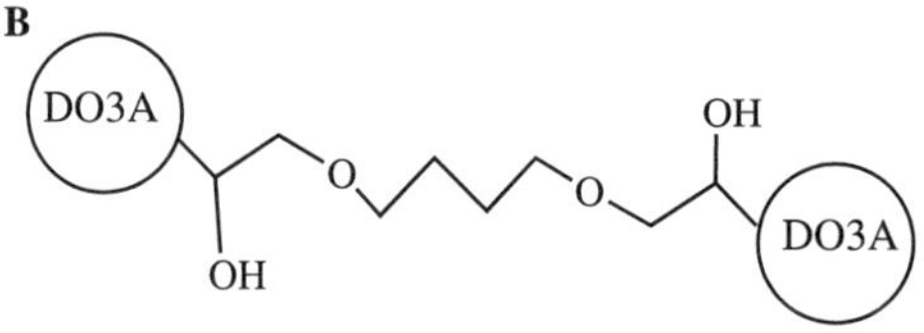

DO3A OH O O DO3A OH

$(DO3A)_2L2$

C

DO3A HO OH O DO3A DO3A O HO OH DO3A

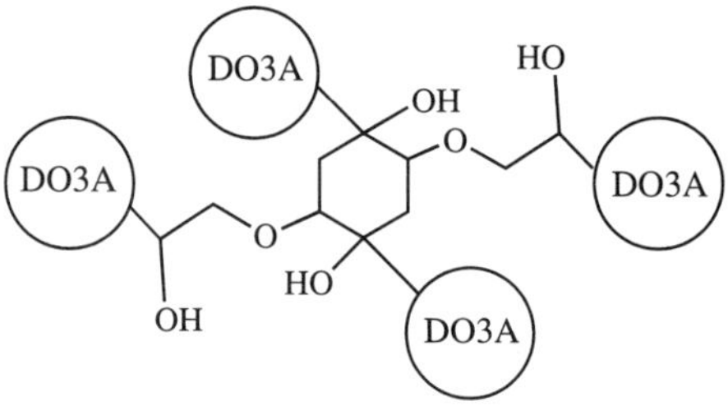

$(DO3A)_4L1$

Figure 7.24 Macrocyclic chelates with multiple Gd(III) ions and large relaxivities. (Adapted with permission from Chart 17 of Caravan, P.; Ellison, J. J.; McMurry, T. J.; Lauffer, R. B. *Chem. Rev.*, 1999, **99**(9), 2293–2352. Copyright 1999, American Chemical Society.)

especially Ca(II). Some K_{sel} data are reported in Table 7.3 (see equation 7.5 for factors affecting K_{sel}). Neutral ligand donors showing the greatest increases in Gd(III) chelate stability are: amide oxygen > pyridyl nitrogen > alcoholic oxygen.[106] However, macrocycles like DO3A (Figure 7.24A) are of continuing interest in Gd(III) imaging agent design. One approach to maximizing relaxivity values has been to increase the molecular weight of the agents along with the numbers of Gd(III) ions incorporated into the molecule. Two examples of the many synthesized and described in reference 101, $(DO3A)_2L2$ (7.24B) and $(DO3A)_4L1$ (Figure 7.24C), are illustrated in Figure 7.24. Their molecular weights and relaxivity values are reported in Table 7.3.

Other high relaxivity Gd imaging agents are macromolecular conjugates prepared by conjugation of the low-molecular-weight chelates such as Gd(III)DTPA to macromolecules such as polylysine and synthetic dendrimers. Figure 7.25 displays the structure of Gadomer-17™, derived from a lysine-

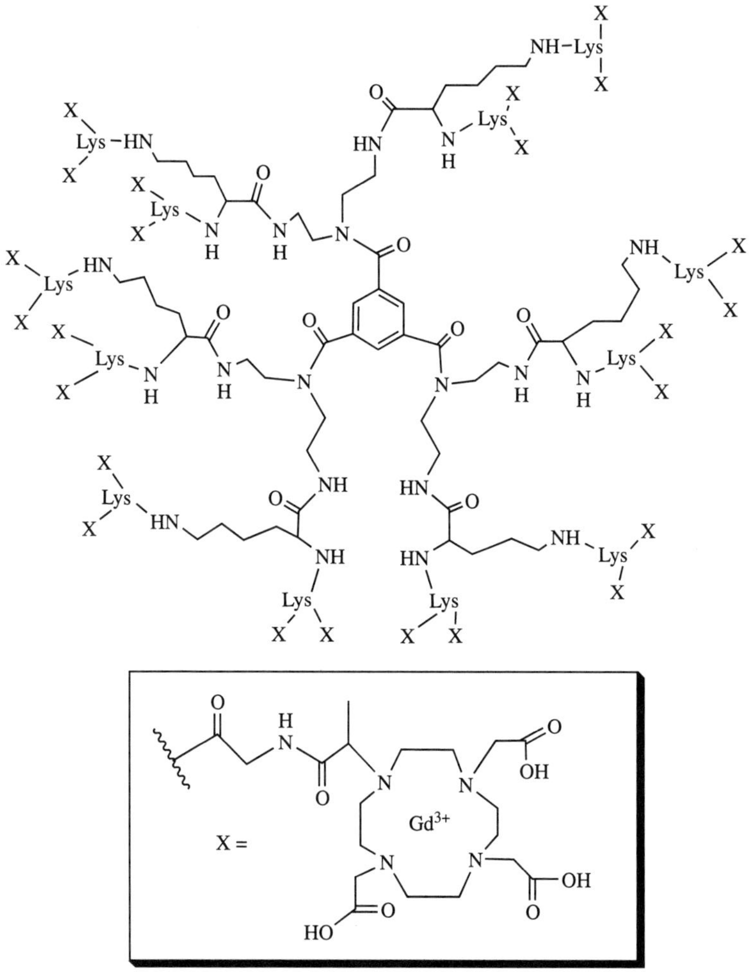

Gadomer-17™

Figure 7.25 Dendrimer-type gadolinium imaging agent Gadomer-17™. (Adapted with permission from Figure 43 of Caravan, P.; Ellison, J. J.; McMurry, T. J.; Lauffer, R. B. *Chem. Rev.*, 1999, **99**(9), 2293–2352. Copyright 1999, American Chemical Society.)

functionalized 1,3,5-benzene tricarboxylic acid core functionalized with 24 GD(III)DTPA chelates.[107] Another approach is the so-called RIME (receptor-induced magnetization enhancement) system. In this approach the Gd(III) contrast agent is targeted to a particular protein or receptor molecule (HSA or human serum albumin being an example for blood pool imaging) as described for technetium reagents in Section 7.3.1.3. Binding causes increased concentration and retention of the Gd(III) complex in the area of the receptor molecule. Binding to the macromolecule also allows the Gd(III) complex to take on the rotational correlation time similar to that of the macromolecule. This can lead to a great increase in relaxivity and consequently a high target-to-background ratio, a desirable goal for imaging agents.[108] MS-325, AngioMARK™, is the prototype MR angiographic agent in this application that may become a noninvasive alternative to X-ray angiography for imaging arterial blockages in the body. See Figure 7.19F for illustration of the ligand system.

7.4 METAL TRANSPORT AND METALLOCHAPERONES

7.4.1 Introduction

Metal transporters are proteins that carry metal ions across cell membranes. One example, human copper transport protein (hCtr1), consists of 190 amino acids. It receives Cu(I) ions on the cell surface and transports them into the cell cytosol. The process is illustrated in Figure 7.26. The metallochaperone concept, that of soluble metal receptor proteins that carry metal ions to specific target proteins, is a relatively new one with most research having been published since 1997. Metallochaperones do not detoxify metals, other proteins such as metallothionein and glutathione carry out this function. Metallochaperones garner metals from their very small homeostatic concentrations inter- or intracellularly and deliver those metals to the metalloenzyme that requires them for activity. The chaperone must be a thermodynamically stable complex as it approaches its target apoprotein, but the metal within the complex must be kinetically labile to allow fast metal ion exchange for its proper transport and distribution.[109] A number of different

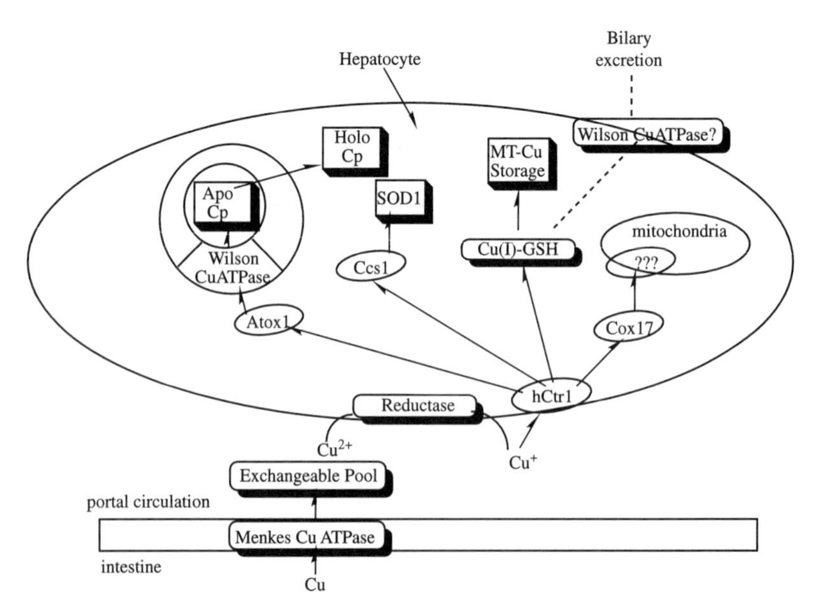

Figure 7.26 Copper transport pathways. (Adapted with permission from Sarkar, B. *Chem. Rev.*, 1999, **99**(9), 2535–2544. Copyright 1999, American Chemical Society.)

metallochaperones may be required to move an entering metal ion from its initial position within the cell to the target apoenzyme. At this time, most information exists concerning copper transporters and metallochaperones, and those are the examples that will be discussed here. It should be noted, however, that homeostasis, the conservation of proper concentrations of species in a living system, is required for all metals and that the concentrations of copper and iron in living cells are intimately related and in many instances co-dependent.

In biological terms, copper is absorbed from the gastrointestinal tract and enters an inter- and intracellular exchangeable pool. The known pathways are described in Figure 7.26 as adapted from references 109 and 110. During uptake, copper is reduced to Cu(I) and absorbed by the cell via a copper transporter (Ctr). The human copper transporter, hCtr, has been described by Zhou and Gitschier.[111] These authors have described the human gene CTR1 that encodes a high-affinity copper-uptake protein and hypothesized that this protein is required for copper delivery to mammalian cells. This hypothesis was tested by inactivating the CTR1 gene in mice by targeted mutagenesis. Early embryonic lethality in mutant embryos was traced to lack of activity in the collagen cross-linking cupro-enzyme lysyl oxidase. This fact plus other mouse abnormalities in less lethal mutations led the authors to conclude that the gene CTR1 encodes a protein essential for copper uptake in mammalian cells. After transport by the Ctr protein, copper ions are stored in biomolecules such as glutathione, metallothionein, or the multicopper oxidase ceruloplasmin, CP, before uptake by metallochaperones for transport to enzymes.

Cytoplasmic Cu(I)-glutathione (GSH) (Figure 7.27A) donates copper to intracellular proteins such as metallothionein (Figure 7.27B). From there, various

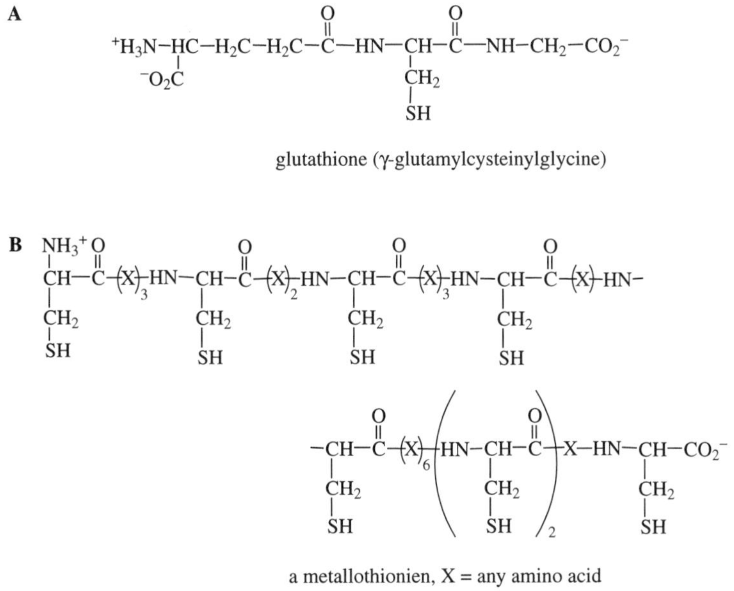

A

$$^{+}H_3N{-}HC(CO_2^-){-}H_2C{-}H_2C{-}C(=O){-}HN{-}CH(CH_2{-}SH){-}C(=O){-}NH{-}CH_2{-}CO_2^-$$

glutathione (γ-glutamylcysteinylglycine)

B

$$NH_3^+{-}CH(CH_2SH){-}C(=O){-}(X)_3{-}HN{-}CH(CH_2SH){-}C(=O){-}(X)_2{-}HN{-}CH(CH_2SH){-}C(=O){-}(X)_3{-}HN{-}CH(CH_2SH){-}C(=O){-}(X){-}HN{-}$$

$$-CH(CH_2SH){-}C(=O){-}(X)_6{-}\left(HN{-}CH(CH_2SH){-}C(=O)\right)_2{-}X{-}HN{-}CH(CH_2SH){-}CO_2^-$$

a metallothionien, X = any amino acid

Figure 7.27 (A) glutathione and (B) a metallothionein.

copper chaperone proteins known as Atox1, Cox17, and CCS1 deliver copper to copper-transporting ATPases,[112] cytochrome oxidase,[113] and superoxide dismutase (SOD),[114] respectively. See Section 5.2.3 in Chapter 5 for a detailed discussion of superoxide dismutase. Copper is distributed in all cellular organelles—nucleus, mitochondria, lysosomes, endoplasmic reticulum, and cytosol. Superoxide dismutase is found in the cytosol, cytochrome oxidase is found in the mitochondria, and metallothionein is found in the cytosol, nucleus, and lysosome. Further discussion of copper chaperone proteins in this chapter will include those concerned with superoxide dismutase and the disease state familial amyloid lateral sclerosis (FALS) associated with superoxide dismutase mutants, Wilson's disease associated with copper accumulation in organs and tissue, and Menkes disease associated with neurodegeneration caused by copper deficiency in a number of tissues.

7.4.2 The Atx1 Metallochaperone

The yeast Atx1 copper chaperone, a 72-residue protein, receives copper ions from the membrane-based copper transport protein, Ctr, or some other intracellular copper storage molecule and transfers it to a copper-transporting ATPase called Ccc2. The oxidized apoprotein (PDB code 1CC7) and Hg-Atx1 (PDB code 1CC8) have been structurally characterized (X-ray diffraction) by the authors of reference 115. Ccc2 translocates copper across intracellular membranes to the multicopper oxidase Fet3, required for high-affinity iron uptake. The human counterpart of Atx1 is known as Atox1 or Hah1. The yeast and human proteins have high homology and have been proposed to operate in the same manner.[115] Atox1 receives copper from hCtr and shuttles it to the multicopper oxidase ceruloplasmin, CP, a Fet3 analog, via the Menkes and Wilson disease proteins, the human counterparts of Ccc2. (See Figure 7.26.) More will be said later about the Menkes and Wilson disease proteins. Atx1 and Atox1 each contain one repeat of the copper chaperone motif MX**C**XX**C** or MT/H**C**XX**C** (M = methionine, T/H = either threonine or histidine, X = any amino acid, and C = cysteine) that binds a single metal ion via the two cysteine residues (in bold type). Copper chaperones are small polypeptides that adopt a $\beta\alpha\beta\beta\alpha\beta$ fold as shown in Figure 7.28 for the Atx1 copper chaperone. Other copper chaperones described later have more complex tertiary structures, but all known so far contain at least one repeat of the $\beta\alpha\beta\beta\alpha\beta$ fold. The tertiary structure can be described as two α-helices superimposed on a four-stranded β-sheet with a solvent-exposed metal binding site. The Atx1 copper chaperone structure in Figure 7.28 is a Wavefunction, Inc. Spartan '02 for Windows™ visualization of the NMR solution structure data deposited in the protein data base (PDB code: 1FD8).[116] The structure represents the Cu(I)-bound form of Atx1, a 73-amino-acid metallochaperone protein from baker's yeast, *Saccharomyces cerevisiae*. It exhibits the $\beta\alpha\beta\beta\alpha\beta$-fold tertiary structure just described. Following the EXAFS data of Pufahl, et al.,[117] a Cu(I) ion (in space-fill form) is placed between two sulfur atoms of cys15 (to the right of the copper ion) and cys18 (to the left of and behind the copper ion). The aa residues are shown in ball-and-spoke form). Cysteine residues are indicated in yellow ribbon form along the –CXXC– portion of the metal-binding motif. The

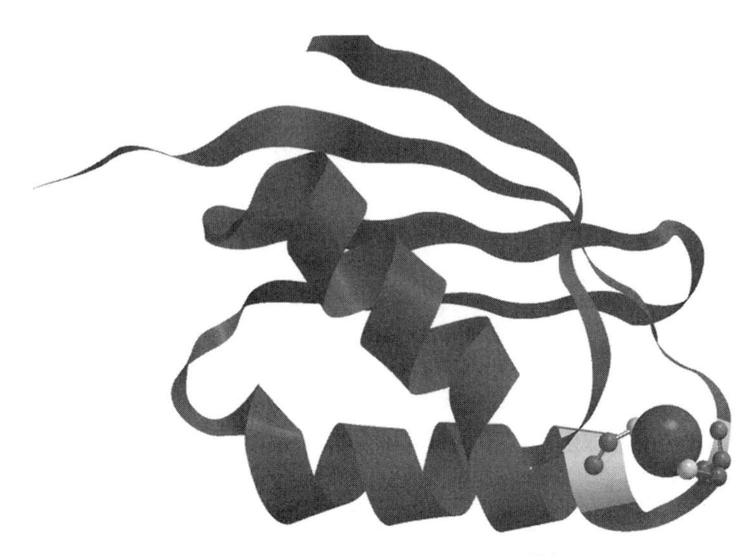

Figure 7.28 Wavefunction, Inc. Spartan '02 for Windows™ representation of the Atx1 copper chaperone protein (PDB code: 1FD8) with data from reference 116. See text for visualization details. Printed with permission of Wavefunction, Inc., Irvine, CA. (See color plate.)

Atx1 chaperone apoprotein (without copper) shows tertiary structural differences in the loop area containing the cysteine ligand residues. Structural data for the apoprotein are deposited with PDB code: 1FES.[116]

The X-ray crystallographic structure of HgAtx1 has been solved by the authors of reference 115 (PDB: 1CC8). The elements of secondary structure are the following: (1) β-strand 1 consisting of residues 5–10; (2) loop area 1 containing part of the chaperone motif met13 thr14 cys15 (Hg ligand), residues 11–15; (3) α-helix 1, residues 16–27 containing the rest of the chaperone motif ser16 gly17 cys18 (Hg ligand); (4) loop 2, residues 28–34; (5) β-strand 2, residues 35–39; (6) loop 3 connecting β-strand 2 to β-strand 3, residues 40–43; (7) β-strand 3, residues 44–49; (8) loop 4 connecting β-strand 3 to α-helix 2, residues 50–52; (9) α-helix 2, residues 53–62; (10) loop 5 connecting α-helix 2 to β-strand 4, residues 63–68; (11) β-strand 4, residues 69–73. The hydrophobic core of the protein, another common feature of copper chaperone proteins, consists of residues phe9, val11, met13, val22, leu26, leu29, val33, ile36, ile38, val45, val47, ile56, ile60, and val67. The βαββαβ fold of Atx1 places it in a structural family of proteins called "ferredoxin-like." Other members of the class include the fourth binding repeat of the Menkes protein (described later in this section), the bacterial mercury ion transporter MerP, ferredoxins, and small DNA- and RNA-binding domains such as that seen in zinc fingers described in Section 2.4 of Chapter 2. In the crystal structure the Hg–cys distances are 2.33 and 2.34 Å and the coordination sphere of the mercury ion is close to linear with an S–Hg–S bond angle of 167°. The crystal structure of apo-Atx1 shows a major change in the metal binding loop with cys15 shifting position to form a disulfide bond with cys18 accompanied by rearrangement of thr14, ser16, and gly17 positions. The rest of the apo-Atx1 molecule is virtually identical to the HgAtx1 form. EXAFS data for CuAtx1 have established that Cu(I) binds to two sulfur ligands at a distance of 2.25 Å with a third possible ligand at 2.40 Å.

The Atx1 mechanism for facile metal transfer has been developed based on the knowledge gained in the preceding research. The Atx1 metal-binding loop favors low metal coordination numbers (two or three coordinate) by providing few side-chain ligating atom possibilities. This factor may also be the basis for facile metal transfer to and from target proteins. The metal-binding site is on the protein's surface, making it accessible for docking with another protein. The apoAtx1 structure shows that the metal-binding site is flexible, rearranging to form a disulfide bond when metal ions are absent. As will be seen in the following text, the same or similar metal-binding site occurs in freely diffusible protein molecules or as part of a larger protein system involved in metal trafficking.[115]

7.4.3 Hah1 or Atox1 Metallochaperone

The human Atox1 or Hah1 proteins and the N-termini of its target proteins (Ccc2 and multicopper oxidase Fet3 or Wilson/Menkes disease proteins) belong to the Atx1-like family of metal binding domains characterized by the conserved MT/HCXXC sequence motif (M = methionine, T = threonine, H = histidine, C = cysteine, X = any amino acid). The crystal structure of Hah1 has been determined in the presence of Cu(I), Hg(II), and Cd(II). The 1.8-Å resolution structure of CuHah1 reveals a copper ion coordinated by cys residues from two adjacent Hah1 molecules. The CuHah1 X-ray crystallographic structure was the first of a copper chaperone bound to copper and provides structural support for direct metal ion exchange between conserved MT/HCXXC motifs in two domains. The structures of HgHah1 and CdHah1, determined to 1.75-Å resolution, also reveal metal ion coordination by two MT/HCXXC motifs. An extended hydrogen-bonding network, unique to the complex of two Hah1 molecules, stabilizes the metal-binding sites and suggests specific roles for several conserved residues. Taken together, the structures provide models for intermediates in metal ion transfer and suggest a detailed molecular mechanism for protein recognition and metal ion exchange between MT/HCXXC-containing domains. The Hah1 metallochaperone protein is implicated in copper delivery to the Menkes/Wilson disease proteins, about which more will be said below. Figure 7.29 is a representation of the Hah1 metallochaperone with the copper ion shown in space fill and the cysteine ligands in monochrome ball-and-spoke form. In this structure the metal-binding sequence begins with met10 and continues with thr11, cys12, gly13, gly14, ending with cys15.[118] Two repeats of the βαββαβ-fold tertiary structure motifs are evident. In the right-hand motif, cys12 lies in front of the copper ion whereas cys15 lies below and behind it. In the left-hand motif, cys15 lies in front and cys12 lies above and behind the copper ion.

Figure 7.29 Wavefunction, Inc. Spartan '02 for Windows™ representation of the Hah1 metallochaperone from reference 118. X-ray data deposited as PDB code 1FEE. See text for visualization details. Printed with permission of Wavefunction, Inc., Irvine, CA. (See color plate.)

7.4.4 Superoxide Dismutase Metallochaperones

The enzyme copper–zinc superoxide dismutase (CuZnSOD) serves as a biological antioxidant, disproportionating the superoxide ion. CuZnSOD or SOD1 is discussed in detail in Section 5.2.3 of Chapter 5. In vitro the copper- and zinc-dependent enzyme superoxide dismutase, SOD1, binds copper ions with extremely high affinity ($K_d \cong 10^{-15}$ M) even though total copper concentration in a living cell reaches the micromolar range.[119] The question of such high affinity in the face of ample Cu(II) or Cu(I) intracellular supply is answered with the knowledge that the total cytoplasmic *free* copper concentration is less than 10^{-18} M or less than one copper ion per cell.[120] In thermodynamic terms, almost all hydrated copper ions are immediately and tightly coordinated by amino acids or biopolymers—peptides, proteins, and other species with free sulfur ligands. In kinetic terms, less than 0.01% of the total cellular copper becomes free in the cytoplasm during the lifetime of the cell. Despite high cellular capacity for copper uptake and chelation, metallochaperones succeed in acquiring copper and delivering it to metalloenzymes that require it.

The sequence for delivery of copper ions to SOD1 passes from the copper transporter (Ctr) by an unknown pathway to the copper chaperone for SOD1 (CCS) and by a studied pathway from CCS to SOD1. The CCS protein has been studied structurally and found to be similar to other copper chaperones such as those discussed above—Atx1 and Atox1 (Hah1). Copper chaperone for superoxide dismutase (CCS) differs from other copper metallochaperones in that it folds into three functionally distinct protein domains with the N-terminal end of domain I including the MXCXXC copper binding site found in Atx1. Surprisingly, it was found that a CCS molecule lacking this domain can still insert copper into SOD1 in vivo, provided that the cell is not starved for copper. It was proposed, therefore, that the Atx1 domain is needed to maximize CCS function under extreme copper limiting conditions.[121] The central domain of CCS (domain II) is homologous to its target, SOD1, and can, under extreme circumstances, turn CCS into a SOD-like molecule that scavenges superoxide ion.[122] Domain II physically interacts with SOD1 and is proposed to secure the enzyme during copper insertion.[123] The C-terminal domain III of CCS is small (~30 amino acids) but is necessary for activating SOD1 in vivo. This peptide is highly conserved among diverse species and includes an invariant CXC motif that can bind copper.[121] It has been proposed that domain III, perhaps in concert with the domain I N-terminal copper binding site, inserts copper into the active site of SOD1.[121,124] It is believed that human CCS (hCCS) may have important relevance in the fatal motor neuron disease, familial amyotrophic lateral sclerosis (FALS), because inherited mutations in SOD1 and toxicity from bound copper ions have been implicated in certain FALS models. FALS is discussed in Section 7.4.5.

The mechanism for copper loading of the human antioxidant enzyme copper–zinc superoxide dismutase (SOD1) by its partner metallochaperone protein is an active research area. It has been found that the copper chaperone for superoxide dismutase (CCS) activates the antioxidant enzyme CuZnSOD (SOD1) by directly inserting the copper cofactor into the apo form of SOD1. Neither the mechanism of protein–protein recognition nor of metal transfer is clear. The metal transfer step has been proposed to occur within a transient copper donor–acceptor complex that is either a heterodimer or a heterotetramer (i.e., a dimer of dimers). To determine the nature of this intermediate, a mutant form of SOD1 was generated by replacing a copper binding residue his48 with phenylalanine (H48F mutant). This protein cannot accept copper from CCS but did form a stable complex with apo- and Cu-CCS. In addition, copper enhanced the stability of the dimer by an order of magnitude. The copper form of the heterodimer was isolated by gel filtration chromatography and was found to contain one copper and one zinc atom per heterodimer. These results support a mechanism for copper transfer in which CCS and SOD1 dock via their highly conserved dimer interfaces in a manner that precisely orients the cys-rich copper donor sites of CCS and the his-rich acceptor sites of SOD1 to form a copper-bridged intermediate.[125] In reference 126 the authors show that the human copper chaperone for CuZnSOD1 (hCCS) activates either human or yeast enzymes in vitro by direct protein to protein transfer of the copper cofactor. Interestingly, when denatured with organic solvents, the apo-form of human SOD1 cannot be reactivated by added copper ion alone as shown in the previous work, suggesting an additional function of hCCS such as facilitation of an active folded state of the enzyme. While hCCS can bind several copper ions, metal binding studies in the presence of excess copper scavengers that mimic the intracellular chelation capacity indicate a limiting stoichiometry of one copper and one zinc per hCCS monomer. This protein is active and, unlike the yeast protein, is a homodimer regardless of copper occupancy. Various studies have indicated that Cu(I) is bound by residues from the first and third domains of hCCS, and no bound copper is detected for the second domain of hCCS in either the full-length or truncated forms of the protein. Copper-induced conformational changes in the essential C-terminal peptide of hCCS are consistent with a "pivot, insert, and release" mechanism that is similar to one proposed for the well-characterized metal handling enzyme, mercuric ion reductase.[127]

In an X-ray crystallographic study of copper chaperone for yeast superoxide dismutase (yCCS) bound to ySOD1, O'Halloran, Rosenzweig, and co-workers found that CCS activates the eukaryotic SOD1.[128] The 2.9-Å resolution structure of yeast SOD1 complexed with yeast CCS (yCCS) reveals that SOD1 interacts with its metallochaperone to form a complex comprising one monomer of each protein. The heterodimer interface is remarkably similar to both the SOD1 and yCCS homodimer interfaces. Striking conformational rearrangements are observed in both the chaperone and target enzyme upon complex formation, and the functionally essential C-terminal domain of yCCS is well-positioned to play a key role in the metal ion transfer mechanism. This domain is linked to SOD1 by an intermolecular disulfide bond that may facilitate or regulate copper delivery.[128]

7.4.5 Copper Toxicity, Disease States, and Treatments

As stated previously, the total normal cytoplasmic free copper concentration is less than 10^{-18} M or less than one copper ion per cell. In thermodynamic terms, almost all hydrated copper ions are immediately and tightly coordinated by amino acids or biopolymers—peptides, proteins, and other species with free sulfur ligands. An excess of copper ions activates metallothionein synthesis for storage or removal of the excess. Copper chaperones mediate transfer of copper ions from extracellular or storage locations to their target proteins. Instability of copper ion concentrations in vivo results in various disease states. Three of these—FALS, Menkes, and Wilson's diseases—are described below.

7.4.5.1 Familial Amyotrophic Lateral Sclerosis (FALS). Familial amyotrophic lateral sclerosis (FALS) is an inherited neurodegenerative disease caused by mutations in CuZn superoxide dismutase (SOD1). FALS comprises approximately 10–15% of the ALS patient population, and about 20–25% of FALS cases are associated with dominantly inherited mutations in the gene that encodes human CuZnSOD. Evidence exists that the disease arises from a gain of toxic property rather than a loss of superoxide scavenging activity. It is known that in addition to its superoxide dismutation activity, SOD1 catalyzes oxidation of substrates by hydrogen peroxide (H_2O_2) at rates comparable with its own inactivation by H_2O_2. Valentine and co-workers studied this SOD1 property using the spin trap reagent 5,5′-dimethyl-1-pyrroline *N*-oxide (DMPO), which reacts with H_2O_2 to give the EPR-detectable hydroxyl adduct DMPO-OH. The reaction is catalyzed by wild-type (WT) CuZnSOD.[129] Because previous studies had indicated that substrate oxidation occurs at the Cu(II) ion bound at the enzyme's active site, the researchers hypothesized that FALS-associated mutant SOD1s might enhance similar oxidative reactions of substrates with H_2O_2. The catalysis might even be enhanced because the SOD1 mutants have more open structures (than WT SOD1) that facilitate substrate access to the active site. In keeping with the prediction of greater substrate oxidation by H_2O_2 for mutants than for WT SOD1, significantly higher concentrations of the EPR-marker DMPO-OH was generated with the FALS-associated mutants (ala4val, A4V and gly37arg, G37R) than found for the WT SOD1 or for Cu(II) added as $CuSO_4$. Addition of Cu(II) chelating agents diethyldithiocarbamate (DDC) or penicillamine (Figure 7.30A) decreased the DMPO-OH generation proportionate to their concentration in mutants contrasting with small increased DMPO-OH concentrations in WT enzyme. The researchers also attempted to find out if the chelators would inhibit neural degeneration processes in a neural cell culture FALS model because it is known that overexpression of WT SOD1 inhibits apoptosis (cell death) whereas FALS-associated mutants enhance apoptosis in the model. The chelator DDC inhibited apoptosis by 30–70% in experimental systems with FALS-associated mutants but had no effect on cells overexpressing WT SOD1. It was concluded that CuZnSOD mutants do indeed catalyze the oxidation of substrates present in motor neuron cells in vitro, although both the possible oxidants and substrates in vivo remain unidentified. The observations indicated that Cu(II) chelators in vitro (1) inhibit DMPO-OH production (and thus may inhibit the ability of mutant enzymes to catalyze harmful substrate oxidations in vivo) and (2) increase the viability of neural cells expressing FALS-associated mutant enzymes. Researchers hope that further testing of these chelators in animal models of FALS may lead to beneficial treatments in human patients with FALS-associated SOD1 mutations.

In other experiments a broad range of FALS-SOD1 mutants (A4V, G37R, G41D, H46R, H48R, G85R, G93C, and I113T) were studied as expressed in in vivo SOD1 assays in yeast *Saccharomyces cerevisiae* lacking CuZn SOD1.[130] Two of the yeast-expressed mutant SOD1s—H46R (his46arg) and H48Q (his48gln)—involve histidine residues that coordinate the catalytic copper ion in SOD1. Two others, A4V and I113T, occur at the SOD1 dimer interface. Three—G37R, G41D, and G93C—occur at turns of the β barrel, a characteristic tertiary structure of the SOD enzymes. G85R occurs at the base of the active site channel. Figure 1A of reference 130 illustrates the position of the point mutations. All mutants bind copper and scavenge free radicals in vivo. Some, but not all, mutants show decreased copper binding, the same being true for decreased superoxide scavenging, activity. H46R and H48Q mutants were least active in superoxide scavenging, while G85R and I113T mutants also exhibited reduced superoxide scavenging ability. All six non-active-site mutants protected yeast cells from high Cu(II) concentrations, whereas the active-site mutants (H46R and H48Q) offered less protection—that is, less copper buffering capacity. When copper concentrations were decreased by addition of increasing amounts of a Cu(II) chelating agent, H46R, H48Q, and G85R (at close proximity to the active site) lost superoxide scavenging ability most rapidly. It was found that all mutants as well as wild-type (WT) SOD1 required the SOD1 copper chaperone (CCS) to bring SOD1 levels to >2% of normal in low Cu(II) concentration conditions. The in vivo experimental results demonstrated that neither loss of SOD1 activity nor loss (or decrease) of copper binding ability is common to the SOD1 mutants. Additionally, the authors find that nitration of tyrosines (thought to be a pathological feature of FALS) increases in SOD1 mutants that more efficiently use the substrate peroxynitrite as a substrate. The researchers also believe that H46R and H48Q mutants release their copper more readily and must be recharged repeatedly, producing nonuniform intracellular SOD1 activity. In agreement with other researchers, these authors believe that copper loading (and perhaps zinc loading and SOD1 polypeptide folding) is chaperone-assisted in vivo and that FALS-SOD1-mediated toxicity arises from aberrant copper-mediated chemistry catalyzed by the less tightly folded (and therefore less constrained in substrate selection) mutant enzymes.[130] Therefore, mutant-specific SOD1 copper chaperone inhibitors (that would preclude use of peroxynitrite as a substrate for instance) may represent a therapeutic avenue for FALS treatment in the future.

7.4.5.2 Wilson and Menkes Diseases. Wilson and Menkes diseases are two copper-transport linked diseases involving mutations in the so-called Wilson and Menkes proteins.[110] As stated above and illustrated in Figure 7.26, copper is transported into the yeast cell via the plasma-membrane protein Ctr1.[115] Copper is then distributed to various cellular locations including metalloenzymes in the secretory pathway that occurs via a P-type ATPase, Ccc2. This enzyme translocates copper across intracellular membranes to the multicopper oxidase Fet3. The human counterparts include hCtr1—the transport protein—to the Atx1-like protein Atox1, then to the multicopper oxidase ceruloplasmin, CP, (a Fet3 analog) via the P-type ATPase human Ccc2 analogs—the Menkes and Wilson disease proteins. Mutations in the latter proteins cause Menkes syndrome and Wilson disease disorders of copper metabolism.[115] Metal binding to the Atx1-like MT/HCXXC motifs within Menkes/Wilson disease proteins have been reported and are discussed further below.[131,132] It is hoped that further characterization of these disease states and the metalloenzymes in their pathways will lead to new therapeutic approaches, including genetic manipulations, to their treatment.

Wilson disease is an autosomal recessive disorder of copper transport involving accumulation of copper in the liver and brain of affected individuals. Treatment methods target the excessive amounts of copper deposited in vital organs (liver, brain, kidney). D-Penicillamine (Figure 7.30A), a copper chelator administered orally, effectively reduces the copper bound to proteins. Toxicity limits its usefulness with side effects involving the immune system or connective tissues. Administration of zinc as a therapeutic agent with low toxicity for Wilson disease began in 1983. Its mechanism of action is to induce metallothionein in intestinal cells. The metallothionein binds copper and holds it until the cells are sloughed off. Zinc thus inhibits absorption of copper from the intestine and also blocks reabsorption of secreted copper from saliva and gastric juices. Menkes disease is a fatal genetic disorder with a widespread defect in intracellular copper transport. Neurodegeneration, caused by copper deficiency in a number of tissues, results in early childhood death in affected individuals. Only one treatment regimen is known that exploits a copper–histidine 1–2 complex found to be the main exchangeable

A penicillamine

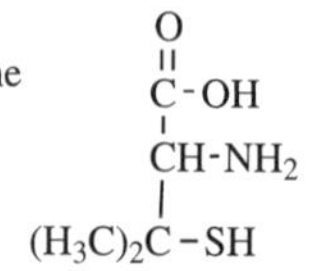

$$\mathrm{(H_3C)_2C{-}SH{-}CH(NH_2){-}C(=O){-}OH}$$

B Possible structures for copper-histidine complexes

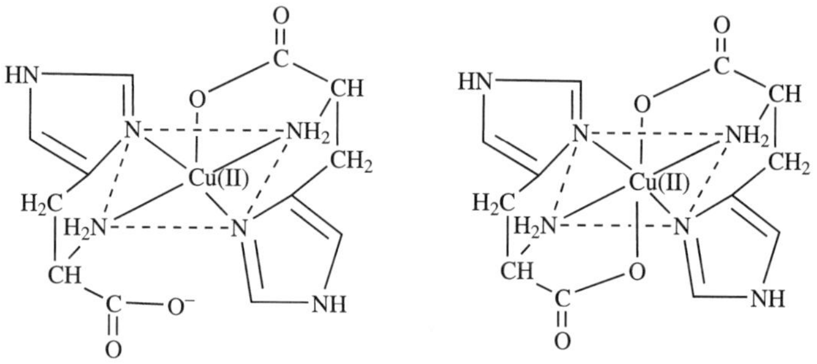

pentacoordinate

hexacoordinate

Figure 7.30 (A) D-Penicillamine used for Wilson's disease treatment. (B) Copper–histidine complex used for Menkes syndrome treatment.

form of copper in human serum. The copper–histidine complex used in treatment have the proposed structure(s) shown in Figure 7.30B.

Isolation of both Wilson and Menkes disease genes has been carried out. The Wilson disease gene (ATP7B) in chromosome 13 spans at least 80 kb of genomic DNA and is composed of 21 exons (see Section 2.3.5 of Chapter 2) encoding the 1411 amino acid residue WND protein that functions as a P-type ATPase involved in copper transport.[110] The N-terminal domain of the copper-transporting ATPase binds six Cu(I) ions. The Menkes disease gene (ATP7A) was mapped to the Xq13 region of the X-chromosome; it encodes a 1500-amino-acid protein, called MNK, also predicted to be a P-type ATPase. The N-terminus contains six metal-binding motifs similar to those in the Wilson disease ATPase.[110] Both disease genes encode very similar proteins (54% of their amino acid residues are identical), and they have significant homology with other cation-transporting P-type ATPases. Both WND and MNK, schematically portrayed in Figure 7.31, contain six repetitive copper-binding domain sequences of approximately 30 amino acid residues at their N-terminal terminus. The copper-binding motifs are separated from the ATP-binding domain by a transmembrane portion of the protein. Each copper-binding repeat features a GMT**C**XX**C**XXXIE sequence motif found earlier in the structure of mercury-binding proteins and bacterial Cd-ATPase.[132] The bold cysteines are believed to be the metal chelation sites. The transmembrane segment contains a cys-pro-cys motif present in the transduction domain of other bacterial heavy metal ATPases.[110] The asp (D) residue in the ATP-binding domain is phosphorylated.

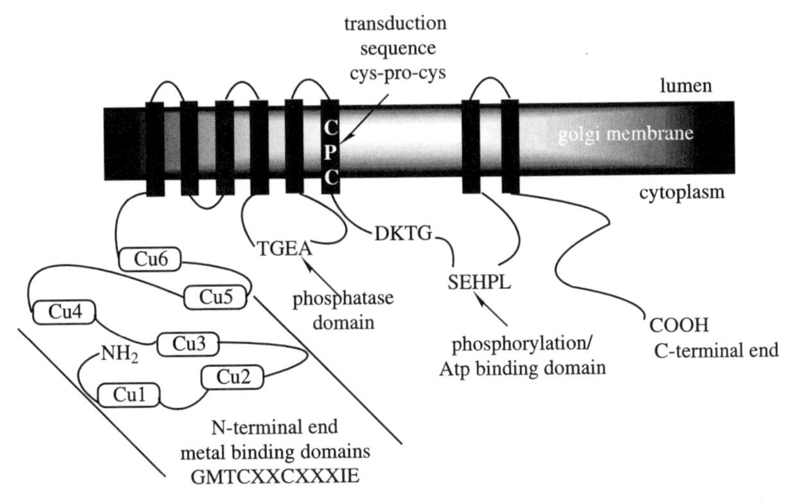

Figure 7.31 Schematic structures for Menkes and Wilson disease copper–transporting P–type ATPases. (Adapted with permission from Figure 2 of Sarkar, B. *Chem. Rev.*, 1999, **99**(9), 2535–2544. Copyright 1999, American Chemical Society.)

Lutsenko et al.[132] studied the N-terminal portions (approximately 600 amino acid residues) of WND and MNK both in vitro and in vivo. The in vitro results indicated that both N-WND and N-MNK bound copper when it was added in a partially chelated Cu(II) form. The uptake increased when Cu(II) was presented in the presence of a reducing agent, and the authors believed that Cu(I) was the actual bound species. In vivo, it was found that one copper ion per metal-binding repeat motif was bound in the Cu(I) form when $CuCl_2$ was added to N-WND and N-MNK proteins; however, without added reducing agent, N-WND released its copper more slowly. Other experiments indicated that cysteine was indeed the metal-binding site in both proteins and that each of the six binding motifs appeared to take up copper ion selectively. When N-WND and N-MNK proteins were denatured and refolded, metal selectivity decreased, indicated by uptake of zinc as well as copper. The authors concluded that the protein fold was important in providing high selectivity for copper.

An NMR solution structure of a copper-binding domain of a Menkes copper transporting ATPase has been carried out.[133] The 72-residue metal-binding domain, called mbd4, has a ferredoxin-like βαββαβ fold as noted for other copper chaperones such as Atx1 and Hah1 in Figures 7.28 and 7.29. In the mbd4 case, structural differences between the apo and metal-complexed forms are limited to the metal-binding loop, which is disordered in the apo structure but well-ordered in the Ag(I)-bound structure. Ag(I) binds in a linear bicoordinate manner to the two cys residues of the conserved GMTCXXC motif; Cu(I) likely coordinates in a similar manner. The Menkes mbd4 was the first bicoordinate copper-binding protein to be characterized structurally. Sequence comparisons with other

Figure 7.32 Apo form of metal binding domain 4 of Menkes copper-transporting ATPase described in reference 133 (PDB: 1AW0). Visualized using Wavefunction, Inc. Spartan '02 for Windows^{TM}. See text for visualization details. Printed with permission of Wavefunction, Inc., Irvine, CA. (See color plate.)

heavy-metal-binding domains reveal both a conserved hydrophobic core and metal-binding motif. Figures 7.32 and 7.33 are Wavefunction, Inc. Spartan '02 for Windows^{TM} visualizations of the metal-binding domain of a Menkes copper transporting ATPase molecule deposited in the Protein Data Bank (PDB codes: 1AW0 and 2AW0). The binding motif is shown in blue/green ribbon form and consists of the sequence GMT**C**NS**C**VQSIE, where the bold lettering indicates metal–binding cysteine residues. Differences in folding can be noted between the apo (Figure 7.32) and metallated forms (Figure 7.33). Metal binding takes place in the loop between the first β-strand and the first α-helix in the 72-amino-acid domain. The mbd4 domain is shown in purple ribbon form, and the GMT**C**NS**C**VQSIE motif is shown in blue/green to distinguish it. Note that the motif's C-terminal end reaches into the first α-helix. Cysteine residues are shown in ball-and-spoke rendering, forming a linear coordination sphere with the Ag(I) ion—in black space-fill form in Figure 7.33.

7.4.6 Conclusions

Knowledge of the intracellular thermodynamics and kinetics of metal metabolism may become useful in the design of compounds that alter intracellular metal ion availability. This in turn may be useful in controlling such biological phenomena as cancer cell proliferation, disorders of metal metabolism, and metal-induced neurotoxicity.

Figure 7.33 Metallated form of metal-binding domain 4 of Menkes copper-transporting ATPase described in reference 133 (PDB: 2AW0). Visualized using Wavefunction, Inc. Spartan '02 for Windows^{TM}. See text for visualization details. Printed with permission of Wavefunction, Inc., Irvine, CA. (See color plate.)

7.5 METALS IN MEDICINE (MIM)

7.5.1 Summary, Goals; Past, Present, and Future MIM Development

The goal of this chapter has been to introduce readers to chemical, biological, structural, developmental, and clinical aspects of therapeutic and diagnostic metal-containing compounds. Some topics have been the subject of extensive and continuing research by university, governmental, pharmaceutical, and medical laboratories. Platinum-containing anticancer agents fit into this category. Other areas, those of SOD mimics and metallochaperones for example, are newer, less well characterized potential medicinal agents. Many other worthy metal-containing drugs or potential diagnostic or treatment protocols involving metals have not been addressed here. While basic research into potential therapeutic or diagnostic agents may take place in university laboratories, drug development cannot take place without the cooperation of pharmaceutical enterprises and governmental evaluation of the drug's safety and efficacy. Ethical questions of who develops what drug for what intended population at what cost or profit cannot be answered (or even properly asked) in the material presented in this chapter.

The interest of United States federal governmental agencies in the topic is illustrated by a "Metals in Medicine: Targets, Diagnostics, and Therapeutics" meeting held at the National Institutes of Health on June 28–29, 2000. This meeting, sponsored by the National Institute of General Medical Sciences (NIGMS), the National Cancer Institute (NCI), the National Institute of Allergy and Infectious Diseases (NIAID), the National Institute of Diabetes and Digestive and Kidney Diseases (NIDDK), the National Institute of Environmental Health Sciences (NIEHS), the Center for Scientific Review (CSR), and the NIH Office of Dietary Supplements, generated substantial interest in the scientific community. Various invited speakers identified ways that basic research in metallobiochemistry and bioinorganic chemistry contributes to the development of future pharmaceuticals and the health benefits therein. Topics presented included (1) a case history outlining discovery, development, and understanding of platinum-containing anticancer agents, (2) molecular and cellular targets of metals, including metal-catalyzed cleavage of nucleic acids, metal complexes and drug transport, and metal complexes as receptor ligands, (3) studies of metalloenzyme structure and function, and of the roles of metalloenzymes in biological systems, (4) the roles of metals in cell regulation and the pathways by which cells regulate metal ion concentrations and delivery to selected metal sites, (5) molecular modeling of metalloenzymes and metalloproteases, (6) radiology, imaging, and photodynamic therapy, (7) topics in metal metabolism including intracellular transporters and metallochaperones, metal chelation therapy, chromium as a dietary supplement, and mechanisms of metal toxicity, (8) topics in metallotherapeutics and disease including vanadium and diabetes, synthetic superoxide dismutase mimetics, and possible polyoxometalate HIV-1 protease inhibitors, and (9) drug development challenges, opportunities, and support mechanisms, including the NCI developmental therapeutics program (DTP) and the U.S. drug approval process.

An objective of the meeting was to build bridges between research communities and between researchers in academia, industry, and government. NIGMS staff used this opportunity to point out several research grant mechanisms that may be useful to foster the collaborative research needed in this area. Much more useful information about the topic is available at the NIGMSS website: http://www.nigms.nih.gov/news/meetings/metals.html. A further outcome of the meeting was the March 2001 program announcement, Metals in Medicine for those interested in federal support of drug research and developmental funding. Information is available at: http://grants.nih.gov/grants/guide/pa-files/PA-01-071.html. The following comments are adapted from the report of the June 2000 MIM meeting. Quotes are taken from the meeting's report as prepared by Peter C. Preusch, PhD., PPBC Division, NIGMS, NIH.

A limiting factor in the development of metallopharmaceuticals is the perception that metals are toxic. This is certainly true of some metals as mentioned previously for mercury, lead, thallium, and Cr(VI) in Section 7.1.2. However, even essential metals are toxic under some circumstances such as oversupply or concentration in inappropriate physiological locations as seen in the discussions of FALS and Wilson and Menkes diseases. However, all metals are clearly not heavy metals, and even the so-called heavy metals are not necessarily toxic under all circumstances as proven unequivocally for the platinum-containing anticancer agents. "To rule out a large fraction of the periodic table, a priori, eliminates any potential benefit that may be discovered among those elements. Understanding and learning to control the actions of metals in vivo is the key to potential beneficial applications of metals, and the key to avoiding metal toxicities. Drug development is a long and nonlinear process." A promising research lead is merely the first small step in producing a successful drug. Issues of pharmacokinetics, efficacy, toxicity, and economics, along with patents, politics, and personalities, all factor into decisions about whether to continue or to drop a drug manufacture and approval project.

"Medicinal inorganic compounds have an appreciable current market impact and significant growth potential." Current sales, including imaging, diagnostics, and therapeutics, are on the order of $2 billion per year. Platinum-based chemotherapeutics are among the most successful of all anticancer drugs, with combined sales of $700 million per year, second only to Taxol, an organic antitumor drug. New compounds are likely to gain a unique market niche by acting through different mechanisms and/or avoiding common clearance mechanisms—that is, multidrug resistance (MDR). Successful or at least promising results have been achieved for various disease states involving a significant number of inorganic elements from the periodic table. Additional exploration should continue to be fruitful. The success rate of metallopharmaceutical advanced clinical leads takes place at about the going rate for all drugs (1 in 10 survive Phase I trials; of these, 1 in 4 survive Phase II). Phase I trials of candidate drugs must establish safety in human patients. These first studies in people evaluate how a new drug should be given (by mouth, injected into the blood, or injected into the muscle), how often, and what dosage is safe. A Phase I trial usually enrolls only a small number of patients, sometimes as few as a dozen. A Phase II trial continues to test the safety and efficacy of a particular drug and begins to evaluate how well the new drug works. Phase II studies usually focus on a particular type of cancer, for instance. Phase III trials should provide definitive evidence of efficacy and safety. These studies test a new drug, a new combination of drugs, or a new surgical procedure in comparison to the current standard. A participant will usually be assigned to the standard group or the new group at random (called randomization). Phase III trials often enroll large numbers of people and may be conducted at many doctors' offices, clinics, and treatment centers nationwide. Phase IV studies may be needed to resolve some postapproval questions.

A limiting factor for interest in inorganic pharmaceuticals may be the relatively limited expertise of most pharmaceutical companies in the area of inorganic chemistry. The reality is that most chemists working for companies are organic chemists by training and naturally look in that direction for leads. There is a perception, right or wrong, that metal-containing agents are more toxic. Metal-containing agents have generally been excluded from the high-throughput library screening programs that have become the major paradigm for drug lead discovery even though metal-based drug discovery may be more amenable to a rational design approach. Industry has been involved in this area in the past and remains very active in some specialty areas.

In the governmental arena the National Cancer Institute (NCI) Developmental Therapeutics Program (DTP) screens metal-containing agents for both cancer and AIDS therapy. "The program, which began in 1955, has gone through several evolutions. Until recently an "empirical drug discovery" paradigm has been used based on screening for antiproliferative effects in animal models (1975–1989) or a 60-cell-line screen (1990–present) with implanted hollow-fiber (1995–present) and human xenograft mouse models (1990–present) as the secondary screens. During this time, 14,900 metal-containing compounds were tested (including a wide range of elements). Of those, 1242 were selected as active in the mouse in vivo screen, and 191 were selected in the 60-cell-line screen. Nine were investigated as clinical candidates. Five resulted in Investigational New Drug (IND) applications. Two resulted in New Drug Applications (NDAs) and are currently used mainstream therapies (cisplatin and carboplatin). For comparison, of the 550,000 total compounds tested by the Developmental Therapeutics Program, 14,475 were active in mouse screen models and 7741 (out of 77,000) were active in the 60-cell-line assay; 59 INDs and 11 NDAs resulted."

An Investigational New Drug (IND) application includes documentation that must be submitted to, and accepted by, the FDA before a new drug or biologic can be shipped interstate for human testing. An IND includes all appropriate evidence that clinical investigations (previously carried out in animal models) can be performed with reasonable safety to a human patient. Following initial filing, the IND becomes a "central file" for information on the drug or biologic. All relevant subsequent information is forwarded to the IND—for example, final study reports, serious adverse experience reports, new protocols (both clinical and nonclinical), and a yearly progress report summarizing the last year's information and proposed plan for the future year. For decades, the regulation and control of new drugs in the United States has been based on the New Drug Application (NDA). Since 1938, every new drug has been the subject of an approved NDA before U.S. commercialization. The NDA application is the vehicle through which drug sponsors formally propose that the FDA approve a new pharmaceutical for sale and marketing in the United States. The data gathered during the animal studies and human clinical trials of an Investigational New Drug (IND) become part of the NDA. The goals of the NDA are to provide enough information to permit an FDA reviewer to reach the following key decisions: (1) The drug is safe and effective in its proposed use(s); (2) the benefits of the drug outweigh its risks; (3) the drug's proposed labeling (package insert) is appropriate and contains the proper information; and (4) the methods used in manufacturing the drug and the controls used to maintain the drug's quality are adequate to preserve the drug's identity, strength, quality, and purity.

Reasons for dropping clinical candidates at any stage of the drug approval process included (1) stability, formulation, or other pharmaceutical development issues, (2) renal toxicity or neurotoxicity, and (3) insufficient advantage over current drugs. Since 1997, the NCI has also operated a screen for compounds active against the cytotoxic effects of HIV in CEM cells. Of 80,000 compounds tested, 4050 (or about 5%) were active. Of the compounds tested, 2291 have included metals. Of those, 136 (about 6%) were active, and two became clinical candidates. Both were dropped due to toxicity problems. One clinical candidate was a polyoxometallate, and therefore about 80 other similar molecules were tested. "These were found to be strongly active in vitro, but too toxic in animal models in vivo. If a way around the toxicity problem can be found, interest in these compounds would be very high." Toxicity obstacles in drug development are not unique to metal-containing agents. In fact the frequencies of success for metal-containing drugs in screens and also of failure in clinical development are not different from the experience with development of non-metal-containing agents.

Recently, "the model for anticancer drug development within the NCI program has changed to focus on mechanism-oriented rational drug discovery based on known molecular targets. Comparisons of compound activity against the 60-cell-line screen have developed correlations between cellular sensitivity and molecular target expression using the COMPARE algorithm. Microarray expression and specific target activity assay programs to characterize the levels of 300–400 targets in the 60-cell-line screen are underway. A yeast assay screen against a large number of mutants has been developed to further help identify targets of compound action. Interestingly, metals as a whole, and even the group of platinum compounds tested, do not cluster well in these studies. This implies that different metals act through different mechanisms of action. In the new drug development model, it is not sufficient to find antiproliferative activity in order to test additional chemically related compounds. It is necessary to demonstrate and optimize activity against a specific target. Furthermore, DNA targeting, per se, is no longer considered a sufficiently good rationale. A more specific cellular effect of DNA modifying agents must be demonstrated. These same constraints, however, operate equally in the selection of non-metal-containing candidates. The rapid access to intervention development (RAID) program has been set up to assist investigators in moving their leads to an IND. Information on this program, data on the library of tested compounds, and the COMPARE program are available at http://dtp.nci.nih.gov."

The following points summarize U.S. federal agency interest in continued development and testing of metal-containing compounds as well as nonmetallic compounds as diagnostic and therapeutic agents: (1) In testing protocols, metal-containing drugs fared about as well as non-metal-containing drugs; (2) metals remain viable candidates for testing and development; however, the rationale for considering all compounds in U.S. governmental drug approval process has become more stringent; (3) future developments must be based on rational targeted design; and (4) the greatest need is for an improved ability to predict human activity from cell culture and animal studies.

REFERENCES

1. Orvig, C.; Abrams, M. J. *Chem. Rev.*, 1999, **99**, 2201–2203.
2. Frausto da Silva, J. R. R.; Williams, R. J. P. *The Biological Chemistry of the Elements: The Inorganic Chemistry of Life*. Clarendon Press, New York, 1991.
3. Sadler, P. J. Adv. *Inorg. Chem.*, 1991, **36**, 1–48.
4. Budavari, S. ed., *The Merck Index*, 13th ed., Merck & Co., Rahway, NJ, 2001.
5. Claussen, C. A.; Long, E. C. *Chem. Rev.*, 1999, **99**, 2797–2816.
6. Birch, N. J. *Chem. Rev.*, 1999, **99**, 2659–2682.

7. Briand, G. G.; Burford, N. *Chem. Rev.*, 1999, **99**, 2601–2658.
8. Shaw, C. F., III. *Chem. Rev.*, 1999, **99**(9), 2589–2600.
9. Ali, H.; van Lier, J. E. *Chem. Rev.*, 1999, **99**, 2379–2450.
10. Macarthur, H.; Westfall, T. C.; Riley, D. P.; Misko, T. P.; Salvemini, D. *Proc. Natl. Acad. Sci. USA*, 2000, **97**(17), 9753–9758.
11. (a) Riley, D. P. *Chem. Rev.*, 1999, 99(9), 2573–2587. (b) Melov, S.; Doctrow, S. R.; Schneider, J. A.; Haberson, J.; Patel, M.; Coskun, P. E.; Huffman, K.; Wallace, D. C.; Malfroy, B. *J. Neurosci.*, 2001, **21**(21) 8348–8353.
12. Salvemini, D.; Wang, Z.-Q.; Zweier, J. L.; Samouilov, A.; Macarthur, H.; Misko, T. P.; Currie, M. G.; Cuzzocrea, S.; Sikorski, J. A.; Riley, D. P. *Science*, 1999, **286**, 304–306.
13. Riley, D. P.; Henke, S. L.; Lennon, P. J.; Aston, K. *Inorg. Chem.*, 1999, **38**, 1908–1917.
14. Aston, K.; Rath, N.; Naik, A.; Slomczynska, U.; Schall, O. F.; Riley, D. P. *Inorg. Chem.*, 2001, **40**(8), 1779–1789.
15. Balasubramanyam, M.; Mohan, V. *J. Biosci.*, 2001, **26**(3), 383–390.
16. Thompson, K. H.; McNeill, J. H.; Orvig, C. *Chem. Rev.*, 1999, **99**(9), 2561–2571.
17. Tracy, A. S.; Crans, D. C., eds., *Vanadium Compounds: Chemistry, Biochemistry, and Therapeutic Applications.*, ACS Symposium Series 711, Oxford University Press, New York, 1998.
18. Wilsky, G. R.; Goldfine, A. B.; Kostyniak, P. J.; McNeill, J. H.; Yang, L. Q.; Khan, H. R.; Crans, D. C. *J. Inorg. Biochem.*, 2001, **85**, 33–42.
19. Reul, B. A.; Amin, S. S.; Buchet, J. P.; Ongemba, L. N.; Crans, D. C.; Brichard, S. M. *Br. J. Pharmacol.*, 1999, **126**(2), 467–477.
20. Amin, S. S.; Cryer, K.; Zhang, B.; Dutta, S. K.; Eaton, S. S.; Anderson, O. P.; Miller, S. M.; Reul, B. A.; Brichard, S. M.; Crans, D. C. *Inorg. Chem.*, 2000, **39**(3), 406–416.
21. Posner, B. I.; Faure, R.; Burgess, J. W.; Bevan, A. P.; Lachance, D.; Zhang-Sun, G.; Fantus, I. G.; Ng, J. B.; Hall, D. A.; Soo Lum, B.; Shaver, A. *J. Biol. Chem.*, 1994, **269**(6), 4596–4604.
22. Bevan, A. P.; Drake, P. G.; Yale, J. F.; Shaver, A.; Posner, B. I. *Mol. Cell Biochem.*, 1995, **153**, 49–58.
23. Yale, J. F.; Vigeant, C.; Nardolillo, C.; Chu, Q.; Yu, J. Z.; Shaver, A.; Posner, B. I. *Mol. Cell Biochem.*, 1995, **153**, 181–190.
24. Shaver, A.; Ng, J. B.; Hynes, R. C.; Posner, B. I. *Acta Crystallogr.*, 1994, **C50**, 1044.
25. Yale, J. F.; Lachance, D.; Bevan, A. P.; Vigeant, C.; Shaver, A.; Posner, B. I. *Diabetes*, 1995, **44**(11), 1274–1279.
26. Melchior, M.; Rettig, S. J.; Liboiron, B. D.; Thompson, K. H.; Yuen, V. G.; McNeill, J. H.; Orvig, C. *Inorg. Chem.*, 2001, **40**(18), 4686–4690.
27. Vincent, J. B. *J. Nutr.*, 2000, **130**, 715–718.
28. Yamamoto, A.; Wada, O.; Ono, T. *Eur. J. Biochem.*, 1987, **165**, 627–631.
29. (a) Davis, C. M.; Vincent, J. B. *Biochemistry*, 1997, **36**, 4382–4385. (b) Davis, C. M.; Sumrall, K. H.; Vincent, J. B. *Biochemistry*, 1996, **35**, 12963–12969. (c) Davis, C. M.; Royer, A. C.; Vincent, J. B. *Inorg. Chem.*, 1997, **36**, 5316–5320.
30. Werner, A. *Z. Anorg. Allg. Chem.*, 1893, **3**, 267.

31. (a) Rosenberg, B.; Van Camp, L.; Krigas, T. *Nature*, 1965, **205**, 698. (b) Rosenberg, B. In *Nucleic Acid–Metal Ion Interactions*, T. G. Spiro, ed., John Wiley & Sons, New York, 1980, pp. 1–29.
32. Rosenberg, B.; Van Camp, L.; Trosko, J. E.; Mansoui, V. H. *Nature*, 1969, **222**, 385–387.
33. Higby, D. J.; Wallace, J. H., Jr.; Holland, J. F. *Cancer Chemother. Rep.*, 1973, **57**, 459–463.
34. (a) Giandomenico, C. M. *Inorg. Chem.*, 1995, **34**, 1015; (b) Kelland, L. R. *Cancer Res.*, 1992, **52**, 822.
35. (a) Farrell, N.; van Beusichem, M. *Inorg. Chem.*, 1992, **31**, 634. (b) Farrell, N. *Cancer Investigation*, 1993, **11**, 578–589.
36. Farrell, N.; Appleton, T. G.; Qu, Y.; Roberts, J. D.; Fontes, A. P.; Skov, K. A.; Wu, P.; Zou, Y. *Biochemistry*, 1995, **34**, 15480–15486.
37. Sherman, S. E.; Lippard, S. J. *Chem. Rev.*, 1987, **87**, 1153.
38. Lippert, B., ed. *Cisplatin: Chemistry and Biochemistry of a Leading Anticancer Drug*, John Wiley & Sons, New York, 1999.
39. Harder, H. C.; Rosenberg, B. *Int. J. Cancer*, 1970, **6**, 207.
40. (a) Howe-Grant, M. E.; Lippard, S. J. *Met. Ions Biol. Syst.*, 1980, **11**, 63. (b) Lim, M. C.; Martin, R. B. *J. Inorg. Nucl. Chem.*, 1976, **38**, 1911. (c) Reishus, J. W.; Martin, D. S., Jr. *J. Am. Chem. Soc.*, 1961, **83**, 2457.
41. (a) Fichtinger-Schepman, A. M. J.; van der Veer, J. L.; den Hartog, J. H. J.; Lohman, P. H. M.; Reedijk, J. *Biochemistry*, 1985, **24**, 707–713. (b) Fichtinger-Schepman, A. M. J.; van Dijk-Knijnenburg, H. C.; van der Velde-Visser, S. D.; Berends, F.; Baan, R. A. *Carcinogenesis*, 1995, 2447–2453.
42. (a) Cardonna, J. P.; Lippard, S. J. *Inorg. Chem.*, 1988, **27**, 1454–1466. (b) Gibson, D.; Lippard, S. J. *Inorg. Chem.*, 1987, **26**, 2275–2279. (c) den Hartog, J. H. J.; Altona, C.; van Boom, J. H.; van der Marel, G. A.; Haasnoot, C. A.; Reedijk, J. *J. Am. Chem. Soc.*, 1984, **106**, 1528–1530. (d) den Hartog, J. H. J.; Altona, C; Chottard, J.-C. Girault, J.-P.; Lallemand, J.-Y.; de Leeuw, F. A. A. M.; Marcelis, A. T. M.; Reedijk, J. *Nucl. Acids Res.*, 1982, **10**, 4715. (e) van Hemelryck, B.; Guittet, E.; Chottard, G.; Girault, J.-P.; Huynh-Dinh, T.; Lallemand, J.-Y.; Igolen, J. Chottard, J.-C. *J. Am. Chem. Soc.*, 1984, **106**, 3037–3039. (f) Girault, J.-P.; Chottard, G.; Lallemand, J.-Y.; Chottard, J.-C.; *Biochemistry*, 1982, **21**, 1352–1356. (g) Chottard, J.-C.; Girault, J.-P.; Chottard, G.; Lallemand, J.-Y.; Mansuy, D. J. *J. Am. Chem. Soc.*, 1980, **102**, 5565–5572.
43. den Hartog, J. H. J.; Altona, C; van Boom, J. H.; van der Marel, G. A.; Haasnoot, C. A. G.; Reedijk, J. *Biomol. Struct. Dyn.*, 1985, **2**, 1137.
44. (a) van Boom, S. S. G. E.; Reedijk, J. *J. Chem. Soc. Chem. Commun.*, 1993, 1397. (b) Barnham, K. J.; Djuran, M. I.; del Socorro Murdoch, P.; Sadler, P. J. *J. Chem. Soc. Chem. Commun.*, 1994, 721.
45. Admiraal, G.; van der Veer, J. L.; de Graaff, R. A. G.; den Hartog, J. H. J.; Reedijk, J. *J. Am. Chem. Soc.*, 1987, **109**, 592.
46. Takahara, P. M.; Rosenzweig, A. C.; Frederick, C. A.; Lippard, S. J. *Nature*, 1995, **377**, 649–652.
47. Gelasco, A., Lippard, S. J. *Biochemistry*, 1998, **37**, 9230.
48. Coste, F.; Malinge, J. M.; Serre, L.; Shepard, W.; Roth, M.; Leng, M.; Zelwer, C. *Nucleic Acids Res.*, 1999, **27**(8) 1837–1846. (PDB: 1A2E)

49. Huang, H.; Zhu, L.; Reid, B. R.; Drobny, G. P.; Hopkins, P. B. *Science*, 1995, **270**, 1842–1845.

50. Reedijk, J.; Fichtinger-Schepman, A. M. J.; van Oosterom, A. T.; van de Putte, P. *Struct. Bonding*, 1987, **67**, 53–89.

51. (a) Van Houten, B. *Microbiol. Rev.*, 1990, **54**, 18–51. (b) Page, J. D.; Husain, I.; Sancar, A.; Chaney, S. G. *Biochemistry*, 1990, **29**, 1016–1024.

52. Jamieson, E. R.; Lippard, S. J. *Chem. Rev.*, 1999, **99**(9), 2467–2498.

53. Ishibashi, T.; Lippard, S. J. *Proc. Natl. Acad. Sci. USA*, 1998, **95**, 4219–4223.

54. Villa, R.; Folini, M.; Perego, P.; Supino, R.; Setti, E.; Daidone, M. G.; Zunino, F.; Zaffaroni, N. *Int. J. Oncol.*, 2000, **16**(5), 995–1002.

55. Donahue, B. A.; Augot, M.; Bellon, S. F.; Trieber, D. K.; Toney, J. H.; Lippard, S. J.; Essigmann, J. M. *Biochemistry*, 1990, **29**, 5872–5880.

56. Grosschedl, R.; Giese, K.; Pagel, J. *Trends Genet.*, 1994, **10**, 94–100.

57. (a) McA'Nulty, M. M.; Lippard, S. J. In *Nucleic Acids and Molecular Biology*, Vol. 9, Eckstein, F.; Lilley, D. M. J., eds., Marcel Dekker, New York, 1995, pp. 264–284. (b) Whitehead, J. P.; Lippard, S. J. In *Metal Ions in Biological Systems*, Vol. 32, Sigel, A.; Sigel, H. eds., Marcel Dekker, New York, 1996, pp. 687–726.

58. (a) Read, C. M.; Cary, P. D.; Crane-Robinson, C.; Driscoll,, P. C.; Norman, D. G. *Nucleic Acid Res.*, 1993, **21**, 3427–3436. (b) Weir, H. M.; Kraulis, P. J.; Hill, C. S.; Raine, A. R. C.; Laue, E. D.; Thomas, J. O. *EMBO J.*, 1993, **12**, 1311–1319. (PDB: 1HME) (c) Hardman, C. J.; Broadhurst, R. W.; Raine, A. R. C.; Grasser, K. D.; Thomas, J. O.; Laue, E. D. *Biochemistry*, 1995, **34**, 16596–16607.

59. Ohndorf, U. M.; Rould, M. A.; He, Q.; Pabo, C. O.; Lippard, S. J. *Nature*, 1999; **399**, 708–712. (PDB: 1CKT)

60. Zehnulova, J.; Kasparkova, J.; Farrell, N.; Brabec, V. *J. Biol. Chem.*, 2001, **276**, 22191–22199.

61. (a) Hosking, L. K.; Whelan, R. D.; Shellard, S. A.; Bedford, P.; Hill, B. T. *Biochem. Pharmacol.*, 1990, **40**, 1833. (b) Zhen, W.; Link, J.; O'Connor, M. P.; Reed, E.; Parker, R.; Howell, S. B.; Bohr, V. *Mol. Cell. Biol.*, 1992, **12**, 3689.

62. Reedijk, J. *Chem. Rev.*, 1999, **99**(9), 2499–2510.

63. Hospers, G. A. P.; Mulder, N. H.; de Jong, B.; de Ley, L.; Uges, D. R. A.; Fichtinger-Schepman, A. M. J.; Schepper, R. J.; de Vries, E. G. E. *Cancer Res.*, 1988, **48**, 6803–6807.

64. Batist, G.; Behrens, B. C.; Makuch, R.; Hamilton, T. C.; Katki, A. G.; Louie, K. G.; Myers, C. E.; Ozols, R. F. *Biochem. Pharmacol.*, 1986, **35**, 2257–2259.

65. Andrews, P. A.; Murphy, M. P.; Howell, S. B. *Cancer Res.*, 1985, **45**, 6250–6253.

66. (a) Loh, S. Y.; Mistry, P.; Kelland, L. R.; Abel, G.; Harrap, K. R. *Br. J. Cancer*, 1992, **66**, 1109. (b) Mellish, K. J.; Kelland, L. R.; Harrap, K. R. *Br. J. Cancer*, 1993, **68**, 240.

67. (a) Kelland, L. R.; Mistry, P.; Abel, G.; Friedlos, F.; Loh, S. Y.; Roberts, J. J.; Harrap, K. R. *Cancer Res.*, 1992, **52**, 1710. (b) Kelland, L. R.; Mistry, P.; Abel, G.; Loh, S. Y.; O'Neill, C. F.; Murrer, B. A.; Harrap, K. R. *Cancer Res.*, 1992, **52**, 3857.

68. (a) Johnson, S. W.; Perez, R. P.; Godwin, A. K.; Yeung, A. R.; Handel, L. M.; Ozols, R. F.; Hamilton, T. C. *Biochem. Pharmacol.*, 1994, **47**, 689. (b) Mistry, P.; Kelland, L. R.; Abel, G.; Sidhar, S. Harrap, K. R. *Br. J. Cancer*, 1991, **64**, 215.

69. Versantvoort, C. H. M.; Brosterman, J. J.; Bagrij, T.,R.; Scheper, J. Twentyman, P. R. *Br. J. Cancer*, 1995, **72**, 82.

70. (a) Pendyala, L.; Creaven, P. J.; Perez, R.; Zdanowicz, J. R.; Raghavan, D. *Cancer Chemother. Pharmacol.*, 1995, **36**, 271. (b) Berners-Price, S. J.; Kuchel, P. W. *J. Inorg. Biochem.*, 1990, **38**, 327.

71. Lai, G. M.; Ozols, R. F.; Young, R. C.; Hamilton, T. C. *J. Natl. Cancer Inst.*, 1989, **81**, 535.

72. Farrell, N.; Qu, Y.; van Beusichem, M.; Kelland, L. R. *Anti-Cancer Drug Design*, 1991, **6**, 232.

73. Roberts, J. D.; van Houten, B.; Qu, Y.; Farrell, N. P. *Nucl. Acids Res.*, 1989, **17**, 9719–9733.

74. Bloemink, M. J.; Reedijk, J.; Farrell, N.; Qu, Y.; Stetsenko, A. I. *J. Chem. Soc. Chem. Commun.*, 1992, 1002–1003.

75. Brabec, V.; Kasparkova, J.; Vrana, O.; Novakova, O.; Cox, J. W.; Qu, Y.; Farrell, N. *Biochemistry*, 1999, **38**, 6781–6790.

76. Kloster, M. B. G.; Hannis, J. C.; Muddiman, D. C.; Farrell, N. *Biochemistry*, 1999, **38**, 14731–14737.

77. Lebwohl, D.; Canetta, R. *Eur. J. Cancer*, 1998, **34**, 1522.

78. Rixe, O.; Ortuzar, W.; Alvarez, M.; Parker, R.; Reed, E.; Paull, K.; Fojo, T. *Biochem. Pharmacol.*, 1996, **52**(12), 1855–1865.

79. Wong, E.; Giandomenico, C. M. *Chem. Rev.*, 1999, **99**(9), 2451–2466.

80. Wilkinson, G.; Gillard, R. D.; McCleverty, J. A., eds., *Comprehensive Coordination Chemistry*, Pergamon Press, New York, 1987.

81. (a) Holford, J.; Sharp, S. Y.; Murrer, B. A.; Abrams, M. J.; Kelland, L. R. *Br. J. Cancer*, 1998, **77**, 366. (b) Holford, J.; Beale, P. J.; Boxall, F. E.; Sharp, S. Y.; Kelland, L. R. *Eur. J. Cancer*, 2000, 1984–1990. (c) Holford, J.; Raynaud, F.; Murrer, B. A.; Grimaldi, K.; Hartley, J. A.; Abrams, M.; Kelland, L. R. *Anticancer Drug Des.*, 1998, 1–18.

82. Kelland, L. R.; Murrer, B. A.; Abel, G.; Giandomenico, C. M.; Mistry, P. Harrap, K. R. *Cancer Res.*, 1992, **52**, 822. (b) Kelland, L. R.; Abel, G.; McKeague, M. J.; Jones, M.; Goddard, P. M.; Valenti, M.; Murrer, B. A.; Harrap, K. R. *Cancer Res.*, 1993, **53**, 2581.

83. Hartwig, J. F.; Lippard, S. J. *J. Am. Chem. Soc.*, 1992, **114**, 5646–5654.

84. (a) Cotton, F. A.; Wilkinson, G. *Advanced Inorganic Chemistry*, 4th ed., John Wiley & Sons, New York, 1980, pp. 950ff. (b) Giandomenico, C. M.; Abrams, M. J.; Murrer, B. A.; Vollano, J. F.; Rheinheimer, M. I.; Wyer, S. B.; Bossard, G. E.; Higgins, J. D., III. *Inorg. Chem.*, 1995, **34**, 1015. (c) Roat, R. M.; Jerardi, M. J.; Kopay, C. B.; Heath, D. C.; Clark, J. A.; DeMars, J. A.; Weaver, J. M.; Bezemer, E.; Reedijk, J. *J. Chem. Soc. Dalton Trans.*, 1997, 3615–3621.

85. Clarke, M. J.; Zhu, F.; Frasca, D. R. *Chem. Rev.*, 1999, **99**(9), 2511–2533.

86. Jurisson, S. S.; Lydon, J. D. *Chem. Rev.*, 1999, **99**(9), 2205–2218.

87. Anderson, C. J.; Welch, M. J. *Chem. Rev.*, 1999, **99**(9), 2219–2234.

88. Pinkerton, T. C.; Desilets, C. P.; Hoch, D. J.; Mikelsons, M. R.; Wilson, G. M. *J. Chem. Ed.*, 1985, **62**(11), 965–973.

89. Jurisson, S. S.; Berning, D.; Jia, W.; Ma, D. *Chem. Rev.*, 1993, **93**, 1137–1156.

90. Libson, K.; Deutsch, E.; Barnett, B. L. *J. Am. Chem. Soc.*, 1980, **102**, 2476.

91. (a) Barbarics, E.; Kronauge, J. F.; Davison, A.; Jones, A. G. *Nucl. Med. Biol.*, 1998, **25**, 667. (b) Piwnica-Worms, D.; Holman, B. L. *J. Nucl. Med.*, 1990, **31**, 1166. (c) Piwnica-Worms, D.; Chiu, M. L.; Budding, M.; Kronauge, J. F.; Kramer, R. A.; Croop, J. M. *Cancer Res.*, 1993, **53**, 977.
92. (a) Rao, V. V.; Chiu, M. L.; Kronauge, J. F.; Piwnica-Worms, D. *Nucl. Med.*, 1994, **35**, 510. (b) Herman, L. W.; Sharma, V.; Kronauge, J. F.; Barbarics, E.; Herman, L. A.; Piwnica-Worms, D. *J. Med. Chem.*, 1995, **38**, 2955.
93. Liu, S.; Edwards, D. S. *Chem. Rev.*, 1999, **99**(9), 2235–2268.
94. (a) Lister-James, J.; Knight, L. C.; Maurer, A. H.; Bush, L. R.; Moyer, B. R.; Dean, R. T. *J. Nucl. Med.*, 1996, **37**, 775–781. (b) Pearson, D. A.; Lister-James, J.; McBride, W. J.; Wilson, D. M.; Martel, L. J.; Civitello, E. R.; Dean, R. T. *J. Med. Chem.*, 1996, **39**, 1372–1382. (c) Muto, P.; Lastoria, S.; Varrella, P.; Vergara, E.; Salvatore, M,; Morgano, G.; Lister-James, J.; Bernardy, J. D.; Dean, R. T.; Wencker, D. *J. Nucl. Med.*, 1995, **36**, 1384–1391. (d) Taillefer, R.; Therasse, E.; Turpin, S.; Lambert, R.; Robillard, P.; Soulez, G. *J. Nucl. Med.*, 1999, **40**(12), 2029–2035.
95. Palestro, C. J.; Weiland, F. L.; Seabold, J. E.; Valdivia, S.; Tomas, M. B.; Moyer, B. R.; Baran, Y. M.; Lister-James, J.; Dean, R. T. *Nucl. Med. Commun.*, 2001, **22**, 695–701.
96. Vallabhajosula, S.; Moyer, B. R.; Lister-James, J.; McBride, B. J.; Lipszyc, H.; Lee, H.; Bastidas, D.; Dean, R. T. *J. Nucl. Med.*, 1996, **37**, 1016–1022.
97. Blum, J.; Handmaker, H.; Lister-James, J.; Rinne, N. *Chest*, 2000, **117**, 1232–1238.
98. Volkert, W. A.; Hoffman, T. J. *Chem. Rev.*, 1999, **99**(9), 2269–2292.
99. Lauffer, R. B. *Chem. Rev.*, 1987, **87**, 901–927.
100. Cacheris, W. P.; Quay, S. C.; Rocklage, S. M. *Magn. Reson. Imaging*, 1990, **8**, 467–481.
101. Caravan, P.; Ellison, J. J.; McMurry, T. J.; Lauffer, R. B. *Chem. Rev.*, 1999, **99**(9), 2293–2352.
102. Gries, H.; Miklautz, H. *Physiol. Chem. Phys. Med. NMR*, 1984, **16**, 105–112.
103. Ehnebom, L.; Fjaertoft Pedersen, B. *Acta Chem. Scand.*, 1992, **46**, 126–130.
104. Kumar, K.; Chang, C. A.; Francesconi, L. C.; Dischino, D. D.; Malley, M. R.; Gougoutas, J. Z.; Tweedle, M. R. *Inorg. Chem.*, 1994, **33**, 3567–3575.
105. Shukla, R. B. *J. Magn. Reson., Ser. A*, 1995, **113**, 196–204.
106. (a) Caravan, P.; Mehrkhodavandi, P.; Orvig, C. *Inorg. Chem.*, 1997, **36**, 1321. (b) Thompson, L.; Shafer, B.; Edgar, K.; Mannila, K. *Adv. Chem. Ser.*, 1967, **71**, 169–179.
107. Dong, Q.; Hurst, D. R.; Weinmann, H. J.; Chenevert, T. L.; Londy, F. J.; Prince, M. R. *Invest. Radiol.*, 1998, **33**, 699–708.
108. Lauffer, R. B. *Magn. Reson. Med.*, 1991, **22**, 339.
109. Pena, M. M. O.; Lee, J.; Thiele, D. J. *J. Nutr.*, 1999, **129**, 1251–1260.
110. Sarkar, B. *Chem. Rev.*, 1999, **99**(9), 2535–2544.
111. Zhou, B.; Gitschier, J. *Proc. Natl. Acad. Sci. USA*, 1997, **94**, 7481.
112. Klomp, L. W. J.; Lin, S. J.; Yuan, D. S.; Klausner, R. D.; Culotta, V. C.; Gitlin, J. D. *J. Biol. Chem.*, 1997, **272**, 9221–9226.
113. (a) Amaravadi, R.; Glerum, D. M.; Tzagoloff, A. *Hum. Genet.*, 1997, **99**, 329–333. (b) Glerum, D. M.; Shtanko, A.; Tzagoloff, A. *J. Biol. Chem.*, 1996, **271**, 14504–14509.
114. Culotta, V. C.; Klomp, L. W. J.; Strain, J.; Cassareno, R. L. B.; Krems, B.; Gitlin, J. D. *J. Biol. Chem.*, 1997, **272**, 23469.

115. Rosenzweig, A. C.; Huffman, D. L.; Hou, M. Y.; Wernimont, A. K.; Pufahl, R. A.; O'Halloran, T. V. *Structure*, 1999, **7**, 605–617 (PDB: 1CC7, 1CC8).

116. Arnesano, F.; Banci, L.; Bertini, I.; Huffman, D. L.; O'Halloran, T. V. *Biochemistry*, 2001, **40**, 1528–1539 (PDB: 1FD8, 1FES).

117. Pufahl, R.; Singer, C. P.; Peariso, K. L.; Lin, S.-J.; Schmidt, P. J.; Fahrni, C. J.; Cizewski Culotta, V.; Penner-Hahn, J. E.; O'Halloran, T. V. *Science*, 1997, **278**, 853–856.

118. Wernimont, A. K.; Huffman, D. L.; Lamb, A. L.; O'Halloran, T. V.; Rosenzweig, A. C. *Natl. Struct. Biol.*, 2000, **7**(9), 766–771 (PDB: 1FEO, 1FE4, 1FEE).

119. O'Halloran, T. V.; Culotta, V. C. *J. Biol. Chem.*, 2000, **275**(33), 25057–25060.

120. Rae, T. D.; Schmidt, P. J.; Pufhal, R. A.; Culotta, V. C.; O'Halloran, T. V. *Science*, 1999, **284**, 805–808.

121. Schmidt, P. J.; Rae, T. D.; Pufhal, R. A.; Hamma, T.; Strain, J.; O'Halloran, T. V.; Culotta, V. C. *J. Biol. Chem.*, 1999, **274**, 23719–23725.

122. Schmidt, P. J.; Ramos-Gomez, M.; Culotta, V. C. *J. Biol. Chem.*, 1999, **274**, 36952–36956.

123. Cassareno, R. L.; Waggoner, D.; Gitlin, J. D. *J. Biol. Chem.*, 1998, **273**, 23625–23628.

124. Zhu, H.; Shipp, E.; Sanchez, R. J.; Liba, A.; Stine, J. E.; Hart, P. J.; Gralla, E. B.; Nersissian, A. M.; Valentine, J. S. *Biochemistry*, 2000, **39**, 8125–8132.

125. Torres, A. S.; Petri, V.; Rae, T. D.; O'Halloran, T. V. *J. Biol. Chem.*, 2001 **276**(42), 38410–38416.

126. Rae, T. D.; Torres, A. S.; Pufahl, R. A.; O'Halloran, T. V. *J. Biol. Chem.*, 2001, **276**, 5166–5176.

127. Qian, H.; Sahlman, L.; Eriksson, P. O.; Hambraeus, C.; Edlund, U.; Sethson, I. *Biochemistry*, 1998, **37**, 9316–9322.

128. Lamb, A. L.; Torres, A. S.; O'Halloran, T. V.; Rosenzweig, A. C. *Natl. Struct. Biol.*, 2001, **8**, 751–755.

129. Wiedau-Pazos, M.; Goto, J. J.; Rabizadeh, S.; Gralla, E. B.; Roe, J. A.; Lee, M. K.; Valentine, J. S.; Bredesen, D. E. *Science*, 1996, **271**, 515–518.

130. Corson, L. B.; Strain, J. J.; Culotta, V. C.; Cleveland, D. W. *Proc. Natl. Acad. Sci. USA*, 1998, **95**, 6361–6366.

131. Didonato, M.; Narindrasorasak, S.; Forbes, J. R.; Cox, D. W.; Sarka, B. *J. Biol. Chem.*, 1997, **272**, 33279–33282.

132. Lutsenko, S.; Petrukhin, K.; Cooper, M. J.; Gilliam, C. T.; Kaplan, J. H. *J. Biol. Chem.*, 1997, **272**(30), 18939–18944.

133. Gitschier, J.; Moffat, B.; Reilly, D.; Wood, W. I.; Fairbrother, W. J. *Natl. Struct. Biol.*, 1998 **5**(1), 47–54 (PDB: 1AW0, 2AW0).

INDEX